# ALGORITHMIC AND HIGH-FREQUENCY TRADING

The design of trading algorithms requires sophisticated mathematical models, a solid analysis of financial data, and a deep understanding of how markets and exchanges function. In this textbook the authors develop models for algorithmic trading in contexts such as: executing large orders, market making, targeting VWAP and other schedules, trading pairs or collection of assets, and executing in dark pools. These models are grounded on how the exchanges work, whether the algorithm is trading with better informed traders (adverse selection), and the type of information available to market participants at both ultra-high and low frequency.

*Algorithmic and High-Frequency Trading* is the first book that combines sophisticated mathematical modelling, empirical facts and financial economics, taking the reader from basic ideas to the cutting edge of research and practice.

If you need to understand how modern electronic markets operate, what information provides a trading edge, and how other market participants may affect the profitability of the algorithms, then this is the book for you.

ÁLVARO CARTEA is a Reader in Financial Mathematics at University College London. Before joining UCL he was Associate Professor of Finance at Universidad Carlos III, Madrid-Spain (2009–2012) and from 2002 until 2009 he was a Lecturer (with tenure) in the School of Economics, Mathematics and Statistics at Birkbeck – University of London. He was previously JP Morgan Lecturer in Financial Mathematics at Exeter College, University of Oxford.

SEBASTIAN JAIMUNGAL is an Associate Professor and Chair, Graduate Studies in the Department of Statistical Sciences at the University of Toronto where he teaches in the PhD and Masters of Mathematical Finance programs. He consults for major banks and hedge funds focusing on implementing advance derivative valuation engines and algorithmic trading strategies. He is also an associate editor for the SIAM Journal on Financial Mathematics, the International Journal of Theoretical and Applied Finance, the journal Risks and the Argo newsletter. Jaimungal is the Vice Chair for the Financial Engineering & Mathematics activity group of SIAM and his research is widely published in academic and practitioner journals. His recent interests include High-Frequency and Algorithmic trading, applied stochastic control, mean-field games, real options, and commodity models and derivative pricing.

JOSÉ PENALVA is an Associate Professor at the Universidad Carlos III in Madrid where he teaches in the PhD and Master in Finance programmes, as well as at the undergraduate level. He is currently working on information models and market microstructure and his research has been published in Econometrica and other top academic journals.

# ALGORITHMIC AND HIGH-FREQUENCY TRADING

**ÁLVARO CARTEA**
*University College London*

**SEBASTIAN JAIMUNGAL**
*University of Toronto*

**JOSÉ PENALVA**
*Universidad Carlos III de Madrid*

# CAMBRIDGE
## UNIVERSITY PRESS

University Printing House, Cambridge CB2 8BS, United Kingdom

One Liberty Plaza, 20th Floor, New York, NY 10006, USA

477 Williamstown Road, Port Melbourne, VIC 3207, Australia

314-321, 3rd Floor, Plot 3, Splendor Forum, Jasola District Centre, New Delhi - 110025, India

79 Anson Road, #06-04/06, Singapore 079906

Cambridge University Press is part of the University of Cambridge.

It furthers the University's mission by disseminating knowledge in the pursuit of
education, learning and research at the highest international levels of excellence.

www.cambridge.org
Information on this title: www.cambridge.org/9781107091146

First published 2015
4th printing 2018

A catalogue record for this publication is available from the British Library

Library of Congress Cataloging in Publication data
Cartea, Álvaro.
Algorithmic and high-frequency trading / Álvaro Cartea, Sebastian Jaimungal, José Penalva.
pages   cm
Includes bibliographical references and index.
ISBN 978-1-107-09114-6 (Hardback : alk. paper)
1. Electronic trading of securities–Mathematical models.   2. Finance–Mathematical models.
3. Speculation–Mathematical models.   I. Title.
HG4515.95.C387 2015
332.64–dc23   2015018946

ISBN 978-1-107-09114-6 Hardback

Additional resources for this publication at www.cambridge.org/9781107091146

To my girls, in order of appearance, Victoria, Amaya, Carlota, and Penélope.

— Á.C.


To my parents, Korisha and Paul, and my siblings Shelly, Cristina and especially my brother Curt for his constant injection of excitement and encouragement along the way.

— S.J.


To Nuria, Daniel, Jose María and Adelina.
For their patience and encouragement every step of the way, and for never losing faith.

— J.P.

# Contents

# Preface

We have written this book because we feel that existing ones do not provide a sufficiently broad view to address the rich variety of issues that arise when trying to understand and design a successful trading algorithm. This book puts together the diverse perspectives, and backgrounds, of the three authors in a manner that ties together the basic economics, the empirical foundations of high-frequency data, and the mathematical tools and models to create a balanced perspective of algorithmic and high-frequency trading.

This book has grown out of the authors' interest in the field of algorithmic and high-frequency finance and from graduate courses taught at University College London, University of Toronto, Universidad Carlos III de Madrid, IMPA, and University of Oxford. Readers are expected to have basic knowledge of continuous-time finance, but it assumes that they have no knowledge of stochastic optimal control and stopping. To keep the book self-contained, we include an appendix with the main stochastic calculus tools and results that are needed. The treatment of the material should appeal to a wide audience and it is ideal for a graduate course on Algorithmic Trading at a Master's or PhD level. It is also ideal for those already working in the finance sector who wish to combine their industry knowledge and expertise with robust mathematical models for algorithmic trading. We welcome comments! Please send them to `algo.trading.book@gmail.com`.

## Brief guide to the contents

This book is organised into three parts that take the reader from the workings of electronic exchanges to the economics behind them, then to the relevant mathematics, and finally to models and problems of algorithmic trading.

Part I starts with a description of the basic elements of electronic markets and the main ways in which people participate in the market: as active traders exploiting an informational advantage to profit from possibly fleeting profit opportunities, or as market makers, simultaneously offering to buy and sell at advantageous prices.

A textbook on algorithmic trading would be incomplete if the development of strategies was not motivated by the information that market participants see in electronic markets. Thus it is necessary to devote space to a discussion of data and empirical implications. The data allow us to present the context which determines the ultimate fate of an algorithm. By looking at prices, volumes, and the details of the limit order book, the reader will get a basic overview of some of the key issues that any algorithm needs to account for, such as the information in trades, properties of price movements, regularities in the intraday dynamics of volume, volatility, spreads, etc.

Part II develops the mathematical tools for the analysis of trading algorithms. The chapter on stochastic optimal control and stopping provides a pragmatic approach to material which is less standard in financial mathematics textbooks. It is also written so that readers without previous exposure to these techniques equip themselves with the necessary tools to understand the mathematical models behind some algorithmic trading strategies.

Part III of the book delves into the modelling of algorithmic trading strategies. The first two chapters are concerned with optimal execution strategies where the agent must liquidate or acquire a large position over a pre-specified window and trades continuously using only market orders. Chapter 6 covers the classical execution problem when the investor's trades impact the price of the asset and also adjusts the level of urgency with which she desires to execute the programme. In Chapter 7 we develop three execution models where the investor: i) carries out the execution programme as long as the price of the asset does not breach a critical boundary, ii) incorporates order flow in her strategy to take advantage of trends in the midprice which are caused by one-sided pressure in the buy or sell side of the market, and iii) trades in both a lit venue and a dark pool.

In Chapter 8 we assume that the investor's objective is to execute a large position over a trading window, but employs only limit orders, or uses both limit and market orders. Moreover, we show execution strategies where the investor also tracks a particular schedule as part of the liquidation programme.

Chapter 9 is concerned with execution algorithms that target volume-based schedules. We develop strategies for investors who wish to track the overall volume traded in the market by targeting: Percentage of Volume, Percentage of Cumulative Volume, and Volume Weighted Average Price, also known as VWAP.

The final three chapters cover various topics in algorithmic trading. Chapter 10 shows how market makers choose where to post limit orders in the book. The models that are developed look at how the strategies depend on different factors including the market maker's aversion to inventory risk, adverse selection, and short-term lived trends in the dynamics of the midprice.

Finally, Chapter 11 is devoted to statistical arbitrage and pairs trading, and Chapter 12 shows how information on the volume supplied in the limit order book is employed to improve execution algorithms.

## Style of the book

In choosing the content and presentation of the book we have tried to provide a rigorous yet accessible overview of the main foundational issues in market microstructure, and of some of the empirical themes of electronic trading, using the US equities market as the one most familiar to readers. These provide the basis for a thorough mathematical analysis of models of trade execution, volume-based algorithms, market making, statistical arbitrage, pairs trading, and strategies based on order flow information. Most chapters in Part III end with exercises of varying levels of difficulty. Some exercises closely follow the material covered in the chapter and require the reader to: solve some of the problems by looking at them from a different perspective; fill in the gaps of some of the derivations; see it as an invitation to experiment further. We have set up a website, `http://www.algorithmic-trading.org`, from which readers can download datasets and MATLAB code to assist in such experimentation.

This book does not cover any of the information technology aspects of algorithmic trading. Nor does it cover in detail certain aspects of market quality or discuss regulation issues.

## Acknowledgements

We are thankful to those who took the time to read parts of the manuscript and gave us very useful feedback: Ali Al-Aradi, Gene Amromin, Robert Almgren, Ryan Francis Donnelly, Luhui Gan, John Goodacre, Hui Gong, Tianyi Jia, Hoi Kong, Tim Leung, Siyuan Li, Eddie Ng, Zhen Qin, Jason Ricci, Anton Rubisov, Mark Stevenson, Mikel Tapia and Jamie Walton. We also thank the students who have taken our courses at University College London, University of Toronto, University of Oxford, IMPA and Universidad Carlos III de Madrid.

Álvaro is grateful for the hospitality and generosity of the Finance Group at Saïd Business School, University of Oxford, with special thanks to Tim Jenkinson and Colin Mayer, and the Department of Statistical Sciences, University of Toronto, where a great deal of this book was written.

Sebastian is grateful for the hospitality of the Mathematical Institute, University of Oxford and the Department of Mathematics, University College London, where parts of this book were written.

José is grateful for the hospitality of the Department of Mathematics, University College London and the Department of Finance at Cass Business School where parts of this book were written, as well as his home institution, the Business Department of the Universidad Carlos III for allowing him to make these visits. He also wishes to thank Artem Novikov of TradingPhysics for his availability and help in accessing the data and clarifying specific issues faced by traders and technicians in high-frequency trading environments.

May 2015, Oxford, London, Toronto, Madrid, Mallorca

# How to Read this Book

This book is aimed at those who want to learn how to develop the mathematical aspects of Algorithmic Trading. It is ideal for a graduate course on Algorithmic Trading at a Master's or PhD level, and is also ideal for those already working in the finance sector who wish to combine their industry knowledge and expertise with robust mathematical models for algorithmic trading.

Much of this book can be covered in an intensive one semester/term course as part of a Graduate course in Financial Mathematics/Engineering, Computational Finance, and Applied Mathematics. A typical student at this stage will be learning stochastic calculus as part of other courses, but will not be taught stochastic optimal control, or be proficient in the way modern electronic markets operate. Thus, they are strongly encouraged to read Part I of the book to: gain a good understanding of how electronic markets operate; understand basic concepts of microstructure theory that underpin how the market reaches equilibrium prices in the presence of different types of risks; and, study stylised statistical issues of the dynamics of the prices of stocks in modern electronic markets. And to read Part II to learn the stochastic optimal control tools which are essential to Part III where we develop sophisticated mathematical models for Algorithmic and High-Frequency trading.

Those with a solid understanding of stochastic calculus and optimal control, may skip Part II of the book and cover in detail Part III. However, we still encourage them to read Part I to gain an understanding of the stylised statistical features of the market, and to develop a better intuition of why algorithmic models are designed in particular ways or with specific objectives in mind.

For a shorter and more compact course on algorithmic trading, students should focus on learning about the limit order book, Chapter 1, then optimal control in Part II, and then concentrate on selected Chapters in Part III, for instance Chapters 6, 8 and 10.

Readers in the financial industry who have some knowledge of how electronic markets are organised may want to skip Chapter 1 but are encouraged to read the other chapters which cover microstructure theory and the empirical and statistical evidence of stock prices before delving into the details of the mathematical models in Part III.

# Part I

## Microstructure and Empirical Facts

# Introduction to Part I

In the first part of the book we give an overview of the way basic electronic markets operate. Chapter 1 looks at the main practical issues when trading: what are the main assets traded and the main types of participants, what drives them to trade, and how do they interact. It also looks at the basic functioning of an electronic exchange: limit orders, market orders, and other types of orders, as well as the limit order book, and basic fee structures. It concludes by looking at the way the limit order book is organised and the basic experience of executing a trade.

Chapter 2 provides an overview of the theoretical economics of trading: what are the economic forces driving the competitive advantage of market makers and other traders and how do they interact. It covers the basic market making models that describe how liquidity is affected by inventory risk or the presence of better informed traders. It also looks at the market maker's trade-off between execution frequency and expected profit per trade, and how informed traders optimally exploit their informational advantage by trading gradually to limit the information leakage of their impact on order flow.

Chapters 3 and 4 look at equity market data to provide an overview of some of the basic empirical regularities that can be observed. Chapter 3 focuses on the time series properties of prices and returns, at daily and intraday frequency. It considers such issues as latency and the effects of limitations on price movements, as well as the dynamic structure of price changes, market fragmentation in US markets, and the comovement of asset prices that drives trading in pairs of assets. Chapter 4 focuses on volume and market quality. It looks at the relationship between volume and volatility, as well as known patterns in volume and prices. This is followed by an overview of different measures of liquidity and market quality: spreads, volatility, depth and trade size, and price impact. The chapter concludes by looking at other issues related to trading such as patterns in messages, order cancellations, executions and hidden orders.

# 1  Electronic Markets and the Limit Order Book

To understand how electronic markets work we must first understand the context in which trading in financial markets occurs. In this chapter, Section 1.1, we provide an overview of how electronic markets function, including short discussions on stocks, preferred stocks, mutual funds and hedge funds. We also discuss types of market participants (noise traders, informed traders/arbitrageurs, market makers) and in Section 1.2 how they interact. Next, in Section 1.3 we describe how electronic exchanges are structured, what limit and market orders are (as well as other order types), how exchanges collect orders in the limit order book (LOB), and the fees charged to market participants. Finally, Section 1.4 provides details of how the LOB is constructed and how market orders interact with it.

## 1.1  Electronic markets and how they function

Many types of financial contracts are traded in electronic markets today, so let us briefly and very superficially consider the main ones. The most familiar of these are shares or company stocks. Shares are claims of ownership on corporations. These claims are used by corporations to raise money. In the US, for these shares to be traded in an electronic exchange they have to be 'listed' by an exchange, and this implies fulfilling certain requirements in terms of the number of shareholders, price, etc. The listing process is usually tied to the first issuance of the public shares (initial public offering, or IPO). The fundamental value of these shares is derived from the nature of the contract it represents. In its simplest form, it is a claim of ownership on the company that gives the owner the right to receive an equal share of the corporation's profits (hence the name, 'share') and to intervene in the corporate decision process via the right to vote in the corporation's annual general (shareholders') meetings. Such shares are called **ordinary shares** (or **common stock**) and are the most common type of shares.

The other primary instrument used by large corporations to raise capital is **bonds**. Bonds are contracts by which the corporation commits to paying the holder a regular income (interest) but gives them no decision rights. The differences between stocks and bonds are quite clear: shareholders have no guarantees on the magnitude and frequency of dividends but have voting rights, bondholders have guarantees of regular, pre-determined payments and no voting rights. There are other instruments with characteristics from both these contracts, the most familiar of which is **preferred stock**. Preferred stock represents a hybrid of stocks and bonds: they are like bonds in that holders have no voting rights and receive a pre-arranged income, but the income they receive has fewer guarantees: its legal treatment is that of equity, rather than debt. This difference is especially relevant when the corporation is in financial distress, as debt is senior to all equity, so that in case of liquidation, debt holders' claims have priority over the corporation's assets –they get paid first. Equity holders, if they get paid, are paid only after all debtholders' claims are settled.

The universe of financial contracts is separated into different asset classes or categories according to the characteristics of the underlying assets. Shares and preferred stock belong to Equities. Bonds belong to their own asset class and are usually differentiated from cash (investments characterised by short-term investment horizons and usually with very heavy guarantees and low returns, such as money market accounts, savings deposits, Treasury bills, etc). There are also more exotic asset classes such as Foreign Exchange (FX), Commodities, Real Estate or Property. An investor will find these different types of assets in electronic exchanges, usually in the form of specialised securities such as mutual funds and exchange-traded funds (ETFs), which allow investors to invest in these asset classes in a familiar, equity-like market which simplifies the process of diversification and is associated with greater liquidity.

A **mutual fund** is an investment product that acts as a delegated investment manager. That is, when an investor buys a mutual fund, the investor gives her cash to a financial management company that will use the cash to build a portfolio of assets according to the fund's investment objective. This objective includes the fund's assets and investment strategy, and, of course, its management fees. The fund's assets can belong to a large number of possible asset classes, including all those described above: equities, bonds, cash, FX, real estate, etc. The fund's investment strategy refers to the style of investment, primarily whether the fund is actively managed or passively tracks an index.

An investor who puts money in a fund participates in both the appreciation and depreciation of the assets as allocated by the fund manager. In order to redeem her investment, i.e. to convert her investment into cash, the investor's options depend on the type of fund she purchased. There are two main types of mutual funds: open-end and closed-end funds. Closed-end funds are mutual funds that are not redeemable: the fund issues a fixed number of shares usually only once, at inception, and investors cannot sell the shares back to the fund. The fund sells the shares initially through an IPO and these shares are listed on an exchange where investors buy and sell these shares to each other.

Open-end funds are funds with a varying number of shares. Shares can be created to meet the demand of new investors, or destroyed (bought back by the fund) as investors seek to redeem theirs. This process takes place once a day, as the value of the fund's (net) assets (its Net Asset Value, NAV) is determined after the market close. Thus, closed-end funds, that do not have to adjust their holdings in response to investor demand, have different liquidity requirements than open-end funds and thus may trade at prices different from their NAV.

A very popular type of fund that, like closed-end funds, are traded in electronic exchanges, are ETFs. Like mutual funds, ETFs act as delegated investment managers, but they differ in two key respects. First, ETFs tend to have very specific investment strategies, usually geared towards generating the same return as a particular market index (e.g., the S&P500). Second, they are not obligated to purchase investors' shares back. Rather, if an investor wants to return their share to the fund, the fund can transfer to the investor a basket of securities that mirrors that of the ETF. This is possible because the ETF sells shares in very large units (Creation Units) which are then broken up and resold as individual shares in the exchange. A Creation Unit can be as large as 50,000 shares. Overall, the general perception one gets is that investors who are looking to reduce their trading costs and find diversified investments prefer ETFs, while investors who are looking for managers with stock-picking or similar unusual skills and who aim to beat the market will prefer mutual funds.

Some investment firms feel that the regulation that is imposed on mutual fund managers to ensure they fulfill their fiduciary duties to investors are too constraining. In response to this they have created **hedge-funds**, funds that pursue more aggressive trading strategies and have fewer regulatory and transparency requirements. Because of the softer regulatory oversight, access to these investment vehicles is largely limited to accredited investors, who are expected to be better informed and able to deal with the fund's managers. Although these funds are not traded on exchanges, their managers are active participants in those markets.

There are also other securities traded in electronic exchanges; in particular, there is a great deal of electronic trading in derivative markets, especially futures, swaps and options, and these contracts are written on a wide variety of assets (bonds, FX, commodities, equities, indices). The concepts and techniques we develop in this book apply to the trading of any of these assets, although we primarily focus our examples and applications on equities. However, when designing algorithms and strategies one must always take into account the specific issues associated with the types of assets one is trading in, as well as the specifics of the particular electronic exchange(s) and the trading objectives of other investors one is likely to meet there.

## 1.2     Classifying Market Participants

When designing trading strategies and algorithms, it is important to understand the different types of trading behaviour one will probably encounter in these exchanges. For instance, one must consider who trades in these exchanges and why. Everyone's motivation is clear, they want to make money, but it is essential to consider what drives them to trade – how it is that they may be looking to make money – because in many cases this will interact with our algorithm design choices and affect whether and how different algorithms achieve the desired trading objective.

Let us start from the creation of the objects of trade we have just discussed. The most familiar of these are shares. We have seen that corporations, or rather, their managers, issue stocks or equity in order to raise capital. These stocks are one of the primary objects of trade which are created when a company goes public and goes through the process of having them listed on an exchange, usually via an IPO. A corporation issues shares to raise capital for diverse economic activities, ranging from manufacturing electronic music players to mining ores in remote places. It is important to remember that these shares are claims on a corporation and as such are subject to the decisions of the company. Hence, one type of participant is corporate managers who create some of the assets that are traded in the exchanges, and who will, at times, actively participate in the market by increasing or reducing the supply of their corporation's shares, e.g., through secondary share offerings (SSOs), share buybacks, stock dividends, conversion of bonds into shares (and vice versa), etc.

We have also seen that there are other objects traded in exchanges. In equity markets we find funds (mutual funds, ETFs) created by financial management companies to commercialise their services. These funds manage large numbers of financial contracts, are very active participants in electronic exchanges, and originate a substantial fraction of the trading observed in exchanges. These 'supply-side' traders can have long-term investment goals (e.g., funds which focus on 'value investing', the kind of strategies epitomised by Warren Buffett) or focus on very immediate strategies (e.g., ETFs that replicate the returns of the S&P500). There are also **proprietary traders** who trade on a (sometimes real, sometimes illusory) trading advantage, which range from the large hedge funds we saw earlier, to small individual 'day-traders' moving in and out of asset positions from their home-offices. Proprietary traders trade on their competitive advantage: be it identifying fundamentally mispriced assets, identifying price momentum or sentiment-based price changes, having special technical abilities to process market information and identify patterns (technical traders), being able to time price movements based on news (be it the announcement of government economic figures or processing Twitter feeds), or identifying fleeting unjustified price discrepancies between equivalent assets (arbitrageurs).

Another, very important, group of market participants are 'regular investors' and '**fundamental traders**'. These are investors who have a direct use for the assets being traded. They may be individuals who buy stocks in the hope of being able to share in their growth as the corporation increases its economic value-creation and its shares appreciate in value. Or, they may want to rebalance their investments because of a change in circumstances (in response to a sudden need for cash, a change in their taste for risk or their outlook for the future). They may be corporations that use financial contracts to hedge risks such as changes in the prices of inputs and outputs from their production activity. Traders in Brent, copper, or electricity futures worry about non-financial issues such as the number of refineries going offline for repairs, the discovery of new methods for safely transmitting electricity, or whether that tropical storm off the coast of Florida is going to turn into a hurricane and make landfall near Miami or Dade. And, one cannot ignore that governments also have a stake in market outcomes. They may want to manage their currency, issue debt to finance public expenditures, or repurchase assets to increase liquidity or maintain market stability.

The effects of the interaction amongst all these traders is one of the key issues studied in the field of market microstructure, which we will familiarise ourselves with in Chapter 2, and which helps us structure the concepts and issues behind our approach to trading. We differentiate three primary classes of traders (or trading strategies) below.

1. **Fundamental** (or **noise** or liquidity) **traders**: those who are driven by economic fundamentals outside the exchange.
2. **Informed traders**: traders who profit from leveraging information not reflected in market prices by trading assets in anticipation of their appreciation or depreciation.
3. **Market makers**: professional traders who profit from facilitating exchange in a particular asset and exploit their skills in executing trades.

Usually, one may consider arbitrageurs as a fourth type of trader, though, for our purposes we subsume arbitrageurs into informed traders moving in anticipation of price changes. Also, although it is not unusual to bundle noise and liquidity traders together, it is unusual to put them together with fundamental traders. The term 'Noise traders' is frequently employed to describe trading that is orthogonal to any events driving market prices, and 'Liquidity traders' is used for traders driven by the need to liquidate or accumulate a position for liquidity reasons orthogonal to market events.

'Fundamental traders', on the other hand, is a term usually reserved for traders that have medium- and long-term investment strategies based on detailed analysis of the actual business activity that underlies the asset being traded. This would naturally classify them as informed traders. However, a large fraction of their trading strategy arises from portfolio management and risk-return trade-offs that have very little short-term price information beyond that contained in the sheer size of their positions. Thus, from the point of view of a high-frequency trading algorithm, it is reasonable to consider them as 'noise' trades relative to the specific market events within the algorithm's horizon. Having said this, as long as a fundamental trader is trading on information with a short-term price impact (such as knowledge of the volume of a substantial change in positions) they may also be included in the Informed trader category.

We can think of market maker types as 'passive' or 'reactive' trading. This is trading that profits from detailed knowledge of the trading process and adapts to 'the market' as circumstances change, while the first two types represent more 'active' or 'aggressive' trading that only takes place to exploit specific informational advantages gained outside of the trading environment noise and fundamental traders having only a fleeting effect on short-run movements, while informed traders anticipate short-run price movements. This distinction is useful when setting up a trading strategy, although the boundary between the two is not always clear. Professional traders often leverage informational advantages gained from trading practice into the trading strategies they use for market making.

A common error is to equate market making with liquidity provision and informed trading with the taking of liquidity. Market making activity generally favours the provision of liquidity but a particular market making strategy may at times provide liquidity while at others demand it. Similarly, informed trading does not always occur via aggressive orders, and may at times be better implemented via passive orders that add liquidity. In Chapter 10 we develop algorithms for market makers who always provide liquidity to the market. These algorithms can be extended to show how market making changes when the market maker may take liquidity from the market. Moreover, in Chapter 8 we develop models of optimal execution where the agent's strategies both take and provide liquidity.

## 1.3    Trading in Electronic Markets

After the who and what of electronic markets, let us look at the how. There are many ways to implement an electronic market, though essentially they all amount to having a way for people to signal their willingness to trade, and a matching engine to match those wanting to buy with those wanting to sell.

### 1.3.1    Orders and the Exchange

In the basic setup, an electronic market has two types of orders: **Market Orders** (MOs), and **Limit Orders** (LOs). MOs are usually considered aggressive orders that seek to execute a trade immediately. By sending an MO, a trader indicates that she wants to buy or sell a certain quantity of shares at the best available price, and this will (usually) result in an immediate trade (execution). On the other hand, LOs are considered passive orders, as a trader sending in an LO indicates her desire to buy or sell at a given price up to a certain, maximum, quantity of shares. As the price offered in the LO is usually worse than the current market price (higher than the best buy price for sell LOs, and lower than the best sell price for buy LOs), it will not result in an immediate trade, and will thus have to wait until either it is matched with a new order that wants to trade at the offered price (and executed) or it is withdrawn (cancelled).

Orders are managed by a matching engine and a limit order book (LOB). The LOB keeps track of incoming and outgoing orders. The matching engine uses a well-defined algorithm that establishes when a possible trade can occur, and if so, which criterion is going to be used to select the orders that will be executed. Most

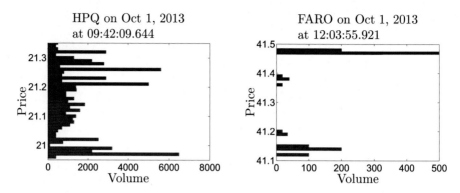

**Figure 1.1** Snapshots of the NASDAQ LOB after the 10,000th event of the day. Blue bars represent the available sell LOs, red bars represent the available buy LOs.

markets prioritise MOs over LOs and then use a price-time priority whereby, if an MO to buy comes in, the buy order will be matched with the standing LOs to sell in the following way: first, the incoming order will be matched with the LOs that offer the best price (for buy orders, the sell LOs with the lowest price), then, if the quantity demanded is less than what is on offer at the best price, the matching algorithm selects the oldest LOs, the ones that were posted earliest, and executes them in order until the quantity of the MO is executed completely. If the MO demands more quantity than that offered at the best price, after executing all standing LOs at the best price, the matching algorithm will proceed by executing against the LOs at the second-best price, then the third-best and so on until the whole order is executed. LOs that have increasingly worse prices are referred to as LOs that are deeper in the LOB, and the process whereby an entering market order executes against standing LOs deeper in the LOB is called 'walking the book'. Section 1.4 provides a more detailed view on how the LOB is built, and how MOs walk the book.

Figure 1.1 shows a snapshot of the limit order book (LOB) on NASDAQ after the 10,000th event of the day for two stocks, FARO and HPQ, on Oct 1, 2013 (see subsection 3.1.1 for a description of how this is constructed from the raw event data). The two are quite different. The one in the left panel corresponds to HPQ, a frequently traded and liquid asset. HPQ's LOB has LOs posted at every tick out to (at least) 20 ticks away from the midprice. In the right panel, we have FARO's LOB. FARO is a seldom traded, illiquid asset. This asset has thinly posted bids and offers and irregular gaps in the LOB. We discuss further details of this example in Section 1.4.

## 1.3.2    Alternate Exchange Structures

The above approach is not the only possible way to organise an exchange. For example, one could use an alternative matching algorithm, such as the **prorata** rules used in some money markets. With a prorata rule, MOs are matched against the posted LOs available at the best price, in proportion to the *quantities* posted – there is no time-priority rule. There are also markets, e.g. in futures, that mix the two, pro-rata and time-priority.

In addition to this basic setup there are a number of variations in the way exchanges organise offers and trades. For example, some markets introduce an additional priority to orders coming from a certain type of trader (either a designated market maker, or, in some markets, a designated supply-side trader). Many exchanges also use auctions at particular points in time. It is quite typical to have an initial and/or a closing auction, that is an auction at the start of the trading day and/or an auction to close the market. In addition, an exchange will use an auction after a market trading halt (e.g., after a volatility limit has been triggered) so as to smooth the transition back to active trading.

Another dimension of importance when characterising an exchange is the degree (and cost) of transparency. In the US there is clear (legal) distinction between regulated exchanges (such as NASDAQ and NYSE) which have specific obligations to publish information regarding the status of their LOBs, and other electronic markets (electronic crossing networks (ECNs), dark pools, and broker-dealer internalisation). Beyond the legal definitions, we generically distinguish lit (open order book) from dark markets based on whether limit book information is publicly available or not. Within lit markets there are many differences on how and at what price information is available. For example, NASDAQ has an order-based book reporting mechanism whereby the exchange records every message, and each LO is assigned an order identification number which can then be used to match the order with subsequent events, such as cancellations or executions. Other markets (NYSE and NYSE MKT/AMEX in particular) use the level-book method, whereby the market receives a message every time there is an event that impacts the order book, but does not keep tabs on posted orders so they cannot be matched with subsequent cancellations or executions. Throughout this book, most of the algorithms that we develop assume that the agent is trading in a lit market where she can observe the LOB. However, in Chapter 7 we discuss dark pools and develop algorithms for optimal execution when the agent simultaneously trades in a lit and dark market.

## 1.3.3    Colocation

Exchanges also control the amount and degree of granularity of the information you receive (e.g., you can use the consolidated/public feed at a low cost or pay a relatively much larger cost for direct/proprietary feeds from the exchanges). They also monetise the need for speed by renting out computer/server space next to their matching engines, a process called **colocation**. Through colocation, exchanges can provide uniform service to trading clients at competitive rates. Having the traders' trading engines at a common location owned by the exchange simplifies the exchange's ability to provide uniform service as it can control the hardware connecting each client to the trading engine, the cable (so all have the same cable of the same length), and the network. This ensures that all traders in colocation have the same fast access, and are not disadvantaged (at least in terms of exchange-provided hardware). Naturally, this imposes a clear distinction between traders who are colocated and those who are not. Those not colocated will always have a speed disadvantage. It then becomes an issue for regulators who have to ensure that exchanges keep access to colocation sufficiently competitive.

The issue of distance from the trading engine brings us to another key dimension of trading nowadays, especially in US equity markets, namely fragmentation, which we discuss in greater detail in Section 3.6. A trader in US equities markets has to be aware that there are up to 13 lit electronic exchanges and more than 40 dark ones. Together with this wide range of trading options, there is also specific regulation (the so-called 'trade-through' rules) which affects what happens to market orders sent to one exchange if there are better execution prices at other exchanges. The interaction of multiple trading venues, latency when moving between these venues, and regulation introduces additional dimensions to keep in mind when designing successful trading strategies.

## 1.3.4    Extended Order Types

The role of time is fundamental in the usual price-time priority electronic exchange, and in a fragmented market, the issue becomes even more important. Traders need to be able to adjust their trading positions fast in response to or in anticipation of changes in market circumstances, not just at the local exchange but at other markets as well. The race to be the first in or out of a certain position is one of the focal points of the debate on the benefits and costs of 'high-frequency trading'.

The importance of speed permeates the whole process of designing trading algorithms, from the actual code, to the choice of programming language, to the hardware it is implemented on, to the characteristics of the connection to the matching engine, and the way orders are routed within an exchange and between exchanges. Exchanges, being aware of the importance of speed, have adapted and, amongst other things, moved well beyond the basic two types of orders (MOs and LOs). Any trader should be very well-informed regarding all the different order types available at the exchanges, what they are and how they may be used. Some examples of the types of orders that you may find are:

- **Day Orders**: orders for trading during regular trading with options to extend to pre- or post-market sessions;
- **Non-routable**: there are a number of orders that by choice or design avoid the default re-routing to other exchanges, such as 'book only', 'post only', 'midpoint peg', ...;
- **Pegged, Hide-not-Slide**: orders that move with the midpoint or the national best price;

- **Hidden**: orders that do not display their quantity;
- **Iceberg**: orders that partially display their quantity (some have options so that the visible portion will automatically be replenished when it is depleted by less than one round lot);
- **Immediate-or-Cancel**: orders that execute as much as possible at the best price and the rest are cancelled (such orders are not re-routed to another exchange nor do they walk the book);
- **Fill-or-Kill**: orders sent to be executed at the best price in their entirety or not at all;
- **Good-Till-Time**: orders with a fixed lifetime built into them so that they will be cancelled if not executed by its expiration time;
- **Discretionary**: orders display one price (the limit price) but may be executed at more aggressive (hidden) prices;

and there are a myriad other variations on the classic MOs and LOs.

When coding an algorithm one should be very aware of all the possible types of orders allowed, not just in one exchange, but in all competing exchanges where one's asset of interest is traded. Being uninformed about the variety of order types can lead to significant losses. Since some of these order types allow changes and adjustments at the trading engine level, they cannot be beaten in terms of latency by the trader's engine, regardless of how efficiently your algorithms are coded and hardwired.[1] Later, when developing the mathematical algorithms in Part III of the book, we assume that the agents employ MOs and LOs and that when LOs are cancelled this is done in full.

## 1.3.5    Exchange Fees

Another important issue to be aware of is that trading in an exchange is not free, but the cost is not the same for all traders. For example, many exchanges run what is referred to as a maker-taker system of fees whereby a trader sending an MO (and hence taking liquidity away from the market) pays a trading fee, while a trader whose posted LO is filled by the MO (that is, the LO with which the MO is matched) will a pay much lower trading fee, or even receive a payment (a rebate) from the exchange for providing liquidity (making the market). On the other hand, there are markets with an inverted fee schedule, a taker-maker system where the fee structure is the reverse: those providing liquidity pay a higher fee than those taking liquidity (who may even get a rebate). The issue of exchange fees is quite important as fees distort observed market prices (when you make a transaction the relevant price for you is the net price you pay/receive,

---

[1] The importance of order types, their use, and the transparency with which they are documented is a key issue. The trader Haim Bodek has made a number of public statements in the last few years that illustrate this.

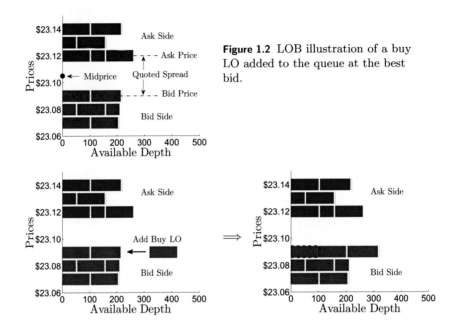

**Figure 1.2** LOB illustration of a buy LO added to the queue at the best bid.

which is the published price net of fees), and their effect in a fragmented market is strongly debated.[2]

## 1.4     The Limit Order Book

Having seen how complex things can get, let us start from the most basic description of the LOB and illustrate it first using an artificial LOB, and later (Figure 1.4) with some actual examples, using detailed message data from two assets, HPQ and FARO, on the NASDAQ stock exchange.

**Addition of LO to LOB.** As mentioned above, electronic exchanges are, at their most basic, described by an LOB and a matching algorithm. We discussed how price-time priority works: an incoming LO joins the LOB at the order's price and is placed last in the execution queue at that price. This is illustrated in an artificial LOB, in Figure 1.2. In this figure, LOs are displayed as blocks of length equal to their quantities. LOs are ordered in terms of time priority from right to left, so that when a new buy LO comes in at $23.09 (the purple block in the bottom panel of Figure 1.2) it will be added to the line of blocks already resting at that price. This new LO joins the queue at the point closest to the y-axis, becoming the third LO waiting to be executed at $23.09.

**MO walks the LOB or is re-routed.** Suppose we are looking at the venue with the LOB depicted at the top of Figure 1.2. Assume that this venue's best

---

2  Colliard & Foucault (2012) provide a very clear theoretical overview of the role of trading fees and their effects on the relationship between quoted and underlying prices.

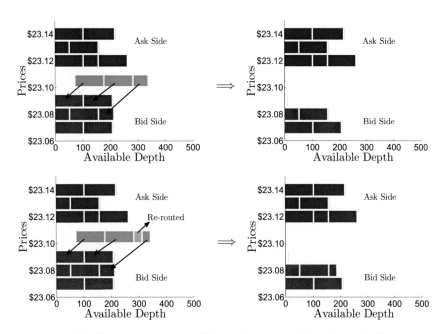

**Figure 1.3** LOB illustration of a sell MO walking the LOB with and without re-routing.

bid is the best buy quote that the market, across all venues, currently displays. A new MO (to sell) 250 shares enters this market as depicted by the sum of the green blocks in the top 'panel of Figure 1.3. The matching engine goes through the LOB, matching existing (posted) LOs (to buy on the bid side) with the entering MO following the rules in the matching algorithm. In the LOB there are two LOs at the best bid $23.09, represented by the two red blocks, both for 100 units, totalling 200 units. These 200 units are executed at the best bid.

What happens to the final 50 units depends on the order type and the market it is operating in. In a standard market, the remaining 50 units will be executed against the LOs standing at $23.08 ordered in terms of time-priority (the MO will 'walk the book'). This is captured by the top panels in Figure 1.3: the left panel shows that the MO coming in is split into three blocks, the first two are matched with LOs at $23.09 and the last with the LOs at $23.08. After the MO is fully executed the remaining LOB is shown in the top right panel of Figure 1.3.

As we mentioned in subsection 1.3.3, in the US, there are order protection rules to ensure MOs get the best possible execution, and which (depending on the order type) may require the exchange to re-route the remaining 50 units to another exchange that is also displaying a best bid price of $23.09. In this case, as shown in the bottom left panel of Figure 1.3, part of the remaining 50 units (the light blue block) is re-routed to another venue(s) with liquidity posted at $23.09. Only once all liquidity at $23.09 in all exchanges is exhausted, can the remaining shares of the MO return and be executed in this venue against any LO resting at (the worse price of) $23.08. In this example, 25 units were re-routed to alternate exchanges, and 25 units returned to this venue and walked the book.

The MO could in principle be an Immediate-or-Cancel (IOC) order, which specifies that the remaining 50 shares that cannot be executed at the best bid should be cancelled entirely.

Because of these order protection rules (trade-through rules – there is no such rule in European markets), you will very seldom observe in the US an MO walking the book straight away. Rather, you may see a large MO being chopped up and executed sequentially in several markets in a very short span of time. This also implies that as depth disappears (as during the Flash Crash of May 6th, 2010) an MO at the end of a sequence of other orders may be executed against very poor prices, and, in the worst circumstances it may be matched with **stub quotes** – LOs at prices so ridiculous that clearly indicate they are not expected to be executed (such trades were observed during the Flash Crash in the following assets: JKE, RSP, Excelon, Accenture, amongst others). Thus, the LOB serves to keep track of LOs and apply the algorithm that matches incoming orders to existing LOs.

The LOB is defined on a fixed discrete grid of prices (the price levels). The size of the step (the difference between one price level and the next) is called the **tick**, and in the US the minimum tick size is 1 cent for all stocks with a price above one dollar. In other markets several different tick sizes coexist. For example, in the Paris Bourse or the Bolsa de Madrid, tick sizes can range from 0.001 to 0.05 euros depending on the price the stock is trading at.

Figure 1.1 shows a sample plot of the limit order book (LOB) on NASDAQ after the 10,000th event of the day for two stocks, FARO and HPQ, on Oct 1, 2013. In blue you find the sell LOs –traders willing to wait to be able to sell at a high price. The best sell price, the ask, is $21.16, while the best buy price, the bid, is $21.15. The difference between the ask and the bid price, the **quoted spread** is

$$\text{Quoted Spread}_t = P_t^a - P_t^b,$$

(where $P_t^b$ and $P_t^a$ are the best bid and ask prices), which in this case, is one cent – the minimum quoted spread. However, some times the bid is equal to the ask and the spread is zero. In that case, the market becomes **locked**, but if this happens, it tends not to last long – although for some very liquid assets it is becoming an increasingly more frequent event. Another common object used when describing the LOB is the **midprice**. The midprice is the arithmetic average of the bid and the ask:

$$\text{Midprice}_t = \tfrac{1}{2}(P_t^a + P_t^b).$$

It is often used to proxy for the true underlying price of the asset – the price for the asset if there were no explicit or implicit trading costs (and hence no spread).

As pointed out earlier, the two LOBs shown in Figure 1.1 are quite different.

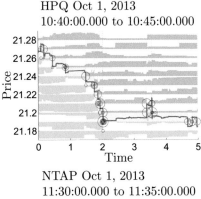

HPQ Oct 1, 2013
10:40:00.000 to 10:45:00.000

**Figure 1.4** Time series of the changes in the LOB for the three assets HPQ, NTAP, and ORCL.

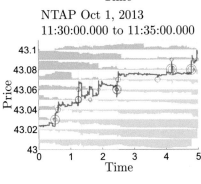

NTAP Oct 1, 2013
11:30:00.000 to 11:35:00.000

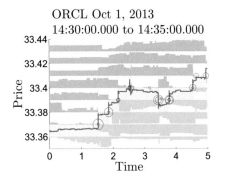

ORCL Oct 1, 2013
14:30:00.000 to 14:35:00.000

The one in the left panel corresponds to HPQ, a frequently traded and liquid asset. HPQ's LOB has LOs posted at every tick out to (at least) 20 ticks away from the midprice and the spread is the minimum spread of 1 tick. In the right panel, we have FARO's LOB. FARO is a seldom traded, illiquid asset. This asset has thinly posted bids and offers and irregular gaps in the LOB. The spread is 20 ticks (20 cents) on a (approximately) \$41 priced asset. The difference in liquidity between these assets is also noticeable from the time at which the 10,000th event of the day takes place for these assets. For HPQ, the 10,000th event corresponds to a timestamp of about 9:42 a.m. (less than 15 minutes after the market opened), while for FARO the 10,000th event did not occur until about 12:04 p.m. (more than two and a half hours after market open). Also note that there are less than 100 units posted if we sum together the depth at the best two price levels on the bid and ask for FARO, while for HPQ there are more than 1,000 shares offered in those first two levels of the LOB – HPQ thus has much greater depth. If one takes into account that FARO trades at a price which is twice as high as that of HPQ, the depth in terms of dollar value of shares posted at those prices is also much greater for HPQ.

The snapshot shown in Figure 1.1 only illustrates a static version of the LOB; however, its dynamics are quite interesting and informative. In Figure 1.4, we show how the LOB evolves through time (over 5 minutes) for three different stocks, HPQ, NTAP and ORCL. On the $x$-axis is time in minutes, and on the $y$-axis are prices in dollars. The static picture we saw in Figure 1.1 is captured by the shaded blue and red regions – the blue regions on top represent the ask side of the LOB, the posted sell volume, while the bid side is below in red, showing the posted buy volume. The best prices, the bid and ask are identified by the edges of the intermediate light shaded beige region, which identifies the bid-ask spread. Volume at each price level, which was captured in Figure 1.1 by horizontal bars, is now illustrated by the size of the shaded region just above/below each price level, although the height of these regions is no longer linear, but a monotonic non-linear transformation that is visually more illustrative.

In addition, Figure 1.4 identifies when incoming orders were executed. The red/blue circles indicate the time, price and size (indicated by the size of the circle) of an aggressive MO which is executed against the LOs sitting in the LOB. When a sell MO executes against a buy LO, it is said to **hit the bid**; analogously, when a buy MO executes against a sell LO, it is said the **lift the offer**. The brown solid line depicts a variation of the asset known as the **microprice** defined as

$$\text{Microprice}_t = \frac{V_t^b}{V_t^b + V_t^a} P_t^a + \frac{V_t^a}{V_t^b + V_t^a} P_t^b,$$

where $V_t^b$ and $V_t^a$ are the volumes posted at the best bid and ask, and $P_t^b$ and $P_t^a$ are the bid and ask prices. The microprice is used as a more subtle proxy for the asset's transaction cost-free price, as it measures the tendency that the price has to move either towards the bid or ask side as captured by number of shares posted, and hence indicates the buy (sell) pressure in the market. If there are a lot of buyers (sellers), then the microprice is pushed toward the best ask/bid price to reflect the likelihood that prices are going to increase (decrease). We explore the microprice and the effect of the relative volumes on the bid and ask side in more depth in Chapter 12 when developing algorithms that take into account volume imbalances in the LOB.

## 1.5    Bibliography and Selected Readings

O'Hara (1995), de Jong & Rindi (2009), Lehalle (2009), Colliard & Foucault (2012), Abergel, Anane, Chakraborti, Jedidi & Toke (2015).

# 2 A Primer on the Microstructure of Financial Markets

To understand the issues and problems faced in the design and implementation of trading strategies, we must consider the economics that drive these trading strategies. To do this we look to the market microstructure literature. Section 2.1 considers the basic market making model that focuses on inventory and inventory risk, as well as the trade-off between execution frequency and profit per trade. It also looks at the conceptual basis for the basic measures of liquidity. The last two sections look at trading when there are informational differences between traders. Section 2.2 from the point of view of the better informed trader, and Section 2.3 from that of the less informed market maker.

For the ecomomics of trading we look to market microstructure, as it is the subfield of finance which focuses on how trading takes place in very specific settings: it "is the study of the process and outcomes of exchanging assets under explicit trading rules" (O'Hara (1995)). Thus, it encompasses the subject of this book, algorithmic and high-frequency trading. It is within the microstructure literature that we find studies of the process of exchanging assets: trading strategies, and their outcomes: asset prices, volume, risk transfers, etc.

A key dimension of the trading and price setting process is that of information. Who has what information, how does that information affect trading strategies, and how do those trading strategies affect trading outcomes in general, and asset prices in particular. Forty years ago finance theory introduced the tools to explicitly incorporate and evaluate the notion of price efficiency, the idea that "market prices are an efficient way of transmitting the information required to arrive at a Pareto optimal allocation of resources" (Grossman & Stiglitz (1976)). This dimension naturally appears in microstructure studies which look into the details of how different trading rules and trading conditions incorporate or hinder price efficiency. What differentiates microstructure studies from more general asset pricing ones is that they focus on two aspects that are key to trading: liquidity and price discovery, and these are the two primary aspects that drive the questions and issues behind the design of effective algorithmic and high-frequency trading.[1]

Trading can take place in a number of possible ways: via personal deals settled over a handshake in a club, via decentralised chat rooms where traders engage

---

[1] Abergel, Bouchaud, Foucault, Lehalle & Rosenbaum (2012) provides a general overview of the determinants and effects of liquidity in security markets and related policy issues.

each other in bilateral personal transactions, via broker-intermediated over-the-counter (OTC) deals, via specialised broker-dealer networks, on open electronic markets, etc. Our focus is on trading and trading algorithms that take place in large electronic markets, whether they be open exchanges, such as the NASDAQ stock market, or in electronic private exchanges (run by a broker-dealer, a bank, or a consortium of buy-side investors).

## 2.1     Market Making

As we saw in Chapter 1, an important type of market participant is the 'passive' market maker (MM), who facilitates trade and profits from making the spread and from her execution skills, and must be quick to adapt to changing market conditions. Another type is the 'active' trader, who exploits her ability to anticipate price movements and must identify the optimal timing for her market intervention. We start with the first group, the 'passive' traders.

Because we are focusing on trading in active exchanges, it is natural to assume that there are many market makers (MMs) in competition. Naturally, trading in a market dominated by a few MMs would need to additionally incorporate how the MMs exercise their market power and how it affects the market as a whole.

MMs play a crucial role in markets where they are responsible for providing liquidity to market participants by quoting prices to buy and sell the assets being traded, whether they be equities, financial derivatives, commodities, currencies, or others. A key dimension of liquidity as provided by MMs is immediacy: the ability of investors to buy (or sell) an asset at a particular point in time without having to wait to find a counterparty with an offsetting position to sell (or buy). By quoting buy and sell prices (or posting limit orders (LOs) on both sides of the book), the MM is willing to provide liquidity to the market, but in order to make this a sustainable business the MM quotes a buy price lower than her quoted sell price. For example an MM is willing to purchase shares of company XYZ at $99 and willing to sell at $101 per share. Note that by posting LOs, the MM is providing liquidity to other traders who may be looking to execute a trade quickly, e.g. by entering a market order (MO). Hence, we have the usual dichotomy that separates MMs as liquidity providers from other traders, considered as liquidity takers.

If our MM is the one offering the best prices, so that the ask is $101 and the bid $99, then the quoted spread is $2. There are a number of theories that explain what determines the spread in a competitive market. Before delving into some of these theories, we consider the issues faced by someone willing to provide liquidity.

## 2.1.1    Grossman–Miller Market Making Model

The first issue faced by an MM when providing liquidity is that by accepting one side of a trade (say buying from someone who wants to sell), the MM will hold an asset for an uncertain period of time, the time it takes for another person to come to the market with a matching demand for liquidity (wanting to buy the asset the MM bought in the previous trade). During that time, the MM is exposed to the risk that the price moves against her (in our example, as she bought the asset, she is exposed to a price decline and hence having to sell the asset at a loss in the next trade).

Recall that the MM has no intrinsic need or desire to hold any inventory, so she will only buy (sell) in anticipation of a subsequent sale (purchase). Grossman & Miller (1988) provide a model that captures this problem and describes how MMs obtain a liquidity premium from liquidity traders that exactly compensates MMs for the price risk of holding an inventory of the asset until they can unload it later to another liquidity trader.

Let us consider a simplified version of their model, with a finite number, $n$, of identical MMs for some given asset and three dates $t \in \{1, 2, 3\}$. To simplify the situation, there is no uncertainty about the arrival of matching orders: if at date $t = 1$ a liquidity trader, denoted by LT1, comes to the market to sell $i$ units of the asset, there will be (for sure) another liquidity trader (LT2) who will arrive at the market to purchase $i$ units (or more generally, to trade $-i$ units, so that LT1's trade (of $i$ units) could be negative or positive (LT1 could be buying or selling). However, LT2 does not arrive to the market until $t = 2$. Let all agents start with an initial cash amount equal to $W_0$, MMs hold no assets, LT1 holds $i$ units and LT2 $-i$ units.

There are no trading costs or direct costs for holding inventory. The focus is on price changes: the asset will have a cash value at $t = 3$ of $S_3 = \mu + \epsilon_2 + \epsilon_3$, where $\mu$ is constant, $\epsilon_2$ and $\epsilon_3$ are independent, normally distributed random variables with mean zero and variance $\sigma^2$. These will be publicly announced between dates $t - 1$ and $t$, that is $\epsilon_3$ is announced between $t = 2$ and $t = 3$, and $\epsilon_2$ is announced between $t = 1$ and $t = 2$. Hence, the realised cash value of the asset can increase or decrease (ignore the fact that there are realisations of $\epsilon_2$ and $\epsilon_3$ that could make the asset value negative – the model serves to illustrate a point). Because the shocks to the value of the asset are on average zero a risk-neutral trader has no cost at all from holding the asset. The model becomes interesting if we assume that all traders, MMs and liquidity traders, are risk-averse. To be more specific, suppose they have the following expected utility for the future random cash value of the asset $(X_3)$: $\mathbb{E}[U(X_3)]$ where $U(X) = -\exp(-\gamma X)$, and where $\gamma > 0$ is a parameter capturing the utility penalty for taking risks (the risk aversion parameter).

Solving the model backwards we obtain a description of trading behaviour and prices. At $t = 3$ the cash value of the asset is realised, $S_3 = \mu + \epsilon_2 + \epsilon_3$. At $t = 2$, the $n$ MMs and LT1 come into the period with asset holdings $q_1^{MM}$ and $q_1^{LT1}$

respectively. LT2 comes in with $-i$ and they all exit with asset holdings $q_2^j$, where $j \in \{MM, LT1, LT2\}$. Note that if, for example, $q_t^j = 2$ this denotes that agent $j$ is holding 2 units when exiting date $t$, so that the agent will be long (that is, has an inventory of) two units. Given the problem as described so far, at $t = 2$ agent $j$ chooses $q_2^j$ to maximise his expected utility knowing the realisation of $\epsilon_2$ that was made public before $t = 2$:

$$\max_{q_2^j} \mathbb{E}\left[U\left(X_3^j\right) | \epsilon_2\right]$$

subject to

$$X_3^j = X_2^j + q_2^j S_3, \qquad X_2^j + q_2^j S_2 = X_1^j + q_1^j S_2.$$

These two constraints capture:

(i) the fact that the cash value of the agent's assets at $t = 3$, $X_3$, is equal to the agent's cash holdings at $t = 3$, which is equal to $X_2$ plus the cash value of the agent's asset inventory $q_2^j$, and

(ii) the fact that the cash value of the agent's assets when exiting date $t = 2$ ($X_2$, and the inventory $q_2^j$) was equal to the cash value of the agent's assets when entering date $t = 2$ ($X_1$, and the inventory $q_1^j$).

Given the normality assumption and the properties of the expected utility function it is straightforward to show that

$$\mathbb{E}\left[U\left(X_3^j\right) | \epsilon_2\right] = - \exp\left\{-\gamma\left(X_2^j + q_2^j \mathbb{E}[S_3 | \epsilon_2]\right) + \frac{1}{2}\gamma^2 \left(q_2^j\right)^2 \sigma^2\right\}.$$

Thus, the problem is concave and the solution is characterised by

$$q_2^{j,*} = \frac{\mathbb{E}[S_3 | \epsilon_2] - S_2}{\gamma \sigma^2},$$

for all agents: the $n$ MMs, LT1, and LT2.

As at date $t = 2$ demand and supply for the asset have to be equal to each other, we can solve for the equilibrium price $S_2$:

$$n q_1^{MM} + q_1^{LT1} + q_1^{LT2} = n q_2^{MM} + q_2^{LT1} + q_2^{LT2}, \tag{2.1}$$

where we use the convention that $q_1^{LT2}$, the assets LT2 came into period 2 with, is equal to his desired trade, $-i$. As we have established above, all $q_2^j$ are equal, so that the right-hand side of the above equation is equal to

$$n q_2^{MM} + q_2^{LT1} + q_2^{LT2} = (n+2) \frac{\mathbb{E}[S_3 | \epsilon_2] - S_2}{\gamma \sigma^2}. \tag{2.2}$$

We also know that at date 1 the total quantity of the asset available was equal to the quantity of assets LT1 brought to the market, so that the LHS of (2.1) is

$$n q_1^{MM} + q_1^{LT1} + q_1^{LT2} = i + q_1^{LT2} = i - i = 0.$$

Hence, substituting into (2.2) and solving, we obtain that in equilibrium, at date

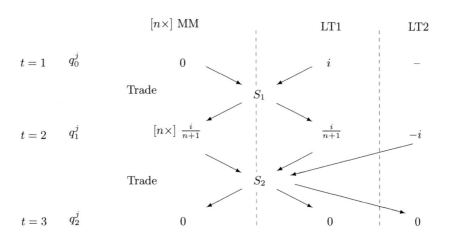

**Figure 2.1** Trading and price setting in the Grossman–Miller model.

$t = 2$, $S_2 = \mathbb{E}[S_3] = \mu + \epsilon_2 + \mathbb{E}[\epsilon_3] = \mu + \epsilon_2$, and therefore, $q_2^j = 0$. This makes sense, as at $t = 2$ there are no asset imbalances, the price of the asset reflects its 'fundamental value' (efficient price) and no one will want to hold a non-zero amount of the risky asset. This analysis is captured in the bottom half of Figure 2.1, where we see the asset holdings of the three types of participants as they enter $t = 2$, $q_1^j$, $j \in \{MM, LT1, LT2\}$, and how after trading at a price equal to $S_2$ they end up with holdings, $q_2^j$, equal to zero.

Consider now what happens at date $t = 1$. Participating agents (the $n$ MMs and LT1 – recall that LT2 will not appear until $t = 2$) anticipate that whatever they do, the future market price will be efficient and they will end up exiting date $t = 2$ with no inventories, so that $X_3 = X_2$. Thus, their portfolio decision is given by

$$\max_q \mathbb{E}\left[U\left(X_2^j\right)\right],$$

subject to

$$X_2^j = X_1^j + q_1^j S_1, \qquad X_1^j + q_1^j S_1 = X_0^j + q_0^j S_1.$$

Repeating the analysis of $t = 2$ at $t = 1$, we obtain that the optimal portfolio solution is

$$q_1^{j,*} = \frac{\mathbb{E}[S_2] - S_1}{\gamma \sigma^2},$$

for all agents, $j$, that are present: the $n$ MMs, and LT1. Also, at date $t = 1$ demand and supply for the asset have to be equal to each other, so that

$$n\, q_0^{MM} + q_0^{LT1} = n\, q_1^{MM} + q_1^{LT1},$$

where $q_0^{LT1} = i$ (recall that if $i > 0$, LT1 is holding $i$ shares he wants to sell), and $q_0^{MM} = 0$. This gives us the following equation:

$$i = (n+1)\frac{\mu - S_1}{\gamma\sigma^2} \iff S_1 = \mu - \gamma\sigma^2\frac{i}{n+1}.$$

The top half of Figure 2.1 reflects how the MMs and LT1 enter the market with asset holdings $q_0^j$ and after trading at $S_1$ they exit date one and enter date $t = 2$ with $q_1^j$.

With this expression we can interpret how the market reaches a solution for LT1's liquidity needs: LT1, a trader who wants to sell a total of $i > 0$ units at $t = 1$, finds that there is no one currently in the market with a balancing liquidity need. There are traders in the market, but they will not accept trading at the efficient price of $\mu$ because if they do, they will be taking on risky shares (they are exposed to the price risk from the realisation of $\epsilon_2$) without compensation. But, if they receive adequate compensation (which we call a liquidity discount, as for $i > 0$, $S_1 < \mathbb{E}[S_2] = \mu$), the $n$ MMs will accept the LT1's shares. However, LT1 is price-sensitive, so if he has to accept a discount on the shares, he will not sell all the $i$ shares at once. In equilibrium, both the $n$ MMs and LT1 end up holding $q_1^{j,*} = \frac{i}{n+1}$ units of the asset each, that is LT1 sells $\frac{n}{n+1}i$ units and holds on to $\frac{i}{n+1}$ units to be sold later. Trading occurs at a price below the efficient price, $S_1 = \mu - \gamma\sigma^2\frac{i}{n+1}$. The difference between the trading price and the efficient price, namely $|S_1 - \mu| = \left|\gamma\sigma^2\frac{i}{n+1}\right|$, represents the (liquidity) discount the MMs receive in order to hold LT1's shares. This size of the discount is influenced by the variables in the model: the size of the liquidity demand ($|i|$), the amount of competition amongst MMs (captured by $n$), the market's risk aversion ($\gamma$), and the risk/volatility of the underlying asset ($\sigma^2$). These variables all affect the discount in an intuitive way: the size of the liquidity shock, risk aversion, and volatility all increase the discount, while competition reduces it. This occurs when LT1 wants to sell, i.e. $i > 0$. If LT1 wanted to buy, $i < 0$, then the solution would be the same except that instead of a discount, the MMs would receive a premium equal to $|S_1 - \mu|$ per share when selling to LT1.

From this analysis we can also see that as competition ($n$) increases, the liquidity premium goes to zero, the price converges to the efficient level, $S_1 = \mu$, and LT1's optimal initial net trade, $q_1^{LT1,*} - q_0^{LT1}$, converges to his liquidity need ($i$).

## 2.1.2    Trading Costs

We have seen how the Grossman & Miller (1988) framework helps to understand how the cost of holding assets (in this case, via the uncertainty it generates to the risk-averse MMs) affects liquidity via the cost of trading ($|S_1 - \mu|$) and the demand for immediacy (as at $t = 1$ LT1 only executes $\frac{n}{n+1}i$ rather than her desired $i$). Also, competition between MMs is crucial in determining these trading costs. But what drives $n$? A natural answer is that $n$ is driven by the trading costs borne by the MMs. In this case, we must distinguish between participation costs, which are needed to be present in the market and do not depend on trading activity, and trading costs that do depend on trading activity, such as trading fees (which we ignored in the previous analysis).

Grossman & Miller (1988) link competition, $n$, with participation costs. They do this by introducing an earlier stage to the model in which potential MMs decide whether they want to actively participate in the market and provide liquidity or prefer to do something else. The decision is determined as a function of a participation cost parameter $c$ which proxies for the time and investments needed to keep a constant, active and competitive presence in the market, as well as the opportunity cost the MM gives up by being in the market and not doing something else. The conclusion, which can be obtained without going into the details of the analysis, is that the level of competition decreases monotonically with supplier's participation costs. Thus, participation costs, proxied by the cost parameter $c$, increase the size of the liquidity premium (via its effect on competition, $n$).

The parameter $c$ captures the fixed costs of participating in the market, but we could also consider introducing into the model a cost of trading that depends on the level of activity in the market. In particular, we introduce trading costs that depend on the quantity traded, like actual exchange trading fees. Exchange trading fees are usually proportional to dollar-volume but here, for simplicity, we use fees proportional to number of shares traded parameterised by $\eta$. Given that fees are known, these fees act like a participation cost for liquidity traders.

The first effect of having $\eta > 0$ is that liquidity traders with a desired trade ($|i|$) that is small relative to trading fees, will find trading too expensive and refrain from trading (we invite the interested reader to compute the minimum desired trade size $\hat{i}$ as a function of $\eta$). For sufficiently large desired trades (so that trading is preferred to not trading by all participants) the model gives us the following solution. Suppose every trader pays $\eta$ per share regardless of whether they are buying or selling the asset. To simplify, assume that any remaining inventories after $t = 2$ are liquidated at $t = 3$. Also, assume LT1 wants to sell $|i|$ units ($i > 0$), and LT2 wants to buy the same amount (the reverse case looks the same but the trading fees enter the problem with the opposite sign).

At $t = 2$, since the MMs and LT1 enter the period with a positive inventory (and will be wanting to sell now or at $t = 3$) their optimal final period holdings are

$$q_2^j = \frac{\mathbb{E}[S_3 - \eta \,|\, \epsilon_2] - (S_2 - \eta)}{\gamma \sigma^2}, \quad j \in \{MM, LT1\},$$

while the demand for shares by LT2 is

$$q_2^{LT2} = \frac{\mathbb{E}[S_3 + \eta \,|\, \epsilon_2] - (S_2 + \eta)}{\gamma \sigma^2}.$$

As everyone anticipates that their trading positions need to be liquidated anyway, the trading fees do not affect the price at $t = 2$, and we obtain $S_2 = \mathbb{E}[S_3 \,|\, \epsilon_2] = \mu + \epsilon_2$ (as before when there were no fees, $\eta = 0$).

At $t = 1$, LT1 has a similar position to that at $t = 2$, as any quantities he doesn't sell now he will have to sell later, so that $\eta$ disappears from the solution and his supply will be given by:

$$q_1^{LT1} = \frac{\mathbb{E}[S_2 - \eta] - (S_1 - \eta)}{\gamma\sigma^2}.$$

On the other hand, MMs anticipate that whatever they buy, they will have to sell later, which changes their asset demand functions to

$$q_1^{MM} = \frac{\mathbb{E}[S_2 - \eta] - (S_1 + \eta)}{\gamma\sigma^2}.$$

The resulting market equilibrium condition is now

$$i = n\, q_1^{MM} + q_1^{LT1} = \frac{\mu - S_1}{\gamma\sigma^2} + n\frac{\mu - S_1 - 2\eta}{\gamma\sigma^2}.$$

This gives us the following equation:

$$i = (n + 1)\frac{\mu - S_1}{\gamma\sigma^2} - \frac{2n\eta}{\gamma\sigma^2} \iff S_1 = \mu - \gamma\sigma^2\frac{i}{n + 1} - 2\left(\frac{n}{n + 1}\right)\eta,$$

and recall that for LT1, $i > 0$.

Thus, we conclude that the presence of trading fees introduces an extra liquidity discount to the initial price $S_1$. What the model tells us is that almost all the trading fees are paid by the liquidity trader initiating the transaction: he pays his own trading fee, $\eta$ per share, plus a substantial fraction $(n/(n + 1))$ of the two transaction fees paid by the MMs $(2\eta)$ though indirectly, via a lower sale price, a lower $S_1$. It also affects the immediacy he obtains from the market, as his holdings at the end of $t = 1$ are no longer $q_1^{LT1,*} = -i/(n + 1)$ but

$$q_1^{LT1,*} = \frac{i}{n + 1} + 2\left(\frac{n}{n + 1}\right)\frac{\eta}{\gamma\sigma^2}.$$

If we look at competition, we can see that participation costs and fees have very different effects. Participation costs enter directly through $c$ while trading fees enter through expected future profits, which will be lower as MMs must bear a fraction of the trading fees. In particular, for each trade, the MM pays $2\eta$, but recovers $2n/(n + 1)\eta$ through the liquidity discount. Therefore, an increase in trading fees has a smaller effect on liquidity via competition but a greater direct effect on immediacy and the liquidity discount.

## 2.1.3     Measuring Liquidity

We have seen how in the Grossman & Miller (1988) model, trading costs, whether setup costs or trading fees, are mostly paid by liquidity traders, whether explicitly (as their own trading fees) or implicitly in the price (greater liquidity discount when selling and larger premium when buying). We now consider how these divergences from 'efficient' prices may be observed in electronic exchanges.

The Grossman & Miller (1988) model avoids looking into the details of the trading mechanism by solving for equilibrium prices and demands in a 'Walrasian auctioneer'-type context where all trading takes place at once, and at a single price.[2] In electronic asset markets, decisions are not taken all at the same point in time, but the equilibrium analysis can be easily reinterpreted in the context of an electronic market. For example, suppose liquidity traders are very eager to trade and do so by sending MOs into the exchange. When the liquidity trader's orders hit the market, they meet the LOs that were posted by the patient MMs and are resting in the limit order book (LOB).

Then, the Grossman–Miller model would correspond to the following sequence of events: as LT1's MOs enter the market, they execute against LOs in the LOB which adjusts to the incoming MO. As the execution price changes, so does LT1's strategy and eventually, after selling $i\frac{n}{n+1}$ shares, the price has moved too far and LT1 stops trading. Overall, LT1's market order executes at the average price of $S_1$, either because it was sent as a large order that walked the LOB (or LOBs, if the order is routed to multiple markets), or because it was split up into several small orders that triggered a gradual move of the bid side in the LOB away from the initial starting point. Then, the discount received by LT1 is the difference between the average price received, $S_1$, and the initial midprice when the first MO hit the market (which is the usual proxy for the efficient price, $\mathbb{E}[S_2]$).

We can rewrite $S_1$ as a linear function of the quantity traded, $q^{LT1}$:

$$S_1 = \mu + \lambda q^{LT1},$$

so that in the Grossman–Miller model we would have

$$\lambda = -\frac{1}{n}\gamma\sigma^2, \qquad \text{and} \qquad q^{LT1} = i\,\frac{n}{n+1}.$$

The $\lambda$ parameter captures the market's price reaction to LT1's total order, its price impact. The notion of price impact is very important both for trading and for theoretical work, and the use of a linear structure such as the one described by the parameter $\lambda$ is very common. In particular, $\lambda$ is used to describe the liquidity of the market for this asset – a more liquid market will have a lower $\lambda$, either because of greater competition ($n$), lower risk tolerance ($\gamma$), or lower volatility ($\sigma^2$), and this results in a lower liquidity discount/premium for liquidity traders.

There is a second popular way to measure liquidity based on price changes, and it is quite easy to see how this model works. The measure is based on the autocovariance of the asset's return, though for the Grossman–Miller model it is easier to describe it when looking at the autocovariance in asset price changes rather than returns. To see how this measure is constructed, let us introduce an additional date $t = 0$ prior to LT1's order submission ($t = 1$), and a random public news event, $\epsilon_1$, that affects the asset's final liquidation price, $S_3 = \mu +$

---

[2] The notion of a Walrasian auctioneer comes from the work of Léon Walras who describes the prices that arise under perfect competition as the result of a simultaneous auction in which supply is equated to demand. The Walrasian auctioneer is the abstract manager of this auction.

$\epsilon_1 + \epsilon_2 + \epsilon_3$. The public news is announced between $t = 0$ and $t = 1$. Define the following constants:

$$\mu_0 = \mathbb{E}[S_3], \qquad \mu_1 = \mathbb{E}[S_3 \,|\, \epsilon_1], \qquad \mu_2 = \mathbb{E}[S_3 \,|\, \epsilon_1, \epsilon_2] \qquad \mu_3 = S_3,$$

and let $\epsilon_t$, $t = 1, 2, 3$ be normal, i.i.d. random variables with mean zero and variance $\sigma^2$. The discrete process $\mu_t$ is a martingale, and we refer to it as the efficient market price.

According to the model above, at $t = 0$ there are no liquidity traders and no trade so that $S_0 = \mathbb{E}[S_3] = \mu_0$ will be the equilibrium price. The model shows that the subsequent equilibrium prices at $t = 1$ and $t = 2$ are:

$$S_1 = \mu_1 + \lambda \, q^{LT1}, \qquad \text{and} \qquad S_2 = \mu_2 \,.$$

To construct the autocovariance of price changes, let $\Delta_1 = S_1 - S_0$ and $\Delta_2 = S_2 - S_1$, and the autocovariance of price changes be given by the following expression:

$$\begin{aligned} \text{Cov}\,[\Delta_1, \Delta_2] &= \text{Cov}\left[\mu_1 + \lambda \, q^{LT1} - \mu_0, \mu_2 - \mu_1 - \lambda \, q^{LT1}\right] \\ &= \text{Cov}\left[\epsilon_1 + \lambda \, q^{LT1}, \epsilon_2 - \lambda \, q^{LT1}\right] = -\lambda^2 \, \text{Var}\left[q^{LT1}\right] < 0 \,. \end{aligned}$$

In this simple (essentially static) model, where all the action takes place at $t = 1$, the autocovariance of price changes captures market liquidity just like price impact does. An interesting effect that we see here is that as liquidity increases and $\lambda$ goes to zero, so the autocovariance of price changes, and the price process converges to the underlying ('efficient price') martingale process $\mu_t$.

The two measures, price impact and the autocovariance of price changes (or returns), become distinct in richer dynamic settings, and capture different dimensions of the market's reaction to incoming MOs. For example, in the continuous-time models of later chapters, the average growth of the efficient price is affected by the rate at which MOs arrive to the market and this effect decays at an exponential rate. This permanent effect of the efficient price of the asset affects all market participants and is different from the temporary effect that each trader sees in their execution prices, which is captured by the parameter ($\lambda$) and does not affect the dynamics of the efficient price.

### 2.1.4     Market Making using Limit Orders

In the transition from the Walrasian auctioneer in the Grossman–Miller model to the measurement of price impact, we have proposed that MMs participate through the posting of LOs. We now consider why an MM would behave in this way and the simplest solution to how she does it.

The usual first reference for this is the model of Ho & Stoll (1981), but working with the original model requires familiarity with the techniques for solving stochastic dynamic programming problems which we see in Part II. Instead, we set up a static version of the model that captures some of the basic elements of the MM's problem. As in the Grossman & Miller (1988) model, the MM is a professional trader who profits from intermediating between different liquidity traders. In this case, we consider a small risk-neutral trader with costless inventory management and infinite patience. She does not require compensation for her services, but makes a profit from optimally choosing how to provide liquidity in an uncertain environment populated by other MMs who do not react to our MM's decisions.

Uncertainty in this context comes from the timing and size of large incoming MOs, and there are no information problems: all information is public so that everyone agrees what the current value of the asset is, which we denote by $S_t$ and refer to as the midprice. Our trader is one of many MMs. We take other MMs' behaviour as given, and this behaviour is represented by a fixed LOB, unaffected by our MM's decisions. Our MM makes money by adding her LOs to the book and clearing the resulting inventory at later dates. Because our MM has no inventory costs, incurs no trading costs, is risk-neutral and infinitely patient, we can assume that she liquidates her inventory at the midprice at no cost.

Then, the MM's problem is to choose where on the LOB to place her LOs so as to maximise her profit per trade, optimally balancing the increase in the price per trade received as she increases the distance of the LO from the midprice, with the frequency with which she will trade, which decreases with that distance from the midprice. Formally, the MM's problem is to choose the distance from the midprice, the depths $\delta^{\pm}$. Then, she will post her sell LO at $S_t + \delta^+$ and her buy LO at $S_t - \delta^-$. The uncertainty from MOs comes from the probability that an MO arrives ($p_{\pm}$) and the probability that once it arrives it walks the book up to where the MM's LOs are resting ($\delta^{\pm}$ away from the midprice), which is described by the cdf $P_{\pm}$. Thus, the probability that the buy LO will be filled is $p_- P_-(\delta^-)$. If we assume that the distribution of other LOs in the LOB is described by an exponential distribution with parameter $\kappa^-$, we have $p_- P_-(\delta^-) = p_- e^{-\kappa^- \delta^-}$. Similarly, the probability that the sell LO is filled is $p_+ e^{-\kappa^+ \delta^+}$. Clearly, as the MM posts her LOs deeper in the LOB, the probability that her order (once an MO arrives) decreases, though her profit per trade ($\delta^{\pm}$) increases.

The left panel of Figure 2.2 illustrates a hypothetical LOB around a midprice of $S_t$ and two possible limit orders: a sell LO on the ask side at $S_t + \delta^+$, and a buy LO on the bid side at $S_t - \delta^-$. The right panel describes the corresponding probability distribution, $P^+$ ($P^-$), of execution of the order posted at a distance $\delta^+$ ($\delta^-$) from the midprice, conditional on the arrival of a buy (sell) MO.

Using $\Pi$ to denote the MM's profit per trade, the MM's optimisation problem is given by the following expression:

$$\max_{\delta^+, \delta^-} \mathbb{E}\left[\Pi(\delta^+, \delta^-)\right] = \max_{\delta^+, \delta^-} \left\{ p^+ e^{-\kappa^+ \delta^+} \delta^+ + p^- e^{-\kappa^- \delta^-} \delta^- \right\}. \qquad (2.3)$$

It is straightforward to see that the solution is to post LOs at the following depths:

$$\delta^{\pm,*} = \frac{1}{\kappa^{\pm}}.$$

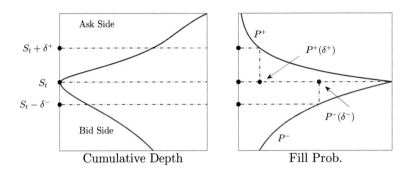

**Figure 2.2** The LOB and the probability of execution.

Given our parametric choice of $P_{\pm}$, the optimal depth is equal to the mean depth in the LOB.

This model captures in a simple way the trade-off between the probability of execution and margin per trade. But, it is very unrealistic in several dimensions: the functional form of all the stochastic components of the model ($P_{\pm}$, and $p^{\pm}$) is very special, constant and exogenous, the MM's decision and that of other traders (as captured by $P(\delta)$) are independent, the MM's objective function is static and very simple. However to address these other issues we need more sophisticated methods and models, so after developing those methods in the following chapters we will revisit some of them. For instance, in Chapter 10 we see how MMs decide how to post limit orders in a fully-fledged dynamic inventory model and how she adjusts her posts if trading with better informed counterparties – a topic that we discuss next.

## 2.2     Trading on an Informational Advantage

So far we have side-stepped one of the main issues in trading: informational differences. Many trades originate not because someone needs cash and sells an asset, or has extra cash and wants to invest, but because one party has (or believes she has) better information about what the price is going to do than is reflected in current prices. So, having seen the basic market making models in the context of public information we turn to the next fundamental issue: how to exploit an informational advantage while taking into account one's price impact. The primary reference in this case is Kyle (1985).

Kyle (1985) looks at the decision problem of a trader who has a strong informational advantage (the case of several competing informed traders is studied in Kyle (1989)) in a context where the price is 'efficient'. The model in Kyle (1985) tells us how the informed trader optimally adjusts his trading strategy to take into account the market reaction, and in particular, the price impact that his trades generate in equilibrium.

To get into the details of the model we first need to define what we mean by 'a strong informational advantage' and price efficiency in this context. To keep things simple we only consider the investor's static decision problem. The same basic idea extends to a dynamic setting. The formal static model is as follows: there is a market for an asset that opens at one point in time. The asset is traded at price $S$, and after trading, the asset has a cash value equal to $v$. The future cash value of the asset, $v$, is uncertain. In particular, $v$ is assumed to be normally distributed with mean $\mu$ and variance $\sigma^2$. In the market, there are three types of traders: an informed trader, an anonymous mass of price-insensitive liquidity traders (traders who need to execute trades whatever the cost), and a large number of MMs that observe and compete for the order flow – that is, the MMs observe and compete for the flow of incoming buy and sell orders from the informed and the liquidity traders.

In contrast to the Grossman & Miller (1988) setting, MMs are risk-neutral, so they do not need a liquidity premium to compensate for the price risk from holding inventory. Therefore, any liquidity premium that arises will come from the need to compensate MMs for their informational disadvantage – and which will be borne by the price-insensitive liquidity traders. These liquidity traders will have, in aggregate, a net demand represented by the random quantity $u$, such that if $u > 0$, on aggregate liquidity traders want to buy $u$ units, while if $u < 0$, these traders want to sell $|u|$ units of the asset. Assume that $u$ is normally distributed with mean zero, variance $\sigma_u^2$, and is independent of $v$. In principle, as liquidity traders are not sensitive to the price ($u$ does not depend on $S$) MMs could charge very large liquidity premia, but competition for order flow between MMs drives the liquidity premium to zero, so that (when there are only MMs and liquidity traders) $S = \mathbb{E}[v]$.

Now consider the possibility that a new trader enters the market, and that this trader (the "insider") knows the exact value of $v$. The insider is the only one who knows $v$ and chooses how much to trade. Let $x(v)$ denote the number of shares traded by the insider. MMs, on the other hand, know that there is an informed trader in the market, but do not know who this trader is.

To make the analysis formal, the model is structured as follows: (i) the insider observes $v$, (ii) on observing $v$ the insider chooses $x(v)$, (iii) $u$ is realised, (iv) the MMs observe the net order flow, $x(v) + u$, (v) based on the net order flow MMs compete to set the asset price, $S$.

To solve the model we use the solution concept of (Bayesian) Nash equilibrium; without going into all the details, this means that all agents optimise given the decisions of all other players, according to their beliefs (which are updated according to Bayes' rule whenever possible). Thus, we require that in equilibrium the insider chooses $x(v)$ to maximise his expected profit, taking into account the dynamics of the game (i.e. that his order will be mixed in with those of the liquidity traders), and anticipating that MMs will set their prices on the basis of what they learn from observing the order flow and what they know about the informed trader's decision problem. Also, we require that MMs choose their prices taking into account the strategy of the insider (in particular, they anticipate the functional form of $x(v)$) and the properties of the uninformed order flow that comes from liquidity traders. In particular, MMs set the market price as a function of net order flow, $S(x + u)$. This is important, as the model naturally tells us that prices are affected by the order flow, so that trading automatically generates a price impact – the average price per unit traded, $S$, moves with the net order flow, $x + u$. We need to look at the equilibrium of the model to see what that price impact function looks like. Nevertheless, in equilibrium, the insider will anticipate the functional form of $S(x + u)$, that is, she will incorporate price impact when choosing $x(v)$.[3] The equilibrium is a fixed point in the optimisation of $x$ given the functional form of $S$, and of $S$ given the functional form of $x$.

Consider what the insider should do. The most natural response is: sell if $v < \mathbb{E}[v] = \mu$ and buy if $v > \mu$, and whether selling or buying, do so as much as possible to leverage his informational advantage. This seems natural, but we must take into account that MMs will adjust their prices to the order flow they observe. Hence, even if $v < \mu$, the insider cannot expect $S = \mu$. In the extreme case where there are no liquidity traders everyone knows that any trade comes from the insider and so the MMs, anticipating the demand as a function of the realisation of $v$, behave optimally and set prices that incorporate all information on $v$ in $x(v)$. Fortunately for the insider, there are liquidity traders that add noise into order flow and allow the insider to camouflage his trade to gain positive expected profits.

So, how do MMs set their prices? The first thing to note is that as MMs compete for order flow, any profits they could extract are competed away. Thus, whatever the price strategy, it will lead to zero expected profits for our (risk-neutral) MMs – though never negative profits as they can always choose not to trade. The zero (expected) profit condition forces prices to have a very specific property: $S = \mathbb{E}[v \mid \mathcal{F}]$, where $\mathcal{F}$ represents all information available to MMs. This property is known as **semi-strong efficiency**: prices reflect all publicly available information (which in our case is order flow which is all the information MMs have).[4] This is why we can readily identify a fundamental property of the MMs' equilibrium strategy:

$$S(x + u) = \mathbb{E}[v \mid x + u].$$

To solve the model we need to find an $x(v)$ that is optimal, i.e. it maximises the insider's expected trading profits, conditional on this pricing rule. Because of the normality of $v$ and $u$, we hypothesise that $S(x + u)$ is linear in net order flow. In particular, let

$$S(x + u) = \mu + \lambda (x + u),$$

---

[3] Formally, liquidity traders are substituted by a "nature" player that executes the random demand $u$.

[4] The notion of price efficiency was introduced by the recent Nobel Laureate, Eugene Fama, see Fama (1970).

where $\lambda$ is an unknown parameter representing the linear sensitivity of the market price to order flow.

Taking this particular functional form as given, consider the insider's problem:

$$\max_x \mathbb{E}\left[x\left(v - S(x + u)\right)\right].$$

Substituting for $S(x+u) = \mu + \lambda(x+u)$ and taking expectations with respect to $u$, we obtain that the objective function is concave and the first-order condition yields

$$x^*(v) = \beta(v - \mu),$$

where $\beta = (2\lambda)^{-1}$.

Because we have hypothesised the functional form of the price function, we must now confirm that the functional form is consistent with the optimal $x(v)$ and at the same time we can characterise $\lambda$. We know that $S = \mathbb{E}[v \mid x + u]$. From the optimal $x$, we know that

$$x + u = \beta(v - \mu) + u = \beta\mu + \beta v - u.$$

As $v$ and $u$ are independent and normal, $x+u$ is normal with mean $\mu(1+\beta)$ and variance $\beta^2\sigma^2 + \sigma_u^2$. We can now compute the joint distribution of $v$ and $x + u$, and from it we can derive $S = \mathbb{E}[v \mid x + u]$, which (using the projection theorem for normal random variables and simplifying) is given by

$$S = \mu + 2(x + u)\frac{\sigma_u}{\sigma},$$

so that the linear sensitivity parameter is $\lambda = 2\sigma_u/\sigma$. This confirms that the hypothesised equilibrium is indeed an equilibrium (for a formal proof, see Kyle (1985)).

Even within the simple, static version of the Kyle model we can clearly see the issues that arise when facing informed trading (also referred to as "toxic order flow"). While in the previous models MMs just needed a liquidity premium (discount) to cover the expected cost from future price uncertainty, the presence of informed traders implies that MMs will be adversely selected, buying when informed traders know it would be better to sell and selling when it would be better to buy. This adverse selection requires a higher premium borne by other (more impatient liquidity) traders. In this model, the additional premium takes the form of price adjustment to order flow (price impact) as described by Kyle's lambda (the $\lambda$ parameter we have just derived). This premium accounts not for the risk that future price movements will be random, as described in Section 2.1.1, but for the adverse selection faced by MMs, as prices will on average move *against* the MMs' position because they trade with better informed traders in the market. The sign of $\lambda$ will be the same as in Grossman & Miller (1988): prices move with the order flow, increasing as buy MOs hit the market and falling as traders sell aggressively.

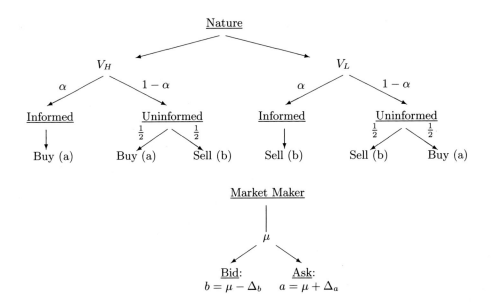

**Figure 2.3** The Glosten-Milgrom model.

## 2.3    Market Making with an Informational Disadvantage

The Kyle model focuses on the informed trader's problem, while using competition to characterise the MM's decisions. As we are very interested in the MM's problem, we now turn to Glosten & Milgrom (1985) for a model that puts the MM at the centre of the problem of trading with counterparties who have superior information.

Again, we look at a simplified and (essentially) static version of the model that allows us to capture the nature of the MM's decision problem. The situation is as before: there are liquidity traders, informed traders, and a competitive group of MMs. The MM is risk-neutral and has no explicit costs from carrying inventory.

Our simple model (described in Figure 2.3) has a future cash value of the asset equal to $v$ which we limit to two possible values: $V_H > V_L$, that is a High, and a Low value. The unconditional probability of $v = V_H$ is $p$. All orders are of one unit, MMs post an LO to sell one unit at price $a$, and a buy LO for one unit at price $b$. We start by assuming that liquidity traders are price insensitive and want to buy with probability $1/2$ and want to sell with probability $1/2$. There are many informed traders, all of whom know $v$ but are limited to trade a single unit, which simplifies their decision: when $v = V_H$ they buy one unit if $a < V_H$, and do nothing otherwise, while when $v = V_L$ they sell one unit if $b > V_L$ and do nothing otherwise. The total population of liquidity and informed traders is normalised to one, and of these, a proportion $\alpha$ are informed and a proportion $(1 - \alpha)$ are uninformed liquidity traders.

Figure 2.3 captures the probabilistic structure of the model: Nature randomly determines whether the underlying state is $V_H$ or $V_L$. Independently of the state, a trader is picked at random from the population, so that with probability $\alpha$ she is informed, and with probability $1 - \alpha$ she is uninformed. An informed trader will always buy at the ask price ($a$) when the asset's value is $V_H$ and sell at the bid ($b$) when the asset's value is $V_L$, while an uninformed trader will buy or sell with equal probability, independent of the true (unknown) value of the asset.

The MM's problem is to choose $a$ and $b$ in this setting. Because liquidity traders are price-insensitive, the optimal solution is trivial: set $a = V_H$ and $b = V_L$, but since MMs compete for business, prices will be set to their (semi-strong) efficient levels – again, this happens because MMs use only public information, *which includes order flow*. Were the MMs to have private information in addition to the order flow, in this setting competition for order flow would incorporate some of that information into prices.

Competition between MMs drives their expected profits to zero. Hence, $a$ and $b$ are determined by the no-profit condition. Rather than solve for $a$ and $b$ directly, define the ask- and bid-halfspreads, $\Delta_a$ and $\Delta_b$ respectively. The sum of the two, $\Delta_a + \Delta_b$, represents the (quoted) spread. Let the expected value of the asset $\mu = \mathbb{E}[v \mid \mathcal{F}]$ where $\mathcal{F}$ represents all public information prior to trading. Then, as described at the bottom of Figure 2.3, MMs will choose $a = \mu + \Delta_a$ and $b = \mu - \Delta_b$ optimally. To determine the effect of choosing $a$ and $b$ on the expected profit and loss, consider what happens when a buy order comes in:

- if it comes from an uninformed liquidity trader she makes an expected profit of $a - \mu = \Delta_a$,
- if it comes from an informed trader she makes an expected loss of $a - V_H = \Delta_a - (V_H - \mu)$.

From the point of view of the MM, the probability that a liquidity trader wants to buy is $1/2$, while the probability that an informed trader wants to buy is $p$ (as all informed traders will buy if the state is $v = V_H$ which occurs with probability $p$). As there are $1 - \alpha$ liquidity traders and $\alpha$ informed ones, the expected profit from posting a price $a = \mu + \Delta_a$ is

$$\frac{(1-\alpha)/2}{\alpha p + (1-\alpha)/2} \Delta_a + \frac{p\alpha}{\alpha p + (1-\alpha)/2} \left( \Delta_a - (V_H - \mu) \right).$$

Setting this expected profit to zero we obtain

$$\Delta_a = \frac{\alpha p}{\alpha p + (1-\alpha)/2} (V_H - \mu) = \frac{1}{1 + \frac{1-\alpha}{\alpha} \frac{1/2}{p}} (V_H - \mu),$$

and following similar reasoning,

$$\Delta_b = \frac{1}{1 + \frac{1-\alpha}{\alpha} \frac{1/2}{1-p}} (\mu - V_L).$$

To interpret these equations let us label the variables. If we think of asymmetric information as 'toxicity' then we can think of $\alpha$ as the prevalence of toxicity, $1 - p$

and $V_H - \mu$ as the magnitude of buy-toxicity and $1 - p$ and $\mu - V_L$ that of sell-toxicity. Then, the equations above describe how MMs adjust the ask-half-spread and the bid-half-spread, and increase it with the prevalence and magnitude of buy- and sell-toxicity.

In later chapters we show how trading algorithms are built to either take advantage of informational advantages or to adjust the depth at which LOs are posted so as to recover losses from trading agents to more informed traders. For example, in Section 7.3 we develop trading algorithms that use the information provided in the order flow to adjust acquisition or liquidation rates when the agent seeks to enter or exit a large position. We also show how the strategy of the MM depends on whether she knows detailed high-frequency information about short-term deviations in the drift of the asset she is trading, see for example Section 10.4.2.

## 2.3.1    Price Dynamics

This simple model can be extended in two different and complementary ways: by incorporating a time dimension and by making liquidity traders price-sensitive. The former is straightforward. In order to avoid having to keep track of the interest rate, set it equal to zero. Then index all variables by time $t$ and set the time of the determination of the cash value of the asset to $T$. Moreover, ensure that probabilities and expectations are adjusted to incorporate the accumulation of public information from trade, as captured by the filtration $\mathcal{F}_t$. As MMs observe different sequences of buy and sell orders they adjust (using Bayes' rule) the estimation of the distribution of $v$, and in particular they set $p_t = \mathbb{P}(v = V_H \,|\, \mathcal{F}_t)$, and $\mu_t = \mathbb{E}[v \,|\, \mathcal{F}_t]$. Then, bid and ask prices will adjust in response to the history of trading, so that

$$a_t = \mu_t + \Delta_{a,t} = \mu_t + \frac{1}{1 + \frac{1-\alpha}{\alpha}\frac{1/2}{p_t}}\left(V_H - \mu_t\right),$$

and

$$b_t = \mu_t - \Delta_{b,t} = \mu_t - \frac{1}{1 + \frac{1-\alpha}{\alpha}\frac{1/2}{1-p_t}}\left(\mu_t - V_L\right).$$

The resulting bid-ask prices display dynamic changes that reflect the public information embedded in the order flow. Note also that at every execution, the execution price ($a_t$ if it is the execution of a market buy order, and $b_t$ for a sell) is equal to the expectation of the underlying asset conditional on the history of order flow, $\mathcal{F}_t$, and also on the information in the execution (that is a buy or a sell). Hence, it can be seen that the realised price process (the price process at execution times) is a martingale (with respect to the objective measure).

2.3.2   Price Sensitive Liquidity Traders

An interesting extension of the static model (which can be further extended to include the dynamics we have just seen) is to allow liquidity traders to avoid trading if the half-spread, $\Delta$, is too high. A direct way to do this is to assume that liquidity traders get an additional (exogenous) value from executing their desired trade, so that trader $i$ gets a cash equivalent utility gain of $c_i$ if he manages to execute his desired trade. Thus, if the transaction cost imposed by the half-spread is too high, higher than $c_i$, trader $i$ will prefer not to execute his trade. Assume that the distribution of the parameter $c_i$ in the population of liquidity traders is described by the cumulative distribution function $F$, such that $F(c)$ is the proportion of liquidity traders that have $c_i \leq c$. We refer to $c_i$ as the agent's urgency parameter.

Then, we can recompute the expected profit of the MM from setting an ask price $a = \mu + \Delta_a$ as above, which will now be given by

$$\frac{(1 - F(\Delta_a))(1 - \alpha)/2}{\alpha p + (1 - F(\Delta_a))(1 - \alpha)/2}\Delta_a + \frac{p\alpha}{\alpha p + (1 - F(\Delta_a))(1 - \alpha)/2}(\Delta_a - (V_H - \mu)).$$

In this expression we have incorporated the fact that whenever a liquidity trader wants to buy $(1-\alpha)/2$, only $1 - F(\Delta_a)$ will have sufficient urgency to execute the trade with a buy-half-spread equal to $\Delta_a$. Introducing this parameter increases the half-spreads, which are now implicitly defined by the following expressions:

$$\Delta_a = \frac{1}{1 + \frac{1-\alpha}{\alpha}\frac{(1-F(\Delta_a))/2}{p}}(V_H - \mu),$$

and following similar reasoning,

$$\Delta_b = \frac{1}{1 + \frac{1-\alpha}{\alpha}\frac{(1-F(\Delta_a))/2}{1-p}}(\mu - V_L).$$

A key issue now is that as the MM increases the halfspread, she faces a smaller population of liquidity traders. If the urgency parameters in the population are relatively small, the MM may find that the above expressions have only the extreme solutions $\Delta_a = V_H - \mu$ and $\Delta_b = \mu - V_L$.[5] These extreme solutions correspond to the solutions without liquidity traders and represent market collapse. With those spreads no one gains anything from trade, and any trade that may occur will come from the informed agents who are indifferent to either trading or not trading – though any trade will immediately reveal the underlying value of the asset and the price will be strong-efficient.

## 2.4   Bibliography and Selected Readings

Grossman (1976), Grossman (1977), Grossman (1978), Ho & Stoll (1981), Gross-

---

[5] By small urgency parameters we mean that no one has an urgency parameter higher than the expected value of the asset, that is, there exists $\epsilon > 0$, such that $F(\mu - \epsilon) = 1$.

man & Miller (1988), Glosten & Milgrom (1985), Kyle (1985), Kyle (1989), de Jong & Rindi (2009), O'Hara (1995), Abergel et al. (2012), Easley, López de Prado & O'Hara (2012), Vayanos & Wang (2009), SEC (2013*b*), SEC (2013*a*), O'Hara, Yao & Ye (2014), Foucault, Kadan & Kandel (Winter 2005), Rosu (2009), Easley, Engle, O'Hara & Wu (2008), Easley & O'Hara (1992), Biais, Glosten & Spatt (2005), Cartea & Penalva (2012), Boehmer, Fong & Wu (2014), Pascual & Veredas (2009), Martinez & Rosu (2013), Martinez & Rosu (2014), Hoffmann (2014), Cvitanic & Kirilenko (2010), Vives (1996), Colliard & Foucault (2012), Foucault & Menkveld (2008), Gerig (2008), Farmer, Gerig, Lillo & Waelbroeck (2013), Gerig & Michayluk (2010), Cohen & Szpruch (2012), Jarrow & Li (2013), Moallemi & Saglam (2013).

# 3    Empirical and Statistical Evidence: Prices and Returns

## 3.1    Introduction

The next two chapters contain empirical analysis of different aspects of trading: prices, returns, spreads, volume, etc., using primarily millisecond stamped data, though we start with daily data that will give us a general overview of the main issues. Chapter 3 focuses on prices and returns, while Chapter 4 is dedicated to volume and market quality measures such as spreads, volatility, or depth.

This chapter, first looks at millisecond data. We then turn to look at the properties of returns both at the daily and at much shorter (one second) time intervals, as well as looking at the interarrival times of price changes. Section 3.4 looks at how market conditions may change when facing latency, as well as the issue of tick size. This is followed by a discussion on price dynamics. Section 3.6 provides a glimpse of the issue of market fragmentation in the US, while the last section provides a first look at the empirics of pairs trading.

In addition to the empirical analysis, we also include plausible interpretations and speculation as to what could be behind some of the results of that analysis. These speculations are included to make the chapter more engaging and to encourage the reader to think about the results. However, they should not be interpreted as anything other than speculative theorising, and should be kept separate from the descriptive analysis of the empirical facts that is limited in scope to the data sample we are using.

### 3.1.1    The Data

We use data from several sources. For daily and monthly data we use publicly available aggregated data from Yahoo! Finance, and data from the Center for Research in Security Prices (CRSP). We also use millisecond timestamped **ITCH** data (publicly available industry standard data, similar to the direct data feed, recently timestamps go to nanosecond resolution). Our data have been converted into table format for easier processing and is in binary for speed and storage reasons. For illustration purposes we convert these to more human-readable form. The data are made up of the following fields (we drop two fields that are irrelevant here):

- Timestamp: number of milliseconds after midnight

- Order ID: Unique order ID
- Message Type:
  - "B" – Add buy order
  - "S" – Add sell order
  - "E" – Execute outstanding order in part
  - "C" – Cancel outstanding order in part
  - "F" – Execute outstanding order in full
  - "D" – Delete outstanding order in full
  - "X" – Bulk volume for the cross event
  - "T" – Execute non-displayed order
- Shares: order quantity (Zero for "F" and "D" messages)
- Price: zero for cancellations and executions
- Ticker : the ticker associated with the asset in question
- MPID: Market Participant ID associated with the transaction[1]
- Exchange: ID of the current market (NASDAQ = 1)

These messages record events that affect the limit order book (LOB), so essentially, they capture what happens to limit orders (LOs). LOs are posted (B,S) and later on they are cancelled (C,D) or executed (E,F). So, market orders (MOs) are not recorded but must be deduced from observing how they are executed against standing LOs (or against non-displayed/hidden orders, T).

Consider the following example (the row numbers have been added to facilitate the discussion and we have dropped the MPID column):

```
1: 33219784  4889087  B  1900  345800  TZA   1
2: 33219784  4887036  C   200       0  FMS   1
3: 33219784  4879129  D     0       0  QQQQ  1
4: 33219784  4889088  S  2000  454800  QQQQ  1
5: 33219784  4879130  D     0       0  QQQQ  1
6: 33219784  4889089  S   500  454800  QQQQ  1
7: 33219785  4882599  D     0       0  QQQQ  1
8: 33219785  4888889  F     0       0  STD   1
```

These messages are sent to the market between 33219784 and 33219785 ms from midnight (July 13th, 2010), that is between 09:13:39.784 and 09:13:39.785. We see several messages for the ETF QQQQ, and one each for the ETF TZA, and the stocks FMS and STD (STD has since changed its ticker to SAN).

The first line is for the TZA ETF and should be read as follows: message recorded at 33219784 ms from midnight (09:13:39.784), with order ID number 4889087, the LO is a posted LO to buy (B) for $1,900$ shares at a price of $34.58 (all prices are in dollars $\times$ $10,000$). The number 1 in the final column represents the market code for NASDAQ.

For QQQQ we observe an LO being cancelled (row 3), followed by the posting of a sell LO (4), another LO cancellation (5), a second sell LO posted (6)

---

[1] This information is usually missing from the public feed.

and a third LO being cancelled (7). The posted sell orders include the quantity and price for the orders (2,000 at $45.48 and 500 also at $45.48), while cancelled orders must be matched with their original posted orders (ID 4879130 and 4882599) in order to identify the corresponding prices and quantities. We see the same pattern for the full execution of order ID 4888889 for STD (8) – i.e. no price or quantity – while for FMS (2) we see a partial cancellation of 200 units from order ID 4887036 (the price needs to be read off the original posted order).

From this data, one can reconstruct the complete order book at any point in time, and study how the market changes over time using different variables and methods. We now proceed to give a brief overview of some of the features we observe.

## 3.1.2    Daily Asset Prices and Returns

When trading, the first variable of interest is the price level. If we have to acquire/liquidate a position we want to know what price we can get if we aggressively execute it, and if we are providing liquidity we want to know at what prices shares are being bought and sold.

As we discussed in Section 1.2, each investor is in the market to meet some objective, and will participate for as long as she feels that she is not losing too much money in pursuit of her objective (e.g., if the transaction costs do not consume the expected price gains, or if the market will adjust prices in reaction to her order, eliminating the original mispricing she wanted to profit from – we discuss these below). The observed price process is the outcome of the interaction between these investors. In electronic markets, we see these prices continuously as traders change their positions to meet their objectives in response to changes in market conditions and information flows. Market efficiency theories tell us that the resulting price process is not predictable and any positive expected return you can predict, is there as compensation for bearing risk. Thus, long-term investors receive a compensation for risk, be this market risk, risk from monetary policy changes, or just compensation for future price fluctuations and dividend uncertainty. Liquidity providers also require compensation: they require compensation for leaving offers at the bid and ask, and will continue to post orders while their trades are sufficiently profitable. Other traders pursue strategies aimed at exploiting deviations from market efficiency, such as keeping prices of similar assets close to each other.

Whether one believes in market efficiency or not, the properties of the price process are amenable to analysis and in this chapter we look at some of the methods and results obtained from detailed message data for specific assets.

We analyse the properties of the price process for a selection of assets from equity markets. Our primary focus is on 2013 prices for AAPL (Apple Inc.), as representative of a highly liquid, very highly traded asset. To illustrate differences across assets, we look at three other assets with tickers ISNS, FARO and MENT: ISNS is the company Image Sensing Systems, Inc. Industry (Application Software); FARO is FARO Technologies Inc. (Scientific & Technical Instruments); and MENT is Mentor Graphics Corp. Industry (Technical & System Software). These assets are all in the technology sector and represent different levels of trading activity (although depending on your definition, you can argue about whether AAPL is a technology or a consumer goods firm).

### 3.1.3    Daily Trading Activity

In Table 3.1 we can see different measures of trading activity for these assets: average number of transactions per day on NASDAQ ($N$), average total daily dollar value of shares traded on NASDAQ ($V(\$)$, in 000s), average number of shares traded daily on NASDAQ ($V(Q)$, in 000s), total average number of shares traded in all markets (Total $V(Q)$, in 000s), and **share turnover** (Turnover). Share turnover represents the total number of shares traded during 2013 divided by the number of **outstanding shares** –also included in Table 3.1 (ShrOut, in millions, as of Dec 30th, 2012). From the column with the number of transactions ($N$) and using the fact that the regular market is open for 6.5 hours (from 9:30 to 16:00) we can conclude that ISNS is a very rarely traded asset (traded about once every half hour in 2013), while FARO and MENT are regular small assets (with on average 1 to 3 trades per minute in 2013), and AAPL is one of the most highly traded equity stocks (around 1 trade per second – note that we are using 2013 data, and these numbers are not rescaled to account for the AAPL June 2, 2014 7-for-1 split).

| Asset | $N$ | $V$ ($\$$) $(\times10^3)$ | $V(Q)$ $(\times10^3)$ | Total $V(Q)$ $(\times10^3)$ | ShrOut $(\times10^6)$ | Turnover |
|-------|-----|------|------|------|------|------|
| ISNS | 14 | 18 | 3 | 12 | 5 | 0.62 |
| FARO | 315 | 1,396 | 34 | 137 | 17 | 2.04 |
| MENT | 908 | 3,964 | 204 | 694 | 112 | 1.56 |
| AAPL | 24,582 | 1,505,175 | 3,208 | 14,516 | 941 | 3.89 |

**Table 3.1** Daily Average Volume in 2013 for selected assets.

This pattern is repeated regardless of which measures of volume in Table 3.1 you look at, and whether measured only for the NASDAQ market or for all markets together.

### 3.1.4    Daily Price Predictability

We first look at the properties of the price process by considering returns constructed from changes in prices from market open to market close for each day in 2013. According to the **efficient market hypothesis**, daily asset returns should be close to unpredictable and reflect information in the market. To investigate this we run ordinary least squares (**OLS**) regressions for intra-day (market open to market close) returns for our four assets. We include a number of variables related to market efficiency and market forces as follows.

The first of these variables is the return on the SPY: the SPY is an **exchange-traded fund** (ETF) that tracks the S&P500 index. In subsection 1.1 we discussed ETFs in the context of the different types of asset classes in the market, and saw that the SPY is an asset traded on the exchanges, just like ISNS, FARO, MENT and AAPL. When we buy the SPY we buy a fund (similar to a mutual fund or a pension fund) whose objective is to track the S&P500 at the lowest possible cost. Thus, many investors who just want the value of their investments to move with "the market" (as represented by the S&P500) prefer to buy the SPY rather than invest in an equity-based mutual fund. Moreover, traders would rather purchase the SPY than acquiring all the 500 assets in the index, since it is (much) cheaper to do so, and removes the costs associated with constantly rebalancing one's portfolio to match changes in the weights the different assets represent in the S&P500. The cost of doing so, but doing so efficiently, is already incorporated into the SPY.

Another variable is the volatility index **VIX**: the VIX is an index continuously published by the Chicago Board of Options Exchange which is designed to measure the market's expectation on future short-term volatility in the S&P500 index – it is computed by taking a certain weighted average of short-term options on the S&P500 index. It is used as a proxy for market uncertainty, investor sentiment, the market taste for risk (market risk aversion) and other related concepts. There are ETFs that try to track VIX, there are futures backed by it, and there are options based on the index.

A third variable of importance is **order flow**. By order flow we mean the difference between the number of shares aggressively bought and shares aggressively sold. Naturally, in a market, for every transaction there is a buyer and a seller. But, in electronic markets we can differentiate between posted limit orders (LOs) and executed market orders (MOs). Thus, if a transaction is the result of a passive limit sell (buy) order being lifted (hit) by an aggressive market buy (sell) order, we refer to it as an aggressive buy (sell) order. An aggressive buy (sell) order is driven by some trader's desire for a rapid buy (sell) and indicates her demand for (supply of) shares of this asset to the overall demand/supply in the market. Thus, the order flow is a proxy for the net demand for the asset which, as we saw in Chapter 2, can incorporate information relevant to market making strategies and future price movements.

In Table 3.2 we show the regression coefficients from the OLS regressions for the following two models.

$$r_{t,j} = \alpha + \beta_{1,j}\, r_{t-1,j} + \beta_{2,j}\, \text{SPY}_t + \beta_{3,j}\, \text{VIX}_t + \beta_{4,j}\, \log(1 + Q_t) + \beta_{5,j}\, \text{OF}_t + \epsilon_{t,j}$$

(3.1)

$$r_{t,j} = \alpha + \beta_{1,j}\, r_{t-1,j} + \beta_{2,j}\, \text{SPY}_t + \beta_{3,j}\, \text{VIX}_t + \beta_{4,j}\, \log(1 + Q_t) + \beta_{5,j}\, \text{OF}_t$$
$$+ \beta_{6,j}\, \text{SPY}_t\, \mathbf{1}_{\text{SPY}_t < 0} + \beta_{7,j}\, \text{VIX}_t\, \mathbf{1}_{\text{VIX}_t < 0} + \epsilon_{t,j}.$$

(3.2)

The first model (3.1) (M1 in Table 3.2) is an OLS regression for the intraday return for each of our four stocks, where $r_{t,j}$, $j \in \{$ ISNS, FARO, MENT, AAPL$\}$. The return is computed as the price of the stock at the close of the market minus the price of the stock at the opening of the market, divided by the price of the stock at the opening of the market: $r = \frac{P_{\text{close}} - P_{\text{open}}}{P_{\text{open}}}$. The model includes as right-hand side variables:

(i) $\alpha$: a constant which captures the mean-daily return and should be close to zero,

(ii) $r_{t-1,j}$: the previous day's intra-day return (should be insignificant as returns should not be predictable),

(iii) $\text{SPY}_t$: the contemporaneous intraday return on the SPY – computed like the stock's intra-day return,

(iv) $\text{VIX}_t$: the contemporaneous intraday 'return' on the VIX – computed like the stock's intra-day return,

(v) $\log(1 + Q_t)$: the log of one plus the number of shares of the stock traded in all markets that day (we add one so that when no shares are traded we do not have to deal with $\log(0)$),

(vi) $\text{OF}_t$: order flow for the stock on NASDAQ that day.

In addition, the second model (3.2) (M2 in Table 3.2) includes the variables in the first model (M1), plus the returns on the SPY and the VIX multiplied by indicators which equal 1 on days in which the VIX or the NASDAQ moved down, and 0 otherwise. These two variables allow us to verify if there is an asymmetric reaction of the asset's return to any 'good' or 'bad' news on the market (or its volatility).

The fitted models in Table 3.2 represent **robust OLS** estimates. Robust OLS estimation is very similar to the standard OLS minimisation of the sum squared residuals, that is, the minimisation of the sum of the squared distance between the observations and the fitted values. The main difference between standard and robust OLS is that with robust OLS the errors in the estimation are weighted so as to reduce the impact of outliers on the estimated parameters. To obtain the estimates in Table 3.2, the weighting is done iteratively and using the Huber's loss function which penalises large errors linearly rather than quadratically – more details on OLS and robust OLS can be found in any standard econometrics textbook, e.g., see Greene (2011) or Cameron & Trivedi (2005).

| Variables | ISNS | | FARO | | MENT | | AAPL | |
|---|---|---|---|---|---|---|---|---|
| | M1 | M2 | M1 | M2 | M1 | M2 | M1 | M2 |
| constant | 0.25 | 0.27 | -2.83 | -2.92 | -2.97 | -3.07 | 1.09 | 1.18 |
| $r_{t-1,j}$ | -0.10 | -0.09 | 0.06 | 0.06 | 0.05 | 0.05 | **−0.12** | **−0.12** |
| SPY (%) | -0.60 | **−1.36** | **1.04** | 1.03 | **1.04** | **1.07** | 0.28 | 0.06 |
| VIX (%) | **−0.08** | -0.01 | -0.03 | -0.05 | 0.00 | 0.01 | -0.03 | -0.02 |
| Log Q | 0.01 | 0.00 | 0.25 | 0.27 | 0.23 | 0.24 | -0.08 | -0.08 |
| Order Flow | 0.03 | 0.02 | **0.05** | **0.05** | **0.03** | **0.03** | **0.06** | **0.05** |
| Negv SPY | — | 1.43 | — | 0.11 | — | -0.08 | — | 0.52 |
| Negv VIX | — | -0.19 | — | 0.06 | — | -0.01 | — | -0.01 |
| Adj R | 0.01 | 0.01 | 0.17 | 0.17 | 0.27 | 0.27 | 0.31 | 0.31 |

**Table 3.2** Robust OLS regression of intraday (open-to-close) return. (Bold: 5% significance)

As Table 3.2 shows, the OLS results for the thinly traded ISNS reflect a lot of noise (in the sense that the **R-squared** ("adj R") is very close to zero) and the significance of the coefficients is not very reliable. For the other assets we find that the OLS regression does pick up some information, and there is one variable that is consistently significant for the other three assets, namely order flow. The coefficient is significant and positive for all three assets indicating that NASDAQ order flow and the asset's return move together, which is consistent with the interpretation of order flow as the market's net demand for the asset, and the models of liquidity of Glosten and Milgrom, and Kyle. We expected the constant and the previous day's return to be insignificant, and we find the first to be the case, and the latter to be true for FARO and MENT. AAPL displays a negative autocorrelation in daily returns, which suggests a significant mean-reverting component in the return process which is not consistent with the efficient market hypothesis, and provides evidence of (negative) short-run momentum, although in the microstructure literature (Roll (1984)) the presence of negative autocorrelation can also be explained in terms of the 'bid-ask bounce' – that is, that trades do not take place at the 'market price' but rather an aggressive buy has to cross a non-zero spread to match with the bid and executes against the LO standing there, at the bid. Finally, we find that during 2013 our assets' returns were not significantly affected by changes in market sentiment (as measured through the VIX). FARO and MENT have significant exposure to market movements, though AAPL (somewhat surprisingly) seems not to. We also find no significant evidence that there is an asymmetry between 'good' and 'bad' news from movements either in the market or in market sentiment.

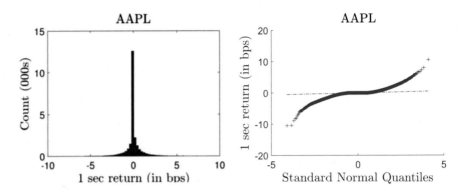

**Figure 3.1** Sample distribution and QQ-plot of the 1-second returns of AAPL on July 30, 2013.

## 3.2      Asset Prices and Returns Intraday

Daily market information is of primary interest only for investors with medium- to long-term investment horizons, but high-frequency trading strategies are executed over very short horizons so we must look at what goes on in much finer detail. To do so, we use millisecond-stamped message-level ITCH data for the NASDAQ market to study prices over several sampling periods.

Focusing on AAPL and on a single day, July 30th 2013, we construct asset returns over one-second intervals. The choice of dates is arbitrary. It was a day with a small positive price gain and positive net order flow: there were 1.45 million shares bought vs 1.24 million shares sold on NASDAQ, and the price increased from market open ($449.96) to market close ($453.32) by $3.36 (+74 bps).

The asset's return is computed using the microprice. The **microprice** $(S_t^*)$, also called the weighted-midprice, is the weighted average of the best bid $(P_t^b)$ and the best ask $(P_t^a)$, weighted by the relative quantities posted at the bid $(V_t^b)$ and ask $(V_t^a)$:

$$S_t^* = \frac{V_t^b}{V_t^b + V_t^a} P_t^a + \frac{V_t^a}{V_t^b + V_t^a} P_t^b. \tag{3.3}$$

The microprice is similar to the midprice, but it incorporates information on order imbalance: e.g., a relatively larger quantity of offers on the bid than on the ask indicates greater buying pressure, and the 'true' price is closer to the ask than to the bid (the microprice was discussed at the end of Chapter 1 – also Chapter 12 is devoted to trading strategies that employ volume imbalance in the LOB as a key variable in trading decisions, the same volume imbalance that moves the microprice towards the ask or the bid).

The choice of sampling frequency is important as it has a very significant effect on the properties of the empirical distribution of the asset's return. If the sampling frequency is very short then many of the observations will be equal to

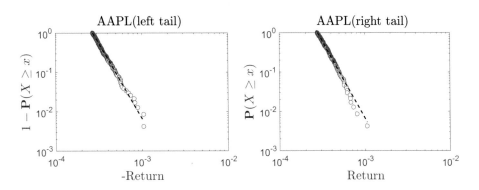

**Figure 3.2** Plot of the tails (log scales) for AAPL on July 30, 2013.

zero. In our example, we sample at the one second frequency, and find that for AAPL on July 30, 2013, 33% of the returns are equal to zero.

We carry out our analysis using returns, and we use basis points (bps) as our unit of analysis (that is, a value of 1 in Figure 3.1 represents a change in the microprice of 1/100th of a percentage point). In our sample, the average microprice is \$454.30, so that a positive return of 1 bps represents an increase in the microprice of 4.5 cents, and 0.22 bps is equivalent to 1 tick, i.e. one cent. Looking at the histogram of the 1-second returns in the left panel of Figure 3.1, we find that the distribution is single-peaked and seems to have fat tails. This is confirmed by the QQ-plot in the right panel of Figure 3.1.

The fat tails exhibited by the asset returns occur often when sampling at short intervals, but may persist at longer frequencies. Figure 3.2 zooms into the right and left tails. Specifically, we use returns above the 95th percentile (1.94 bps) and below the 5th percentile (-1.94 bps) to define these tails. Taking these cutoff values as given, we assume the tails follow a power-law with a probability distribution function given by $f(r)$, where

$$f(r) = \frac{\alpha - 1}{r_{\min}} \left( \frac{r}{r_{\min}} \right)^{-\alpha},$$

and estimate the parameter $\alpha$ using maximum likelihood estimation (MLE). The MLE for $\alpha$ can be shown to be (see, e.g., Clauset, Shalizi & Newman (2009))

$$\hat{\alpha} = 1 + T \left[ \sum_{t=1}^{T} \log \left( \frac{r_t}{r_{\min}} \right) \right]^{-1}.$$

Using the given cutoffs, these estimators give us $\hat{\alpha}_{\text{right}} = 3.35$ and $\hat{\alpha}_{\text{left}} = 3.38$. The model fit of the tails corresponds to the black dashed lines in Figure 3.2, and indicate that the 1-second returns have very heavy tails.

To gain a sense of whether the market behaves in accordance with the efficient market hypothesis, we estimate the autocorrelation function (ACF) from the return data. Recall that the ACF $f(n)$ is given by the correlation of $r_{t-n}$ with $r_t$

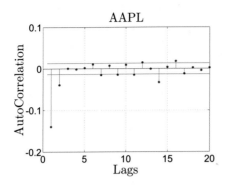

**Figure 3.3** Sample ACF of 1-second returns for AAPL on July 30, 2013.

and can be estimated from the sample correlation. The sample ACF for APPL on July 30, 2013 is shown in the right panel of Figure 3.3. This gives a negative and significant autocorrelation for the first lag, which indicates a significant mean-reversion component in the microprice. We also find that the 12th lag is (weakly) significant and positive, while the 14th is negative and significant. There are no strong theoretical reasons to observe such patterns, so without further research we cannot be sure if this is a spurious pattern that appears on this particular day, or something truly significant.

## 3.3    Interarrival Times

We have seen that even at one second intervals, 33% of the time there are no price changes. As the sampling frequency becomes smaller, it becomes increasingly tenuous to try to model the observed prices as a continuous process, and we need to consider discrete processes. We start by looking at the interarrival times between movements at either the bid or the ask.

Let $\tau_i$ denote the times at which there is change either in the bid or in the ask, and we look at the frequency of interarrival times, $X_i = \tau_{i+1} - \tau_i$. The mean is 10.4 ms and the median 3 ms. Figure 3.4 provides additional insight: the top panels describe the histogram of $X_i$ in absolute and log scales, while the bottom panel contains the QQ-plot relative to the exponential (with the same mean). These graphs indicate that the interarrival times have a power-law distribution with very heavy tails. We estimate the parameter for the right tail (as we did for the 1-second returns above), and we obtain an MLE of $\hat{\alpha} = 3.13$ for the power using the 95th percentile (41 ms) as the cutoff.

We also briefly look at the dynamics of the interarrival process. Figure 3.5 describes the ACF. As shown, changes to the bid and ask are not independent, but have a strong autocorrelated component. This suggests a commonly observed empirical fact about such changes, namely, that they cluster. Rapid changes in the bid/ask are followed by further rapid changes, while a relatively long calm period is also similarly followed by another calm period.

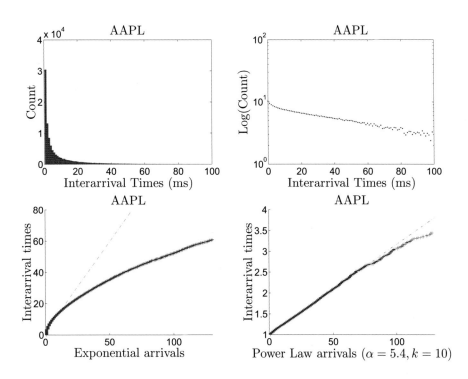

**Figure 3.4** Plot of the frequency of interarrival times $(X)$ in absolute and log scales for AAPL on July 30, 2013. The bottom panel describes the QQ-plot relative to the exponential distribution and the power law distribution: $\mathbb{P}(X_i \leq x) = 1 - (k/x)^{\alpha}$.

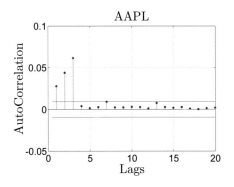

**Figure 3.5** Plot of the autocorrelation function for interarrival times (AAPL, July 30, 2013).

## 3.4    Latency and Tick Size

Suppose we wish to execute a trade, e.g., buy $1,000$ shares of AAPL. The first trade-off we face is immediacy versus cost of execution. If you are concerned about immediacy, the fastest way to execute a trade is to cross the spread and aggressively execute as much as possible using a market order (MO). But, any agent for whom immediacy is very important needs to account for a new issue: message **latency**. Latency refers to the delay between sending a message to the market and it being received and processed by the exchange. Sometimes the time it takes for the exchange to acknowledge receipt of the message is also added in. Latency is random and depends on many factors such as distance between sender and exchange, the structure and type of network, the amount of orders in the network (which can generate congestion), etc. In addition, the word latency is more generally used to express the time it takes for a message to travel from one point to another, such as the latency of a news feed. A classic example of latency is the time it takes for a message to go from the Chicago options exchange (or rather, the CME colocation centre) to New York (or rather, NASDAQ's processing centre in New Jersey). The latency between these two centres is estimated to be between 7.5 and 6.7 ms when travelling by fibre optic cable, or between 4.2 and 5.2 ms when travelling by microwave (on a clear day).

If an agent is trading from home, through a broker, she must be aware of the substantial delay between the moment she asks her broker to execute her order and the time it reaches the market. During that time, market conditions may have changed a great deal. If, however, the agent is trading directly through a broker's feed or she has her own feed into the market, the latency will be much shorter, though still significant relative to centres which are **colocated**. Being colocated, also known as colocation, means that an agent's trading system is physically housed at the electronic exchange's data centre and has a direct connection to the exchange's matching engine. In principle, those who are colocated face similar latency amongst themselves – though there will still be some latency dependent on their software and hardware configurations (colocation was also discussed in Chapter 1).

Latency is an issue for agents for several reasons. An agent who is trading frequently and on narrow margins needs to be aware of the state of the market and be able to adapt her orders, whether to post a new or cancel an existing LO, or submit an MO. Furthermore, when executing an MO, an agent needs to be aware of the relationship between her choice of routing strategy and how other traders may react to the information that may be extracted from observing the strategy's outcomes. A large, poorly routed MO will telegraph its progress through the several exchanges leading to poor execution quality and high execution costs, as other traders reposition themselves to absorb the order at more favourable prices (to them).

For certain assets, circumstances change very fast while for others the market is quite stable. We see this in the movements of prices and spreads in Tables 3.3 and 3.4. The former captures the slow changes in seldom traded assets, such as ISNS, FARO or MENT. The latter looks at more frequently traded stocks, AAPL and ORCL.

Table 3.3 looks at what happens to the bid/ask/midprice/quoted spread from the end of one minute to the next for three assets: ISNS, FARO and MENT. A one minute delay is a very long time for a trader, and one would not expect such a delay unless one is trading through a very slow connection to the exchange (or the order does not go directly to the exchange). We see this in the first column of

| Asset | Var | $\Delta X \neq 0$ (%) | P01 | Q1 | Q2 | Q3 | P99 |
|-------|-----|------|-----|----|----|----|----|
| | | | \multicolumn{5}{c}{Stats (for $\Delta X \neq 0$)} | | | | |
| ISNS | Bid | 4.2 | -43.0 | -4.0 | 1.0 | 5.0 | 40.0 |
| | Ask | 3.4 | -51.0 | -4.0 | -1.0 | 3.0 | 51.0 |
| | Midprice | 6.7 | -25.8 | -2.0 | 0.5 | 2.0 | 26.8 |
| | Quoted Spread | 6.7 | -45.0 | -5.0 | -1.0 | 4.0 | 46.0 |
| FARO | Bid | 48.6 | -21.0 | -2.0 | 1.0 | 3.0 | 19.0 |
| | Ask | 49.0 | -19.0 | -3.0 | -1.0 | 3.0 | 22.0 |
| | Midprice | 63.4 | -16.0 | -1.5 | 0.5 | 2.0 | 16.0 |
| | Quoted Spread | 60.8 | -18.0 | -2.0 | -1.0 | 2.0 | 18.0 |
| MENT | Bid | 44.2 | -5.0 | -1.0 | 1.0 | 1.0 | 6.0 |
| | Ask | 44.1 | -5.0 | -1.0 | -1.0 | 1.0 | 5.0 |
| | Midprice | 52.8 | -4.5 | -1.0 | 0.5 | 1.0 | 5.0 |
| | Quoted Spread | 31.1 | -5.0 | -1.0 | -1.0 | 1.0 | 4.0 |

**Table 3.3** One minute changes in bid, ask, midprice, and quoted spread.

Table 3.3 (labelled $\Delta X \neq 0$) where we find the percentage of minutes for which the bid/ask/midprice/quoted spread is different (not equal to zero). For ISNS we observe that for only 4.2 percent of the time are there changes in the bid from one minute to the next. In the adjacent columns we can see the statistics for those minutes in which the variable of interest changed. From these, we can conclude that for half of the time ISNS's bid price changed, it moved between an increase of 5 cents (Q3) and a drop of 4 cents (Q1). If we look at the midprice for ISNS, we see that it changed only 6.7 percent of the time, and when it did, half of those changes were between an increase of 2 cents and a drop of 2 cents. On the other hand, if one considers assets such as FARO or MENT, a one-minute delay will face changes in the bid (similarly in the ask) around half of the time, as well as (naturally) in the midprice and quoted spread. Note that this is an unconditional analysis for all minutes of 2013, and does not take into account that market participants react to order flow. For example, if one sends a market buy order, then other agents may quickly adjust their quotes upwards in response to this new information. Hence, it is natural that an asset with very little public information (in the form of trades/MOs) like ISNS will see fewer price movements than more frequently traded ones, like FARO or MENT.

This is not the same picture that we find when looking at the most popular stocks (in terms of activity). Assets such as AAPL and ORCL will almost certainly experience changes in prices within one minute, but for these assets a one minute delay is unreasonably long by any standard. Table 3.4 reflects the same information as Table 3.3 but after a 100 ms delay rather than one minute (and for the last three months in 2013 rather than the whole year). Also, rather than report the statistics for the whole sample, we report the median values for the statistics computed at the daily level. That is, after computing the daily percent-

| Asset | Var | $\Delta X \neq 0$ (%) | Stats (for $\Delta X \neq 0$) | | | | |
|-------|-----|------|-------|------|------|------|------|
| | | | P01 | Q1 | Q2 | Q3 | P99 |
| AAPL | Bid | 3.84 | -17.0 | -3.0 | -1.0 | 3.0 | 18.0 |
| | Ask | 4.00 | -18.3 | -3.0 | 1.0 | 3.0 | 17.0 |
| | Midprice | 6.75 | -11.5 | -1.5 | 0.5 | 1.5 | 11.0 |
| | Quoted Spread | 6.69 | -16.0 | -3.0 | -1.0 | 3.0 | 18.0 |
| ORCL | Bid | 0.41 | -2.0 | -1.0 | -0.5 | 1.0 | 2.0 |
| | Ask | 0.40 | -2.0 | -1.0 | 1.0 | 1.0 | 2.0 |
| | Midprice | 0.47 | -2.0 | -1.0 | 0.5 | 1.0 | 2.0 |
| | Quoted Spread | 0.16 | -3.0 | -1.0 | -1.0 | 1.0 | 2.0 |

**Table 3.4** One hundred ms changes in bid, ask, midprice, and quoted spread.

age of 100 ms intervals with non-zero bid price changes for AAPL for each day from October to December 2013, we report that the median of these is 3.84%. Similarly, after computing the first quartile of the non-zero bid price changes for each day, we report the median of these, which is $-3$ cents, while the median of the third quartiles is an increase of 3 cents.

Note that there is a noticeable difference in the frequency and magnitude of price changes for AAPL and ORCL. AAPL sees more frequent changes and of greater magnitude than ORCL. There are two factors that are important here: (i) the volume traded (in dollars) for AAPL is one order of magnitude greater than for ORCL, but also, (ii) the price of AAPL in the last quarter of 2013 was between \$490 and \$560, while that of ORCL was between \$32 and \$38. As both assets have the same **minimum tick size** (of one cent), this means that AAPL can experience much smaller percentage changes in its price (1 cent $= 0.2$ bps of \$500) than ORCL (1 cent $= 2.5$ bps of \$40). Thus, one would expect more frequent price movements for AAPL than for ORCL (even after AAPL's share split of one old share into seven new ones, as after the split and AAPL prices around \$100, a one cent change is equivalent to a 1 bps change, much greater than the 2.5 bps of ORCL).

## 3.5      Non-Markovian Nature of Price Changes

In this section, we investigate how successive price changes are interrelated. For this, we first look at whether the sign of the current price change can predict the sign of the next (non-zero) price change. We continue using AAPL on July 30, 2013 for the analysis, and record price changes every time they occur. In Table 3.5 we see how an increase in the price (bid or ask) is more often than not followed by a reversal, and similarly for a fall in the price.

In Table 3.5 we see that an increase in the ask price (an uptick) is followed by a down tick 57% of the time, while a drop in the ask is followed by an increase

| | Ask | | | Bid | |
|---|---|---|---|---|---|
| | Uptick | Downtick | | Uptick | Downtick |
| $t/t+1$ | (⇑) | (⇓) | | (⇑) | (⇓) |
| Uptick (⇑) | 43.0 | 57.0 | Uptick (⇑) | 36.5 | 63.5 |
| Downtick (⇓) | 61.8 | 38.2 | Downtick (⇓) | 55.3 | 44.7 |

**Table 3.5** Empirical Transition Rates: Single Price Change.

61.8% of the time. For the bid price the numbers are 63.5% and 55.3% of the time. Earlier, in Table 3.4, we saw that a 100 ms latency resulted in no changes in the bid more than 96 percent of the time (in at least half the observation period). But, in Section 3.3 we saw that the median interarrival time of a change in either the bid or the ask is 3 ms (that for the bid is 8 ms). The numbers in Table 3.5 help us reconcile these two seemingly contradictory statements: even though after a 100 ms delay one may not observe a *net* change in the bid, this zero net change is a result of both no changes during the 100 ms period and also of several changes that cancel each other out. So, for latencies greater than 8 ms, if you submit an MO, by the time it hits the market, the price may have moved away but it may also have returned to the price that was there when the MO was submitted. Latency, therefore, introduces execution risk specially for traders who are not colocated.

Table 3.5 also illustrates an asymmetry that appears on this day in the relative frequencies of price reversals. This asymmetry suggests that a shrinking of the quoted spread (a fall in the ask or an increase in the bid) was more likely to be reversed than an increase in the quoted spread (an increase in the ask or a fall in the bid).

So we present Table 3.6 to investigate whether the sign of the current price change can predict the sign of the price changes in the next two periods. This table describes the relative frequency of price reversals ('Reversal', ⇑⇓ or ⇓⇑) relative to consecutive movements in the same direction ('Up'(⇑⇑) and 'Down'(⇓⇓)) conditional on the current price change being up ⇑ or down ⇓. Around 60% of consecutive changes in the bid and in the ask are in opposite directions. Table 3.6 shows not only the conditional probability of the future two price moves, but also the unconditional one. Comparing those transitions we are led to the conclusion that there seems to be little if any difference between the conditional and the unconditional transitions. Thus, even though price changes tend to be reversed, the direction of the current price change (whether it is an uptick or a downtick) does not carry additional information about future price changes.

Finally, we consider a longer sequence of price changes, i.e. we consider whether the signs of the two past price changes can predict the next one. To accomplish this we define four states, one for each possible pair of signs of price changes as follows: $A$ is an uptick followed by another uptick ($A = ⇑⇑$), $B$ is an uptick followed by a downtick ($B = ⇑⇓$), $C$ is a downtick followed by an uptick ($C = ⇓⇑$),

| | Ask | | |
|---|---|---|---|
| | Up$(t+1, t+2)$ | Reversal$(t+1, t+2)$ | Down$(t+1, t+2)$ |
| | (⇑⇑) | (⇑⇓, ⇓⇑) | (⇓⇓) |
| Uptick$(t)$ (⇑) | 17.1 | 59.5 | 23.4 |
| Downtick$(t)$ (⇓) | 19.6 | 59.1 | 21.3 |
| Unconditional | 18.3 | 59.3 | 22.4 |

| | Bid | | |
|---|---|---|---|
| | Up$(t+1, t+2)$ | Reversal$(t+1, t+2)$ | Down$(t+1, t+2)$ |
| | (⇑⇑) | (⇑⇓, ⇓⇑) | (⇓⇓) |
| Uptick$(t)$ (⇑) | 24.0 | 60.3 | 15.7 |
| Downtick$(t)$ (⇓) | 23.7 | 58.1 | 18.2 |
| Unconditional | 23.9 | 59.1 | 17.0 |

**Table 3.6** Empirical Transition Rates: Pairs of Tick Changes.

and $D$ is a downtick followed by another downtick ($D = ⇓⇓$). The price change signs are then used to generate a sequence of $A$, $B$, $C$, $D$ with overlapping observations, so that, e.g., the sequence ⇑⇓⇓⇑ would be represented as $BDC$. Note that there are several transitions that cannot occur. For example, $B$ cannot be followed by $A$, since if the price change was $B$(⇑⇓), then either (i) the next change is ⇑, in which case we transition to the state $C$(⇓⇑), or (ii) the next change is ⇓, in which case we transition to the state $D$(⇓⇓).

The estimated transition frequencies for this Markov chain are provided in Table 3.7. The transition $AA$ should be interpreted as the sequence of ticks ⇑⇑⇑, $BC$ as ⇑⇓⇑, and so on. In the table we can see confirmation of price reversals: the transitions $BC$ and $CB$ have markedly higher probability, suggesting that successive price movements (both for the bid and the ask) tend to go in opposite directions.

## 3.6    Market Fragmentation

Another issue to consider when executing aggressively is that of market fragmentation. We only address this issue superficially here but it is a serious concern for high-frequency traders.

So far we have focused on detailed data from one exchange, NASDAQ. As of October 2014 in the US there were 11 exchanges and around 45 alternative trading venues, most of which were dark pools – dark pools are trading venues that do not publicly display price quotes and in 2014 NASDAQ represented around 20 percent of the trading (later, in section 7.4, we provide execution algorithms were the agent has access to a standard lit market and also to a dark

| | Ask | | | |
|---|---|---|---|---|
| $(t+1)$ | A (⇑⇑) | B (⇑⇓) | C (⇓⇑) | D (⇓⇓) |
| A($t$) (⇑⇑) | 54.4 | 45.6 | - | - |
| B($t$) (⇑⇓) | - | - | 70.0 | 30.0 |
| C($t$) (⇓⇑) | 34.4 | 65.6 | 0.0 | 0.0 |
| D($t$) (⇓⇓) | - | - | 48.6 | 51.4 |

| | Bid | | | |
|---|---|---|---|---|
| $(t+1)$ | A (⇑⇑) | B (⇑⇓) | C (⇓⇑) | D (⇓⇓) |
| A($t$) (⇑⇑) | 43.0 | 57.0 | - | - |
| B($t$) (⇑⇓) | - | - | 62.2 | 37.8 |
| C($t$) (⇓⇑) | 32.8 | 67.2 | - | - |
| D($t$) (⇓⇓) | - | - | 46.9 | 53.1 |

**Table 3.7** Empirical Transition Rates: Pairs of Ticks.

pool). So, we cannot truly talk about 'the market' as a single exchange, but as the aggregation of activity across a large number of venues. To get an idea of the degree of market fragmentation, that is, the extent to which the market for one asset is distributed across different venues, we look at trading during market hours (9:30-16:00) for one asset, AAPL, on July 30, 2013 across all venues using Consolidated Tape data. These data provide information on all transactions and best quotes from all venues.

Here we have reconstructed the bid and ask for the venues for which we have AAPL activity reported during that day. With the reconstructed bid and ask, we compute the percentage of time each exchange's best price (the bid or the ask) coincides with the best price across all venues. Table 3.8 captures this information. We can see that NASDAQ-OMX's bid coincides with the best bid across all venues during 67 percent of regular trading hours, while the same figure is 19 percent for BATS, 43 percent for the NYSE-ARCA, 35 percent for EDGE-X and never for EDGE-A markets.

Thus, optimally executing a trade is not just about timing and prices displayed in one exchange, but also about: how to organise the way an order (or orders) reaches a particular trading venue, what the different laws governing how exchanges should handle orders are and the rules exchanges use to implement them, how to programme the routing of the order, and which order types are best suited to one's particular routing and trading strategy, etc.

In the US the specific regulation, Reg NMS (National Market System), has been set up to facilitate competition between exchanges and to protect investors. In particular, it has specific provisions to protect investors' orders by preventing trade-throughs (i.e. the execution of an order at an inferior price when a better price is available, especially when that price comes from a 'proper' (protected)

|  | Percentage Time at NBBO | |
| --- | --- | --- |
| Exchange | Bid | Ask |
| NASDAQ | 67.8 | 61.3 |
| BATS | 18.8 | 15.7 |
| ARCA-NYSE | 43.4 | 38.3 |
| NSE | 0.0 | 0.0 |
| FINRA | 0.0 | 0.0 |
| CSE | 0.0 | 0.0 |
| CBOE | 1.2 | 0.7 |
| EDGA | 0.0 | 0.0 |
| EDGX | 34.5 | 41.0 |
| NASDAQ-BX | 0.0 | 0.0 |
| NASDAQ-PSX | 0.0 | 0.0 |
| BATS-Y | 4.5 | 0.0 |

**Table 3.8** Percentage of Time that Exchange's Best prices are at the NBBO.

quote in another trading venue). Tables 3.9 and 3.10 look at trade executions in the different venues and compare the price at which the trade was executed relative to the best bid/ask price in the exchange in which the trade is reported (Local) and relative to the best available price across all venues we have reconstructed from the data (NBBO).

| | Local | | | NBBO | | |
| --- | --- | --- | --- | --- | --- | --- |
| Exchange | Bid | Ask | Total | Bid | Ask | Total |
| NASDAQ | 22, 214 | 8, 122 | 30, 336 | 458, 994 | 367, 181 | 824, 675 |
| BATS | 1, 854 | 2, 200 | 4, 054 | 118, 205 | 108, 407 | 225, 512 |
| ARCA-NYSE | 9, 840 | 5, 630 | 15, 470 | 292, 933 | 273, 729 | 566, 361 |
| NSE | 901 | 200 | 1, 101 | 13, 244 | 11, 057 | 24, 301 |
| FINRA | 0 | 0 | 0 | 534, 178 | 406, 346 | 940, 424 |
| CSE | 0 | 0 | 0 | 0 | 0 | 0 |
| CBOE | 0 | 0 | 0 | 5, 005 | 1, 550 | 6, 555 |
| EDGA | 0 | 100 | 100 | 31, 357 | 22, 125 | 53, 482 |
| EDGX | 9, 324 | 2, 300 | 11, 624 | 230, 187 | 207, 005 | 436, 392 |
| NASDAQ-BX | 1, 016 | 1, 100 | 2, 116 | 60, 971 | 48, 365 | 109, 336 |
| NASDAQ-PSX | 0 | 100 | 100 | 600 | 1, 525 | 2, 125 |
| BATS-Y | 100 | 1, 000 | 1, 100 | 16, 519 | 16, 178 | 32, 497 |

**Table 3.9** Number of Shares Executed at Best Prices (Local refers to best price at local exchange if local best price is not NBBO).

Table 3.9 compares for each venue, the executions that occurred at the best price across all venues (NBBO) versus the ones at the best price in that venue (Local), when that venue's best price was not the best across all venues. We can

| Exchange | NBBO | Local Best | Inside Best | Outside Best | Shares (total) |
|----------|------|------------|-------------|--------------|----------------|
| NASDAQ | 39.8 | 1.5 | 57.5 | 1.3 | 2,073,946 |
| BATS | 34.8 | 0.6 | 64.2 | 0.4 | 648,137 |
| ARCA-NYSE | 43.3 | 1.2 | 54.4 | 1.2 | 1,308,771 |
| NSE | 33.1 | 1.5 | 64.3 | 1.1 | 73,486 |
| FINRA | 22.2 | 0.0 | 71.5 | 6.4 | 4,238,247 |
| CSE | 0.0 | 0.0 | 0.0 | 100.0 | 122,250 |
| CBOE | 63.3 | 0.0 | 36.7 | 0.0 | 10,350 |
| EDGA | 27.8 | 0.1 | 71.6 | 0.5 | 192,202 |
| EDGX | 30.6 | 0.8 | 67.4 | 1.1 | 1,425,145 |
| NASDAQ-BX | 41.1 | 0.8 | 55.5 | 3.4 | 266,166 |
| NASDAQ-PSX | 52.1 | 2.5 | 40.5 | 4.9 | 4,075 |
| BATS-Y | 26.0 | 0.9 | 73.2 | 0.0 | 125,220 |

**Table 3.10** Percentage of Shares Executed, by Execution Quality.

see that when trading against a best price, it occurs against the NBBO most of the time. In Table 3.10, we allow for further types of executions, not just against a best price, but also inside the (Local) spread (between the local best bid and ask, while not at the NBBO) and outside the spread (for FINRA and CSE we use the NBBO as reference because no Local bid/ask quotes are reported). The numbers clearly show that almost all trades occur at the computed NBBO or inside the (local) spread: 40 and 58 percent for NASDAQ respectively, 43 and 54 for NYSE-ARCA, 35 and 64 for BATS, and 30 and 67 percent for EDGE-X. The same percentages for trades reported to FINRA are 22 and 72 percent respectively using the NBBO as reference. These numbers suggest quite high execution quality, although the proportion of trades inside the spread seems unusually large if one takes into account that a regular market order should be executing at best prices. A possible explanation is that many of these trades are being executed against hidden orders posted inside the spread and/or via alternative order types which allow aggressive postings inside spread.

## 3.7 Empirics of Pairs Trading

Most traders do not look at one asset at a time, but consider the interactions between different assets. This makes sense when you can extract information from the interaction between different assets. This works best with groups of assets that share common shocks and occurs naturally for assets in the same industry.

In this section we focus on the interaction between two technology stocks (Intel, INTC) and a technology ETF (Merrill Lynch Semiconductor ETF, SMH) on November 1, 2013. These two assets move together for two main reasons. The first is mechanical: around 20% of the ETF holdings are shares of INTC. The second is economic: the ETF is designed to represent the semiconductor industry, and hence its price will move in response to news that affects that industry, and the same news will have a similar effect on the price of INTC. In Chapter 11, devoted to pairs trading and statistical arbitrage, we build on some of the ideas presented here to show how to take advantage of the information provided by a collection of assets. In particular, we present different trading algorithms based on the co-integration in the stock price level or in the drift component of a collection of assets that builds on the following empirical analysis.

Our analysis is based on the following theoretical model: we assume that both INTC and SMH are stocks whose dynamics have a transitory (mean-reverting) component and a permanent (Brownian) component. We express the dynamics of this process in vector form as follows:

$$dS_t = \kappa \left( \theta - S_t \right) dt + \sigma dW_t, \tag{3.4}$$

where $\Sigma = \sigma \sigma'$, and $W_t$ is a Brownian motion.

The presence of a mean-reverting component (the term proportional to $dt$) introduces the opportunity for generating positive expected returns from trading by exploiting that component's predictability. In this case, we use the joint information from the two processes to create a stronger trading signal by constructing a linear combination of the two assets, which is most strongly driven by the mean-reverting component (and which is the basis for the some of the algorithms of Chapter 11).

This is done by transforming the system in Equation (3.4), which has a generic matrix $\kappa$, into an equivalent system,

$$d\tilde{S}_t = \tilde{\kappa} \left( \tilde{\theta} - \tilde{S}_t \right) dt + \tilde{\sigma} dW_t, \tag{3.5}$$

where $\tilde{\kappa}$ is a diagonal matrix; that is, we look for the constants $\{ \alpha_{11}, \alpha_{12}, \alpha_{21}, \alpha_{22} \}$ such that

$$\tilde{S}_{t,1} = \alpha_{11} S_{t,1} + \alpha_{1,2} S_{t,2}$$
$$\tilde{S}_{t,2} = \alpha_{21} S_{t,1} + \alpha_{2,2} S_{t,2} \, ,$$

and

$$\tilde{\kappa} = \begin{bmatrix} \tilde{\kappa}_1 & 0 \\ 0 & \tilde{\kappa}_2 \end{bmatrix} .$$

The resulting $\tilde{\kappa}$ matrix has $\{ \tilde{\kappa}_1, \tilde{\kappa}_2 \}$ equal to the eigenvalues of $\kappa$, and the process $\tilde{S}_{t,j}$ corresponding to the largest of these (in absolute terms), $\max\{ |\tilde{\kappa}_1|, |\tilde{\kappa}_2| \}$, will have the strongest exposure to the mean-reverting process, and hence should contain the most trading-relevant information – i.e. it will generate the best trading signal (see the algorithms developed in Chapter 11).

We illustrate this by a simple estimation of the relationship between INTC and SMH during November 1, 2013. We sample using the midprice and estimate the process at regular intervals (every 5 seconds). We fit the discrete version of

| | **A** | **B** | | | |
| --- | --- | --- | --- | --- | --- |
| | | $\Delta S_{t-1,INTC}$ | | $\Delta S_{t-1,SMH}$ | |
| $\Delta S_{t,INTC}$ | 0.011 | 0.997 | *** | 0.002 | |
| $\Delta S_{t,SMH}$ | 0.035 | 0.003 | | 0.998 | *** |

**Table 3.11** Estimated parameters of VAR (*** significant at 1% level).

the model in (3.4) and use it to compute the values of the transformed model in (3.5) in order to build the trading signal.

To estimate the discrete version of model (3.4) we estimate the vector autoregressive process (VAR)

$$\Delta S_t = A + B \, \Delta S_{t-1} + \varepsilon_t \, ,$$

where $S_t = [S_{t,INTC} \quad S_{t,SMH}]'$ are the asset prices, and $\Delta S_{t,j}$ denotes the change in asset $j$ – $A$ is a vector of constants, $B$ a matrix of constants and $\varepsilon_t$ a vector of white noise. The resulting estimates are described in Table 3.11. From these we can recover the parameters of model (3.4):

$$\kappa = \frac{\mathbb{I} - B}{\Delta t} = \begin{bmatrix} 0.003 & -0.002 \\ -0.003 & 0.002 \end{bmatrix}, \qquad \theta = \kappa^{-1} A \, \Delta t = \begin{bmatrix} 24.30691 \\ 40.91387 \end{bmatrix},$$

and diagonalise $\kappa$ to obtain $\tilde{\kappa}$ and $\tilde{S}$:

$$\kappa = U \cdot \Lambda \cdot U^{-1}, \qquad \tilde{\kappa} = \Lambda = \begin{bmatrix} 0.0047 & 0 \\ 0 & 0.0007 \end{bmatrix},$$

$$\tilde{S}_t = U^{-1} S_t = \begin{bmatrix} 0.682 & 0.547 \\ -0.731 & 0.837 \end{bmatrix} S_t \, .$$

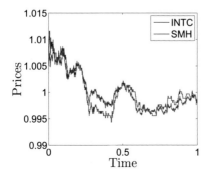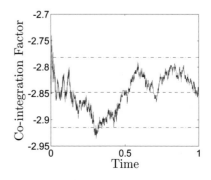

**Figure 3.6** INTC and SMH on November 1, 2013: (left) midprice relative to mean midprice; (right) co-integration factor. The $x$-axis is time in terms of fractions of the trading day. The dashed line indicates the mean-reverting level; the dash-dotted lines indicate the 2 standard deviation bands.

|  | Constant | $r_{t-1}$ | | $r_{t-2}$ | |
|---|---|---|---|---|---|
| $r_{t,INTC}$ | -0.000 | -0.011 | | 0.025 | |
| $r_{t,SMH}$ | -0.000 | -0.057 | *** | 0.014 | |
| $r_{\tilde{S}_{t,1}}$ | 0.000 | -0.195 | *** | -0.079 | *** |
| $r_{\tilde{S}_{t,2}}$ | -0.000 | 0.013 | | 0.044 | *** |

**Table 3.12** Estimated parameters of individual AR(n), *** significant at 1% level.

In Figure 3.6 we display the price process for the two assets in the left panel, and in the right panel the price process for $\tilde{S}_1$, which is called the co-integration factor. Visual inspection (not included here) suggests a much stronger mean-reverting (auto-regressive component) for $\tilde{S}_1$ than for $\tilde{S}_2$. We verify this by running autoregressions on the returns for all four price processes: $r_{INTC}$, $r_{SMH}$, $r_{\tilde{S}_1}$, and $r_{\tilde{S}_2}$. We can see the results in Table 3.12. For the assets, INTC and SMH, the coefficient on lagged returns is only significant for the ETF, SMH. After estimating the model and applying the diagonalisation on $\kappa$, the return on one of the resulting portfolios, $\tilde{S}_1$ (the co-integration factor), has a coefficient on lagged returns that is almost four times larger than the one on SMH (and includes an additional significant coefficient on the returns two periods prior).

## 3.8    Bibliography and Selected Readings

Hasbrouck (1991), Aït-Sahalia, Mykland & Zhang (2005), Bandi & Russell (2008), Barndorff-Nielsen & Shephard (2004), Engle (2000), Hansen & Lunde (2006), Mykland & Zhang (2012), Cartea & Karyampas (2012), Corsi & Renò (2012), Bandi & Renò (2012), Barndorff-Nielsen & Shephard (2006a), Barndorff-Nielsen & Shephard (2006b), Barndorff-Nielsen & Shephard (2004), Corsi (2009), Corsi, Pirino & Renò (2010), Aït-Sahalia et al. (2005), Cartea & Karyampas (2014), Hasbrouck (1995), Hasbrouck (1993), Bauwens & Hautsch (2006), Bauwens & Hautsch (2009), Cameron & Trivedi (2005), Ding, Hanna & Hendershott (2014).

# 4 Empirical and Statistical Evidence: Activity and Market Quality

This chapter continues our overview of empirical matters by looking at volume and market quality. As in the previous chapter we start by looking at daily volume though focusing on its relationship with volatility. We then move on to 'seasonal' patterns observed in the data, both in volume as well as in prices. Section 4.3 turns to market quality. These are variables that affect trade execution, such as spreads, volatility, depth, and price impact. Section 4.4 looks at message activity and the relationship between cancellations, executions and distance from the midprice. The chapter concludes with a look at hidden orders.

## 4.1 Daily Volume and Volatility

So far we have seen that the price level (and the asset's returns) over the course of a whole day are difficult to predict and move with market forces. Over short horizons, these prices have fat tails, are subject to rapid changes (where the speed of change in price levels depends on the frequency with which the asset is traded), these changes tend to cluster in time, and are more likely than not to return to their previous level.

But trading activity, usually measured using volume (either in number of shares or the value of shares traded) has a different dynamic structure that has important ramifications for the way we look at market data. Andersen & Bondarenko (2014) capture this idea very well:

> Since volume and volatility are highly correlated and display strong time series persistence, any variable correlated with volatility will, inevitably, possess non-trivial forecast power for future volatility. This is true for bid-ask spreads, the quote intensity, the transaction count, the (normalized) trading volume ...

This and subsequent sections will consider the empirical aspects of some of these variables associated with volatility.

As a first step, we look at the relationship between volume and volatility using a robust regression for our four main assets (ISNS, FARO, MENT and AAPL) with daily volume as the dependent variable. As in subsection 3.1.4, we estimate two models using robust OLS. The left-hand side variable is the log of the number of shares traded on each trading day of 2013, $\log(1 + Q_t)$ (remember that we must add 1 to $Q_t$ as for some assets, ISNS in particular, there are trading days with no trading and zero volume).

The variables on the right-hand side are the variables we used in the model for intraday returns. In particular, the first model (M1) includes a constant, the lagged values of the left-hand side variable (lagged volume), intraday returns on the VIX and the SPY, the contemporaneous intraday return of the asset, and OF which is the day's net order flow, on NASDAQ, defined as the volume of buy minus the volume of sell orders.

Hence, Model 1 (M1) is

$$\log(1 + Q_{t,j}) = \alpha + \beta_{1,j} \log(1 + Q_{t-1,j}) + \beta_{2,j} \, \text{SPY}_t + \beta_{3,j} \, \text{VIX}_t$$
$$+ \beta_{4,j} \, r_{t,j} + \beta_{5,j} \, \text{OF}_t + \epsilon_j \,,$$

while Model 2 (M2) is

$$\log(1 + Q_{t,j})$$
$$= \alpha + \beta_{1,j} \log(1 + Q_{t-1,j}) + \beta_{2,j} \, \text{SPY}_t + \beta_{3,j} \, \text{VIX}_t + \beta_{4,j} \, r_{t,j} + \beta_{5,j} \, \text{OF}_t$$
$$+ \beta_{6,j} \, (\text{SPY}_t)^2 + \beta_{7,j} \, (\text{VIX}_t)^2 + \beta_{8,j} \, \text{HL}_t + \beta_{9,j} \, (r_t)^2 + \epsilon_j \,.$$

In addition to the variables appearing in M1, M2 also includes

- $(\text{SPY}_t)^2$: the squared value of the intraday return on the SPY ETF, as a proxy for market wide volatility,
- $(\text{VIX}_t)^2$: the squared value of the intraday 'return' on the VIX, as a proxy for the volatility of volatility, or the variation in intraday changes in market sentiment,
- $\text{HL}_t$ (HL–volat): the asset's price range $(\max P_t - \min P_t)$ during the day, as a measure of the day's price volatility, and
- $r_t^2$: the square of the asset's intraday return (another, distinct, measure of intraday volatility)

| Variables | ISNS M1 | ISNS M2 | FARO M1 | FARO M2 | MENT M1 | MENT M2 | AAPL M1 | AAPL M2 |
|---|---|---|---|---|---|---|---|---|
| constant | **6.47** | **5.40** | **4.88** | **5.22** | **7.77** | **7.70** | **5.46** | **9.66** |
| $\log 1 + Q_{t-1}$ | **0.22** | **0.22** | **0.58** | **0.52** | **0.41** | **0.39** | **0.67** | **0.38** |
| SPY (%) | 0.04 | 0.19 | -0.17 | **-0.21** | 0.03 | 0.07 | 0.00 | 0.03 |
| VIX (%) | 0.02 | 0.02 | 0.00 | -0.01 | 0.01 | 0.02 | 0.00 | -0.00 |
| $r_t$ | -0.01 | -0.03 | **0.05** | 0.04 | 0.06 | 0.01 | 0.01 | -0.01 |
| Order Flow | 0.02 | 0.04 | 0.00 | -0.00 | -0.01 | -0.00 | -0.00 | 0.00 |
| SPY$^2$ | — | -0.01 | — | 0.09 | — | -0.11 | — | -0.03 |
| VIX$^2$ | — | 0.00 | — | -0.00 | — | 0.00 | — | 0.00 |
| HL-volat | — | **0.32** | — | **0.10** | — | **0.20** | — | **0.28** |
| $r_t^2$ | — | -0.01 | — | 0.01 | — | -0.02 | — | -0.02 |
| Adj R | 0.03 | 0.18 | 0.30 | 0.50 | 0.17 | 0.24 | 0.38 | 0.65 |

**Table 4.1** Robust OLS regression of intraday volume (Bold: 5% significance)

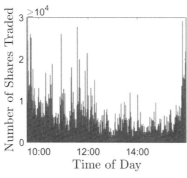

**Figure 4.1** Intraday volume for AAPL on July 30, 2013 at three different scales.

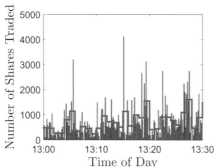

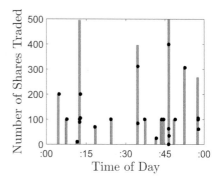

The results in Table 4.1 display no evidence of significant effects from market variables (VIX or SPY) nor from order flow on volume. We also find no effect of the day's intraday return – FARO has a positive and significant effect in M1, but it disappears as we include better proxies for intraday volatility. What we find is substantial support for Andersen and Bondarenko's statement: volume seems to have significant time series persistence as evidenced by the common, positive and significant coefficient on last period's volume (M1 and M2 for all assets), and positive and significant 'correlation' with volatility, as measured by HL-volatility (in M2, all assets). Volatility, as measured by the square of the intraday return, seems statistically insignificant in the presence of HL-volatility.

## 4.2    Intraday Activity

There are other well-known empirical patterns of intraday volume. Figure 4.1 shows the volume (number of shares traded in NASDAQ) at three different time scales over the course of a single trading day for AAPL. The top panel shows the results over the whole day when volume is aggregated in one-minute buckets (although the volumes for the first and last couple of minutes are off the scale). A striking characteristic for this day are the peaks at the beginning of the day and at the end of the day. There is a third peak around noon but (as we see later) it represents a pattern that is specific to this trading day and is not a generic feature observed when trading in this asset.

A second striking feature is the large variability in volume. Computing the descriptive statistics on volume we find a mean of 6,898 shares, a standard deviation of 7,014, and quantiles Q1: 3,299, median: 5,349 and Q3: 8,039. The third and fourth moments give us a measure of skewness of 6.14 and kurtosis 60.83. All these confirm the impression that intraday volume has very large peaks of trading.

The peaks in volume observable at certain minutes, form a pattern that is repeated as one zooms into smaller time intervals. The bottom two pictures in Figure 4.1 look at the pattern of volume during a 30-minute and a one-minute window in the middle of the day. The left panel compares volume aggregated in ten-second buckets with their one-minute average (the thick line). We see substantial variation with large peaks of trading mixed in with periods of relative calm. Zooming in further, the pattern in the right panel is even more striking. For a single minute of the day, the grey columns identify volume aggregated to one second, while the large dots represent volume aggregated at 20 ms (which is almost equivalent to plotting individual transactions). Transactions seem randomly distributed over the minute, and it is not obvious whether the clustering of changes in prices we saw earlier (in subsection 3.3) is also taking place at this time scale. Transactions also appear to happen in multiples of 100 shares.

The fact that transactions occur at round quantities is an institutional feature. The market designers and regulators differentiate between 'odd', 'even', 'mixed' and 'round' lots. A round lot is a message or transaction involving units of even lots (an even lot is 100 shares). Odd lots are trades smaller than an even lot, and can be more expensive to trade (in terms of fees/commissions), while mixed lots are transactions which include both round and odd lots. Odd lots are sometimes aggregated and even not displayed on the consolidated tape (the public ticker that includes 'all' transactions from all exchanges). Also, trading in odd lots is subject to special rules which are in the process of being changed and reviewed (see for example SEC Release No. 34-71057 SEC (2013b), on the reporting of odd lots and changes in the definitions of "market" orders).

There are many algorithms which are based on or linked to volume. For example, some execution algorithms (see Chapters 6, 7 and 8) may require that MOs sent to the exchange do not exceed a percentage of what other market participants are trading at that point in time. In the same vein, some algorithms are designed to trade in a given direction, buy or sell the asset, whilst targeting a given percentage of the market – these are known as POV or percentage of volume algorithms. Moreover, volume plays a very important role in determining execution cost benchmarks. One of the most important of these is VWAP, which stands for volume weighted average price. We devote Chapter 9 to algorithms that target POV and VWAP where we also expand on the discussion of intraday volume to which we now turn.

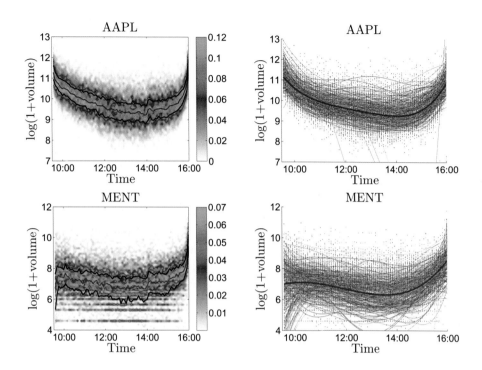

**Figure 4.2** Volume as a function of the time of day.

## 4.2.1   Intraday Volume Patterns

To gain insight into the intraday volume patterns, in Figure 4.2 we show heat-maps (left panels) of daily volume for the year 2013 for AAPL and MENT. The heat-maps are generated by first bucketing traded volume into five-minute windows throughout the day, for every day of the year. Then, for each five-minute bucket, we compute the distribution of volume. The heat-map is a visualisation of the collection of these distributions for each five-minute bucket all at once. In the figures we use coloured lines to represent the first and third quartiles, as well as the median.

We see that volume for AAPL is very large at the beginning of the day, and it gradually slows down until around 14:00, at which time there is a small surge in activity. The 14:00 surge slowly builds up and accelerates during the last half hour of the trading day, peaking at the close.

A reasonable hypothesis for the 14:00 surge is that at that time there are more announcements than is the norm, and these announcements tend to generate greater volume. For example, the monthly Treasury Budget is announced at that time of the day. Earlier we saw this figure for the day of July 30th. There we saw the usual peaks at the beginning and end of the day as well as a peak of trading volume around noon, which we can now show that is atypical for this asset.

To explain the peaks at the beginning and the end of the day, we need to hypothesise about the factors that drive volume. A common hypothesis in the literature, which is also made here, is that new information generates greater volume. In addition to the 14:00 surge, this would also explain why there is such a large volume at and just after the opening, as overnight news is gradually incorporated into prices during that time. But, this hypothesis does not explain the magnitude of the increase in volume at the end of the day, as there is not an unusually large number of announcements at that time. However, in Chapter 7 we will see another possible explanation for this peak in trading at the end of the day: traders who have not been able to meet their liquidation targets will accelerate trade execution as the market approaches its time to close. A second, not unrelated, possible explanation is that traders may prefer to postpone non urgent executions towards the end of the day when execution costs are lower (See for instance Section 4.3.5 where we discuss price impact, in particular Figure 4.11 where we show that the impact of orders (walking the LOB) is lower at the end of the trading day for INTC, and in Chapter 6 we show similar behaviour for SMH, see Figure 6.1). Finally, strategies that target volume (such as the ones developed in Chapter 9) will naturally exacerbate the increase in volume anticipated during this period.

The right panels of Figure 4.2 expand our analysis a bit further. They show a functional data analysis (FDA) approach to viewing the data. In these graphs for every trading day we regress the realised five-minute volume against Legendre polynomials and plot the resulting curve as a thin line in the figure. This generates a smooth volume curve for each day of the year. We then plot the mean of these curves (the solid blue line) which represents the expected (or average) trading volume throughout the day for the corresponding ticker. The conclusions that we drew about the behaviour of volume but we obtain additional insights.

In the right panels we observe four large outliers for AAPL – four curves that disappear off the bottom of the scale. These represent four special days for 2013. Three of these four were predictable while the other was not. The predictable three correspond to July 3 (Independence day), November 29 (Thanksgiving), and December 24 (Christmas) when NASDAQ closed early (13:00) for the holidays. These days are excluded from the calculations for MENT. The fourth outlier corresponds to August 22, 2013. On this day NASDAQ suffered major problems that led to a market shutdown for about three hours. So, in addition to identifying regularities in intraday trading patterns, the FDA curves have helped us identify outliers in the data. Outliers are very important when using historical data for analysis, backtesting, and designing algorithms. Quite often, an outlier will have a disproportionate effect on an algorithm's profitability, whether when backtesting it against historical events or running it live in the market. Thus, it is crucial to keep track of these outliers and account for them in the design and evaluation of algorithms. In our historical analysis of the intraday pattern of volume, AAPL's mean daily volume function drawn in the right panels is lower

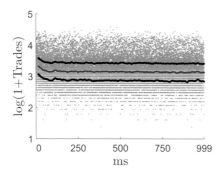

**Figure 4.3** Trading pattern within a second.

than it would have been if we had not included those four particular dates in the estimation.

In order to contrast the intraday behaviour of AAPL with that of a less frequently traded asset, Figure 4.2 displays the same analysis for MENT. We observe that the daily peaks at the beginning and the end of the day are also there, though slightly modified, and distorted by the discreteness of trading lots. For MENT, the initial burst of trading is not as frequent as with AAPL. As we can discern from the pictures on the right, there are a substantial number of days for which trading starts unusually slowly. These slow days balance out the bursts of trading from other days, so that on average volume in early trading does not seem to differ substantially from that during the rest of the day. It is at the end of the day that we see a substantial amount of trading activity in MENT.

We have omitted ISNS and FARO from this analysis, as the frequency of trading for those assets is even lower than for MENT, and thus there are many more zero-volume observations. The resulting figures are qualitatively similar, although there is a lot more noise in the estimation, and a much larger number of zero trading intervals.

### 4.2.2 Intrasecond Volume Patterns

When working at time intervals much finer than a second, a natural question to ask is whether we observe time patterns at such small intervals, like the ones we found over the duration of the day. Focusing only on AAPL we look at the millisecond trading pattern, that is, for each transaction we look at the millisecond in which it occurred. In Figure 4.3 we display the average number of transactions at each millisecond for each day in 2013 (AAPL), as well as the quartiles (Q1, median and Q3) for the daily means for each millisecond (the quartiles have been smoothed using moving averages).[1]

We see that there is hardly a persistent pattern at the millisecond level, although an initial spike is observable at the 000-020 ms range, followed by a subtle

---

[1] To improve the visual presentation the 10 highest realisations in the year have been removed from the figure.

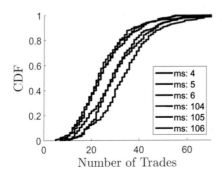

**Figure 4.4** Cumulative distribution function of transaction count for specific milliseconds.

valley that ends around the 100 ms point. To explore this in more detail, in Figure 4.4 we identify the empirical cumulative distribution function of the number of transactions ending at six different milliseconds: three early ones (at 4, 5, and 6 milliseconds), and three later ones (at 104, 105, and 106 ms). From the figure we can see that the early milliseconds stochastically dominate the other three. (The choice of 104-106 is arbitrary and the same pattern is observed if we choose other milliseconds to compare with what happens at 4, 5, and 6 – sometimes even more starkly.)

This pattern suggests that there is an unusual number of transactions that are recorded just after the exact beginning of a second. A plausible explanation is that there may be an unusual number of transactions that are entered (automatically) at the exact end/beginning of a second and what we observe is the latency or clock-asynchronicity of these orders (machines).

## 4.2.3    Price Patterns

It is quite standard to look at volume patterns, but not price patterns. In this section we ask whether executions at prices that end in multiples of 5 cents (round values) are different from those that do not. There is no strong fundamental reason why the price of an asset, which in theory represents a fraction of the income for shareholders generated by a firm, should have a round value, such as $450.25 or $21.00. However, if we look at the frequency with which we observe transactions taking place at different prices grouped by the number of cents in the price, we find the pattern displayed in Figure 4.5.[2] The figure also shows the quartiles Q1, median and Q3 as solid lines.

The patterns are quite evident. There is a very large accumulation of transactions whose prices end in exact dollar-valued prices, a large number of transactions with prices that end in 50 cents, and spikes of larger than usual numbers of transactions at prices that end in units of 10 cents, and even 5 cents (these differences are for the most part statistically significant).

---

[2] To improve the visual presentation the 10 highest realisations in the year have been removed from the picture and the quartile estimation (10 out of 98,280).

Trade by cents in Price (AAPL)

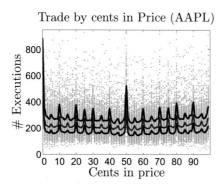

**Figure 4.5** Price pattern: frequency of executions by number of cents in the price.

There is a quite straightforward interpretation for this phenomenon, which is that for some reason (rational or not) there is a preference for providing liquidity at prices that end in round cent values. We use the term 'preference for providing liquidity', as transaction prices are (mostly) determined by an aggressive MO filling a standing posted LO, so it is liquidity providers who decide to accept a larger number of executions at a particular price level.

Why would agents provide liquidity in this way? We can hypothesise that there are a number of stop-loss orders and momentum orders programmed to execute as MOs at round prices. These could be latent, having been programmed by agents who decide to enter/exit when the price moves beyond a certain barrier, which is psychologically or conveniently set at a whole number. This type of reasoning is consistent with chartist ideas such as 'price supports' and 'price ceilings'. If the above reasoning is correct, then the accumulation of executions at those prices may be triggered by psychologically-based demand for liquidity which is then happily provided by agents who do not have such psychological inclinations and expect the unusual demand for liquidity at these prices to be unjustified by economic/market microstructure fundamentals (and hence, a source of profit).

## 4.3 Trading and Market Quality

Financial markets play a key role in helping a market economy to allocate resources over time and uncertainty. Financial markets provide a forum for firms to raise capital, and facilitate investor participation in the general economic progress of the economy. In this context, the stock market provides a forum where equity holders can convert their equity into cash (and vice versa) quickly and at a reasonable price. In this section we look at different ways of measuring the market's effectiveness in this role under the generic heading of 'market quality'.

In subsection 4.2.1 we used two basic arguments to explain intraday patterns: that new information increases trading volume, and that an increased desire to trade (e.g. due to increased trader urgency) interacts with the quality of the market which feeds back to motivate further trading. Market quality enters directly into the second argument: an expected increase in volume generates expectations of better market quality, that is of improved effectiveness of the market in facilitating trade via lower execution costs (spreads), greater price efficiency (less mean-reversion in price levels or lower transitory volatility), etc., which induces greater volume. In the first argument, market quality enters indirectly, as it modulates the relationship between the exogenous forces of new information and traders' desire to execute trades. Either way, if the quality of the market varies, trading activity will vary with it.

So what determines market quality? What determines the effectiveness of the market in facilitating trade? Naturally, the direct cost of trading matters: how much does one pay for shares one wishes to buy; what is the price one pays relative to one's opinion of its market value; how much does one value the information obtained from the market, and how easy is it to complete a transaction? Think of a medieval cattle market. Suppose one lives on a farm which is equidistant from two towns that hold their weekly cattle market on the same day of the month. How does one choose between them? One will probably go to the market that is most likely to offer the best price, and which can be obtained after an easy sale process, and with the best guarantees that the transaction is finalised and one can walk away with the money. In a financial market, where agents are buying and selling, these dimensions along which to evaluate the quality of a market are: having sufficient information to identify the true market value of the asset, being able to buy (or sell) any quantity at prices sufficiently close to the asset's value, and having the confidence that the deals are honoured. Of course, one may have qualms about the existence of such a thing as 'the true market price' of an asset, but regardless, it is a useful concept to work with and is the basis of much of the literature. If one does not believe there is or can ever be such a thing as the true market price, the concept is still useful as a theorectical construct in the study of market microstructure, in the same way as the concept of an ideal gas is useful in physics.

From our short list of dimensions of market quality, the last one (honouring of deals) is usually taken for granted, although when prices suffer large fluctuations we do see some transactions being cancelled by the exchange (usually because they occur at 'ridiculous' prices). The other two issues are captured by measures of market quality such as spreads, **price impact**, volatility, **resilience**, depth, probability of informed trading (referred to as **PIN**), etc. Spreads measure the immediate cost of executing a trade aggressively; price impact measures the cost of executing larger trades via their impact of trading on prices; volatility measures the effectiveness of the price in transmitting information about the market value of an asset; resilience is related to market impact and measures the market's ability to return to equilibrium after a trade; depth measures the amount of visible liquidity in the market, and PIN measures the degree of information asymmetry in the market and hence, like volatility, the ability of the market to transmit information about the market value of an asset. We now look at spreads, volatility, depth and price impact.

## 4.3.1 Spreads

Spreads measure the execution costs of small transactions by measuring how close the price of a trade is to the market price. The first problem, naturally, is to determine what is the 'true market price'. The simplest, and most common approach, is to use the midprice,

$$S_t = \frac{1}{2}(a_t + b_t), \tag{4.1}$$

the simple average of the bid ($b_t$) and the ask ($a_t$) price. This reference is based on the economic concept that the market price is the equilibrium price, the price at which demand equals supply, and in a market with frictions that generate a gap between the best buy price (the ask) and the best sell price (the bid), the equilibrium should lie somewhere in between. The midprice is the simplest way to estimate this market price, although, as we saw in Chapter 2, the spread may arise for different reasons (compensation for inventory risk, or adverse selection from trading against more informed traders) and in some cases the 'true price' may be closer to the bid or the ask.

We saw an alternative estimate of the market price earlier as well, the microprice defined in (3.3). This seems more meaningful economically (and to develop algorithmic trading strategies) as it incorporates the quantities offered to sell and buy at the bid and ask (respectively) to weigh the bid and ask prices, and which may better reflect some of the microstructure issues described above. There are other, more sophisticated models that try to estimate the equilibrium price, e.g. by separating movements in the midprice into a temporary and a permanent component (the permanent being the equilibrium price), but we do not treat them here.

The two most common spread measures are the **quoted** and the **effective spread**, both of which use the midprice as the market price. The quoted spread, QS, is the difference between the ask and the bid prices,

$$QS_t = a_t - b_t,$$

and represents the potential cost of immediacy: the difference in price between posting an LO at the best price and aggressively executing an MO (and hence 'crossing the spread') at any point in time. It also reflects distance from the market price, if one takes the midprice as reference. The direct trading cost of a market sell order would be $S_t - b_t = QS_t/2$, while that of a market buy order would be $a_t - S_t = QS_t/2$ (the quoted half-spread).

In contrast, the effective (half-)spread, ES, measures the realised difference between the price paid and the midprice, which for a market buy order is

$$ES_t = a_t - S_t,$$

while for a market sell order it is

$$\mathrm{ES}_t = S_t - b_t .$$

For an MO executed in full on an exchange against a visible LO, the effective spread is equal to the quoted halfspread (if it does not walk the LOB). Sometimes it will be greater, if it does walk the LOB, or smaller, if it is matched with a **hidden order** inside the spread – it could even be negative, if the hidden order was aggressively posted. (We saw hidden orders in the context of different order types in subsection 1.3.4 and we will discuss this further in subsection 4.5 – a hidden order is an LO that is posted in the LOB but is not visible to market participants.) A negative effective spread reflects that one is buying at a price below or selling at a price above the 'market price' (represented by the midprice). In empirical analysis, these spreads are usually normalised and expressed in bps relative to the midprice.

In Table 4.2 we look at the quoted spread for our four assets during 2013 (ordered from least to most traded). For each asset we compute the time-weighted average quoted spread, tQS, for each minute of the day. This is calculated as follows: for each minute of the day, $t = 1 : 390$, while the market is open (from 9:30-16:00),

$$\mathrm{tQS}_t = \sum_{i=1}^{n-1} (\tau_{i+1} - \tau_i) \, \mathrm{QS}_{t_i} ,$$

where $i \in \{1, \ldots, n\}$ indexes the time (in minutes) at which the quoted spread changes during minute $t$, $\tau_i$. Table 4.2 describes the statistics for the minutes of every day in 2013 (252 trading days), for each asset.

| Asset | Mean | StdDev | P01 | Q1 | Median | Q3 | P99 |
|-------|------|--------|-----|-----|--------|-----|------|
| ISNS  | 33.2 | 270.8  | 2.0 | 11.0 | 22.0  | 40.0 | 129.2 |
| FARO  | 23.9 | 192.0  | 2.4 | 8.9  | 12.0  | 16.6 | 71.0 |
| MENT  | 3.5  | 27.4   | 1.0 | 1.0  | 1.1   | 2.0  | 13.9 |
| AAPL  | 13.6 | 54.7   | 5.4 | 11.0 | 13.8  | 16.9 | 29.3 |

**Table 4.2** Time-average Quoted Spread (in cents).

The descriptive statistics of the resulting sample are provided in Table 4.2. The first thing to note is that the data on the table suggest that more frequently traded assets trade at lower spreads. This positive relationship between volume and market quality can work both ways: volume attracts liquidity and improves market quality, or higher market quality facilitates trade and generates greater volume.

Nevertheless, AAPL seems to have an enormous spread. But, recalling the discussion on tick size earlier, the large spread for AAPL is an illusion as we have not adjusted for the relative tick size. The average midprices (at the end of each minute) for our assets are: ISNS $5.25, FARO $40.62, MENT $19.93, and AAPL $473.00. This implies that the median quoted spreads are: for ISNS 419bps, FARO 29.5bps, MENT 5.5bps and AAPL 2.9bps. By evaluating spreads relative to the midprice we recover the expected relationship between quoted spreads and volume.

Also from Table 4.2 we can compute the interquartile range (as a percentage of the median). From these calculations we find that more frequently traded assets tend to have less volatile quoted spreads (the numbers are: 1.32bps for ISNS, 0.64bps for FARO, 0.91bps for MENT, and 0.42bps for AAPL).

The MENT example illustrates another aspect of the importance of tick sizes which is of great interest, especially to regulators. In the US (for assets with prices greater than one dollar), the minimum tick size is legally fixed at one cent – there are ways to trade in fractions of a cent but the one cent minimum is binding in most cases. Imposing a minimum tick size of one cent may affect trading for some assets, such as MENT. From Table 4.2 we can see that for almost 50 percent of all minutes, the one cent minimum is constraining MENT's quoted spread at that level (one cent). This translates to a possibly significantly large relative minimum quoted spread (around 5bps for MENT) and may be limiting the market quality for this asset.

A final comment on Table 4.2: the numbers in this table are contaminated by an event we mentioned earlier, in subsection 4.2.1, namely that the data is not corrected for trading stops, and in particular for the trading halt during August 22nd. We do not include the corrected table as they are not significantly different and so it is not necessary, but only because our data set includes all the minutes in 2013, almost 100,000 observations per asset, so the overall effect on the statistical aggregates is very small. Nevertheless, we wanted to take this opportunity to point out the importance of knowing the details of your dataset. In the ITCH dataset, all messages are recorded and timestamped, *even when the market is halted and there is no trading*. We mentioned earlier how on August 22nd the NASDAQ halted trading for three hours. During that time, messages kept coming into the exchange and were time-stamped. In particular, many orders were cancelled and the 'ask' and 'bid' moved dramatically. As trading was suspended, these fluctuations led to huge and also negative *artificial* spreads, and they contaminate the data in Table 4.2 (primarily the mean and standard deviation, although the effect is small). If we wanted to use our analysis for trading or designing an algorithm, especially if it involves unsupervised/deep learning, the unfiltered data could generate significant distortions.

We have also computed the effective (half-)spreads. In order to obtain numbers that are comparable to those in Table 4.2 we doubled the effective spreads and included them in Table 4.3. Again we have constructed one-minute buckets, and for each, we have computed the quantity-weighted effective spreads, qES, for our four assets:

| Asset | Mean | StdDev | P01 | Q1 | Median | Q3 | P99 |
|-------|------|--------|-----|-----|--------|-----|-----|
| ISNS  | 12.56 | 45.00 | -42.00 | 3.50 | 9.23 | 19.00 | 65.00 |
| FARO  | 7.63  | 8.61  | -10.00 | 3.33 | 6.50 | 10.76 | 32.84 |
| MENT  | 1.23  | 1.57  | 0.00   | 1.00 | 1.00 | 1.03  | 5.38  |
| AAPL  | 9.32  | 3.85  | 2.67   | 6.61 | 8.83 | 11.47 | 20.20 |

**Table 4.3** Quantity-weighted Effective spread (in cents).

$$qES_t = \sum_{j=1}^{m} \frac{q_j}{\sum_{s=1}^{m} q_s} ES_j ,$$

where $j \in \{1, \ldots, m\}$ indexes the trades that took place during minute $t$, $q_j$ denotes the number of shares in trade $j$, and $ES_j$ is the effective spread for trade $j$.

Effective spreads differ from quoted spreads in several ways. Earlier we saw that the effective spread is equal to the quoted halfspread when a trade executes against a visible LO and does not walk the LOB. In our dataset, trades are recorded via the execution of the posted LO, so we do not have information on the MO that was sent to the market. This implies that none of our executions walk the LOB. This biases our measure of the ES downwards, but the bias is small, as we see very few executions of LOs away from the bid/ask during previous milliseconds (a necessary condition for an MO to walk the LOB at NASDAQ, as the remaining quantity may need rerouting in search of best execution in all markets). Another reason why this bias is small, is that in the fragmented US market, when an MO comes into a market and it is greater than the depth at the bid/ask, the part that is not executed is usually routed to other markets, and only under very special circumstances will it literally walk the LOB. Note, that in general this rerouting makes it virtually impossible to reconstruct the quantity of a large incoming MO without specific information from the agent who sent it.

Thus, our measured ES has to be equal to or lower than the (current) quoted spread. A visible trade will generate ES $=$ QS/2. As not all posted LOs are visible, some trades will be executed at prices better than the bid/ask. This will generate an ES that is strictly smaller than the QS, and may even produce a *negative* ES. One obtains a negative effective spread if an incoming market buy (sell) order meets a hidden sell (buy) order that is below (above) the midprice. That this occurs is evident by looking at the first percentile of the empirical distribution for ES (the P01 (first percentile) statistic in Table 4.3).

There is another difference between ES and QS, namely that ES can only be measured when there is a trade, while quoted spreads are always observable. Therefore, it may be possible that quoted spreads differ from effective spreads if market conditions around trades are systematically different from those without

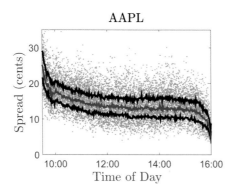

**Figure 4.6** Intraday spread pattern: interquartile range for one-minute returns.

trades. Looking at the interquartile range and the standard deviations of ES and QS in Tables 4.2 and 4.3 we find that ES is less volatile than QS. This would happen if executions tend to be concentrated around times of narrow quoted spreads, something we explore in more detail in subsection 4.4.

We have seen that assets with greater trading frequency have better market quality in the sense that execution costs for small trades (the quoted spread) is smaller. If we look at the intraday pattern of trades we find further evidence that lower execution costs occur when trading is high (as anticipated by the theoretical discussions in Chapter 2, where we saw that if the number of informed traders stays constant and the number of uninformed traders increases, the spread shrinks). In Figure 4.6 we plot the one-minute time-averaged quoted spreads for AAPL in 2013, as well as the first quartile, median and third quartile. As the figure shows, quoted spreads are initially high, decline rapidly during the first half-hour of trading, and are mostly constant throughout the remainder of the day until the last (half) hour of trading, when the spread rapidly declines. This pattern in quoted spreads is also seen in the effective spread. Recall that in Figure 4.2 we saw that the afternoon is associated with increased trading, and hence we find, as hypothesised, that during a period with a constant flow of information more trading and lower spreads occur together.

This connection between trading volume and spreads fails during the morning where the situation is completely reversed: declining volumes go hand-in-hand with declining spreads. This can be explained by appealing to the other factor affecting volume which we discussed earlier, namely information. When the market opens, and during the subsequent hour of trading, the market absorbs all the information that has accumulated since the last market close. This would explain the heavier trading. But a lot of new information is also associated with a great deal of uncertainty. Theoretically, as we saw in Chapter 2, in the presence of greater uncertainty it is optimal to post wider bid-ask spreads, and in the market making algorithms developed in Chapter 10, see for example Section 10.3, we show that greater price uncertainty increases the depths of the quotes that a risk-averse market makers sends to the LOB. Thus, more information at the beginning of the day explains the coincidence of wider spreads with greater volume.

## 4.3.2    Volatility

We now consider another dimension of market quality, namely volatility. Volatility measures price fluctuations and represents a cost (i.e. low market quality) in the sense that a rapidly changing price makes it difficult to determine the actual market price of the asset. Of course, one may observe price changes because the true market value of the asset is changing, and hence the literature differentiates between fundamental volatility and microstructure noise. The first captures the fluctuations in the true market price, while microstructure noise represents extraneous fluctuations due to the way the market operates. There is a large (and growing) number of measures of raw volatility (unconditional volatility which does not distinguish fundamental volatility from microstructure noise) and of microstructure volatility.

For simplicity we use the term volatility to refer to raw volatility, and we measure the volatility of asset returns, rather than of asset prices. We have seen several measures of volatility when studying the relationship between volume and volatility in Section 4.1. The simplest such measure is the realised volatility: the standard deviation of a sample of returns. We have seen volatility measured using the square (or the absolute value) of the return – this is useful if you have very few observations and you are working in a sufficiently small time scale so that the mean return can be safely assumed to be (essentially) equal to zero. Another common alternative is to use the range of the return (or price): e.g. by taking the difference between the maximum (max) and minimum (min) values of the price over a certain interval and normalising it by either the minimum value, the mean/median, the initial value, or the average of the min and the max. Here we look at realised volatility, the range of returns, and the number of times the bid or the ask changes.

| Asset | Mean | StdDev | P01 | Q1 | Median | Q3 | P99 |
|---|---|---|---|---|---|---|---|
| ISNS | 16.6 | 54.8 | 0.0 | 0.0 | 0.0 | 14.4 | 160.3 |
| FARO | 8.3 | 12.7 | 0.0 | 3.8 | 6.6 | 10.3 | 31.3 |
| MENT | 5.6 | 6.6 | 0.0 | 3.2 | 4.6 | 6.5 | 20.1 |
| AAPL | 5.5 | 4.2 | 1.0 | 3.3 | 4.7 | 6.7 | 18.1 |

**Table 4.4** Realised one-min volatility (15 min samples).

Table 4.4 displays the statistical properties of realised volatility measured as $\sigma_t$, the standard deviation of one-minute returns over fifteen minute periods (for every day in 2013), that is, for every 15-minute period (each 15-minute period indexed by $t$),

$$\sigma_t^2 = \sum_{j=1}^{15} \left( r_j - \frac{1}{15} \sum_{s=1}^{15} r_s \right)^2,$$

where $j \in \{1, \ldots, 15\}$ is the index for each of the individual minutes within the 15 minute period $(t)$, and $r_j$ is the realised return for minute $j$ in $t$. What we observe as we go from AAPL to ISNS in Table 4.4 is that the more frequently traded asset also has a higher mean volatility.

| Asset | Mean | StdDev | P01 | Q1 | Median | Q3 | P99 |
|-------|------|--------|-----|-----|--------|-----|-----|
| ISNS | 0.10 | 1.37 | 0.00 | 0.00 | 0.00 | 0.00 | 0.00 |
| FARO | 3.77 | 2.33 | 1.10 | 2.50 | 2.84 | 4.26 | 12.05 |
| MENT | 3.00 | 3.47 | 0.00 | 0.00 | 2.18 | 5.04 | 12.84 |
| AAPL | 6.23 | 2.85 | 3.72 | 4.72 | 5.28 | 6.45 | 19.47 |

**Table 4.5** Interquartile Range of one-minute returns.

Table 4.5 displays an alternative way to look at the same idea, only now we are looking at the statistical properties of a different variable, sampled over smaller time intervals. In Table 4.5 we include the statistics for the interquartile range of one-minute returns. That is, for AAPL, 5.28bps is the median of 252 observations, one for each trading day, of the interquartile range observed for the one-minute returns during that day.

Because the sampling method is distinct we observe some interesting differences. ISNS displays a zero interquartile range for most days. This is natural, as it is an asset that displays very few price movements – the median realised 15-minute volatility is zero as we saw in Table 4.4. By focusing on the median interquartile range for each day, this sampling method misses the very large but relatively rare price movements that are responsible for the high volatility numbers for ISNS in Table 4.4.

A different effect is responsible for the differences between MENT and FARO. Despite MENT having similar trading activity than FARO, it has lower volatility. MENT has more than 25% of days with an interquartile range equal to zero, but it also has lower realised one-minute volatility. The difference between MENT and FARO has probably much more to do with MENT's relative tick size. As we saw above when looking at the quoted spread, the one cent tick size is a binding constraint for MENT most of the time. This leads to an unusual degree of price stickiness, as many small price movements are not sufficient to push the bid or ask a whole cent (5 bps) away from their current levels. Thus, despite having similar activity levels as FARO, its price displays lower volatility.

In subsection 3.5 we looked at the non-Markovian nature of price changes. We found that there is a significant tendency (at least for AAPL on July 30th, 2013) for price movements to reverse themselves. Thus, looking at the volatility of

| Asset | Mean | StdDev | P01 | Q1 | Median | Q3 | P99 |
|-------|------|--------|-----|----|--------|----|----|
| ISNS | 2 | 29 | 0 | 0 | 0 | 0 | 16 |
| FARO | 11 | 25 | 0 | 0 | 3 | 13 | 100 |
| MENT | 6 | 18 | 0 | 0 | 2 | 7 | 75 |
| AAPL | 150 | 149 | 7 | 64 | 109 | 185 | 709 |

**Table 4.6** Number of Changes in the ask or bid.

one-minute returns misses many such price movements. To account for all price changes, we construct yet another measure of volatility: we count the number of changes in the bid and ask within a one-minute period, and report the statistics in Table 4.6.

We find that the less frequently traded assets (ISNS, FARO and MENT) also have more stable prices (bid and ask) – at least 25 percent of the time we see no price changes at all. For ISNS this is even more marked, as it happens at least 75 percent of the time. Nevertheless, the average is 2 per minute, suggesting that price changes occur infrequently but when they do, there are a lot of them. For MENT and FARO we see more price movements than for ISNS which is consistent with what we found earlier. MENT sees fewer price changes than FARO even though there is more trading in the former stock, but this is linked to the issue of minimum tick size discussed above.

The statistics in Table 4.6 indicate that AAPL displays almost two orders of magnitude more price changes than MENT or FARO. However, the realised volatility of its return is lower than that of FARO and similar to that of MENT (in Table 4.4). We interpret this as reflecting the interaction of small relative tick size and large frequency of trading. A one cent price change for AAPL (with an average price of around $500 in 2013) is 0.2 bps. An asset with such a small relative tick size and with such a large trading activity is bound to have a price level that is very sensitive (and hence generates many changes in the bid/ask within a minute), but most of the resulting changes will be rapidly reversed (as we saw in Tables 3.5 and 3.6), generating relatively low realised volatility (Table 4.4).

To conclude our discussion of volatility, in Figure 4.7 we look at how volatility (in terms of the interquartile range of one-minute returns) changes over the course of one trading day. The figure uses dots to displays the interquartile range for each minute of the day, estimated using all trading days in 2013 (390 observations – one per minute). The lines represent fitted quartic curves. The blue line is fitted with standard OLS while the red line is fitted using robust OLS, which controls for outliers by reducing the weight of the more extreme observations at the beginning and the end of the day.

Comparing the intraday volatility pattern in Figure 4.7 to the intraday volume pattern in Figure 4.2, we see a common pattern: high at the beginning of the

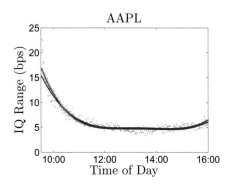

**Figure 4.7** Intraday volatility: interquartile range for one-minute returns.

day, lower as the day progresses until it reaches a plateau between noon and 15:00, followed by an increase until market close. While Figure 4.7 shows a large left-slanted smile for the volatility pattern, Figure 4.2 shows a more symmetric volume smile and, if anything, slanted to the right. This is consistent with trading at the beginning of the day being subject to more uncertainty and also being more informationally driven, while trading at the end of the day is driven less by information, and possibly more by traders rushing to close their positions.

### 4.3.3 Market Depth and Trade Size

| Asset | Mean | StdDev | P01 | Q1 | Median | Q3 | P99 |
|-------|------|--------|-----|-----|--------|-----|------|
| ISNS  | 619  | 787    | 51  | 150 | 300    | 750 | 3,250 |
| FARO  | 142  | 125    | 14  | 86  | 122    | 171 | 484  |
| MENT  | 661  | 694    | 117 | 351 | 527    | 784 | 2,852 |
| AAPL  | 189  | 169    | 64  | 127 | 161    | 210 | 662  |

**Table 4.7** Average Depth at the Bid and Ask (number of shares).

Market quality is not just about the informational content of prices or the cost of executing a small order, it is also about depth. By depth, we mean the volume posted in the LOB and available for immediate execution. In this section we focus mostly on the volume at-the-touch, that is at the bid and ask price levels. Mutual and pension funds manage a large fraction of people's wealth and they need markets to adjust their positions, pay their investors, and evaluate their performance. These funds move large quantities of shares. In any one day a fund may want to buy or sell thousands, tens of thousands, or even more shares of any one company. Table 4.7 illustrates how unreasonable it is to think that the market will match those trades at the published bid/ask prices. The table shows the distribution of the one-minute time-weighted average of the quantity of shares available at the ask and bid on NASDAQ for every day in 2013. This number does not exceed 1,000 shares in 75% of cases for any one of the four

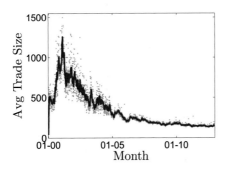

**Figure 4.8** Average monthly trade size, AAPL 2000-2013.

assets (not even AAPL). Recall (see Table 3.1) that in 2013, NASDAQ intraday trading represents around 22 percent of total volume.

Depth and trade size are not independent of one another: the decision of how much to trade depends on the expected availability of shares resting as LOs for immediate execution, and similarly, the decision of how much to offer will depend on the expected flow of incoming MOs. If the depth is thin (few orders resting in the LOB), MOs will be small – which implies that in thin markets, relatively urgent large orders that would walk the LOB need to be broken up into smaller MOs which are then sequentially executed over a period of time.

The institutional, legal, technological and economic changes of the last 15 years have produced a steady decline of the average trade size (the number of shares in a single trade). We can see this for AAPL in Figure 4.8 which displays the monthly average trade size, computed from CRSP data – the dots are the monthly averages and the dark line is a smoothed representation (a moving average). It shows a sharp increase in the early 2000s to a peak of around 1,300 shares per trade, and a steady decline to around 200 shares by the end of 2013. (Nowadays, with the 7:1 split, the numbers should be different, though a similar pattern is observable in most stocks.) Table 4.8 gives the statistics on NASDAQ trade sizes (total number of shares traded, $Q$, divided by the number of trades, $n$) for 2013. As can be seen there, trade sizes in AAPL are smaller than for the other three assets, though not substantially so.

| Asset | Mean | StdDev | P01 | Q1 | Median | Q3 | P99 |
|-------|-------|--------|------|------|--------|-------|--------|
| ISNS | 206.2 | 348.7 | 1.0 | 100.0 | 100.0 | 200.0 | 1600.0 |
| FARO | 99.8 | 76.4 | 1.0 | 64.0 | 100.0 | 100.3 | 374.8 |
| MENT | 199.9 | 172.9 | 6.0 | 100.0 | 150.0 | 240.3 | 867.1 |
| AAPL | 121.1 | 40.5 | 52.2 | 95.7 | 115.4 | 139.0 | 252.8 |

**Table 4.8** Average Size of a Trade ($Q/n$).

When developing algorithms that provide liquidity to the market, the depth, captured by the shape of the LOB, is critical because this dictates where a trader

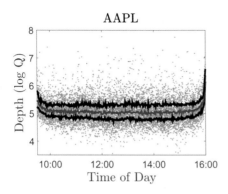

**Figure 4.9** Intraday depth: (log) average quantity of shares posted at the bid and at the ask.

should post her LOs. We already discussed a simple model of market making in Section 2.1.4, and in subsequent chapters we develop these ideas further in the context of optimising market making and optimal execution algorithms. For instance, Chapter 8 looks at optimal execution (buying or selling large positions) when the agent employs LOs and possibly also MOs. In Chapter 10 the shape of the LOB plays a critical role in the optimal posting of liquidity for a market maker, and we consider how different assumptions about the shape of the book affect the optimal posting strategy.

To close this discussion we look at the intraday patterns of depth in Figure 4.9. This figure shows the intraday pattern of the average quantity of shares posted at the bid and ask every minute of 2013. As expected, there is the usual sharp increase at the end of the trading day. This is consistent with greater market quality in the form of narrower spreads (as seen earlier, in Figure 4.6) and the increased desire of traders to close positions at the end of the day. Somewhat, though only marginally, surprising is that depth is also higher at the beginning of the day, where we hypothesised that high price uncertainty leads to wider spreads. The figure suggests that the theoretical trade-off between the benefits of market making from increased order arrival and the cost from higher uncertainty discussed in subsection 2.1.1 is resolved in practice in favour of providing greater liquidity.

### 4.3.4 Price Impact

A main concern for participants that wish to execute a large order is that the order will have an adverse price impact: increasing the price when buying aggressively and lowering it when selling. There are several variables that can be used to measure the price impact of an order. Measuring depth, as we have just done, gives us a measure of price impact, in the sense that the depth at the bid/ask tells us how large a market order can be with a zero price impact, i.e. without walking the LOB.

But, in a single market, a large order would consume all the volume at the best quote and work its way through to the next tick and so on until the order is filled. Thus, whenever an MO walks the LOB, the average price per share is worse than the best quote at the moment the MO was sent to the LOB. In Chapter 6 we have a thorough discussion of how to devise optimal strategies to minimise the market impact of large orders.

So far we have only considered what happens to the MO as it reaches the LOB, but executing a (relatively large) trade can be quite complex. In the US, with 11 public exchanges, regulators have felt a need to regulate what should happen to orders that consume all existing liquidity in one venue at the best bid/ask, in order to protect investors. This regulation and the multiplicity of exchanges raise related issues of time delays between different venues. So, when sending an MO one has to design the routing strategy very carefully: to which exchanges and when to submit the order, and what will happen if that order consumes all available liquidity at some point. So, an agent needs to know what happens as the order is executed, but also what happens in the aftermath of a trade.

With the information we have from NASDAQ we now look at what happens after orders are executed there. Table 4.9 looks at the executions for AAPL on NASDAQ on July 30th, 2013. As mentioned earlier (in Chapter 3), we do not observe MOs, but rather, what happens to existing posted orders. To illustrate the kind of analysis that can be done, we make the simplistic assumption that all orders executed at the same time (same millisecond) and at the same price level, are all part of the same MO, and are aggregated accordingly.

Table 4.9 looks at what happens when different types of buy and sell orders are executed, i.e. we look at the arrival of MOs under different circumstances. The row labelled "Benchmark" captures what happens on average by including every 10 ms interval during the trading day (2.34 million observations). This serves as benchmark with which to compare what happens in the 10 ms time interval AFTER an MO comes in. We compare this benchmark with what happens after the following six events:

- Buy[Sell] Order: an order is executed against orders posted on the ask (sell-side) [bid (buy-side)] of the book;
- Buy[Sell] Order (n-o): same as Buy [Sell] Order but ignoring buy [sell] orders within 10 ms of a previous buy [sell] order;
- Buy[Sell] Order-Large: buy [sell] orders for strictly more than 300 shares.

In Table 4.9 we include statistics on order arrival on the bid and ask side in the benchmark case, as well as after these six events. The columns labelled '$n = 0$' describe the percentage of cases in which we do not observe an MO arriving on the bid/ask side after an execution. In the benchmark case we see that despite the high level of trading activity in AAPL, our interval size is sufficiently small so that in roughly 99.4 percent of cases we do not observe an order arriving. After any type of execution, this proportion falls on both sides of the book. We observe that after a Buy MO there is at least a second MO arriving in 30 percent of cases, and at least one sell MO in 5 percent of cases. A similar pattern is observed after a Sell MO: in at least 25 percent of cases a sell MO is followed by at least one other MO, while in at least 5 percent of cases a sell MK is followed by a buy MO. We also see very similar patterns if we exclude MO arrivals that occur just after other MOs. The number of arrivals, however, increases substantially after a large order: a large buy (sell) is followed by other buys in 43 (38) percent of cases, and by sells in 8 (8) percent of cases. If we look at the quartiles conditional on there being at least one MO (the columns labeled 'Q1', 'Q2' and 'Q3'), the arrival of a market buy (sell) order seems to have no clear effect on the distribution of arriving sell (buy) orders, but we do see evidence suggesting that incoming orders on the same side may be more frequent. So, our very preliminary and limited analysis suggests that order arrival seems to be followed by further order arrival on both sides of the book, and more on its own side than on the other side of the book.

| Event | $n$ | Buy Market Orders | | | | Sell Market Orders | | | |
|---|---|---|---|---|---|---|---|---|---|
| | | $n = 0$ | Q1 | Q2 | Q3 | $n = 0$ | Q1 | Q2 | Q3 |
| Benchmark | 2,340,000 | 99.6 | 1 | 1 | 2 | 99.7 | 1 | 1 | 2 |
| Buy Order | 6,852 | 70.7 | 1 | 2 | 3 | 94.3 | 1 | 1 | 2 |
| Buy Order (n-o) | 5,707 | 71.5 | 1 | 2 | 3 | 94.4 | 1 | 1 | 2 |
| Buy Order-Large | 532 | 57.5 | 1 | 2 | 5 | 92.1 | 1 | 1 | 2 |
| Sell Order | 7,358 | 94.5 | 1 | 1 | 2 | 74.9 | 1 | 2 | 3 |
| Sell Order (n-o) | 6,269 | 94.7 | 1 | 1 | 2 | 75.7 | 1 | 2 | 3 |
| Sell Order-Large | 347 | 92.2 | 1 | 1 | 2 | 62.2 | 1 | 2 | 4 |

**Table 4.9** Market Impact of an execution on MOs (AAPL 20130730).

In Table 4.10 we do a similar analysis where we look at how different events affect the bid and ask prices. For this table we follow the convention that +1 is a one cent move away from the best price, that is: if the ask price is $453.02, a +1 in the ask is a change from $453.02 to $453.03, while on the bid side, with the bid price at $452.96, +1 in the bid is a price drop of one cent, that is a change from $452.96 to $452.95. With this convention, positive price changes represent moves away from the midprice and negative price changes represent moves towards the midprice, which allows us to provide a more streamlined presentation of the different effects of MOs on bid and ask sides.

Returning to Table 4.10, we consider how different orders affect the best price on their own side of the LOB, that is, the left side of the table describes how the ask price reacts to an aggressive buy MO, and the right decribes how the bid side reacts to an aggressive sell MO.

We consider two benchmark cases: the column 'Ask' ('Bid') is the benchmark case that looks at average changes in the ask (bid), that is after every 10 ms interval. The first row tells us the percentage of time for which there was no change in the bid (99.5 percent) and no change in the ask (also 99.5 percent). We also look at what happens to the ask (bid) after a 'Buy' ('Sell') order comes in, and the percentage of times when the ask (bid) stays the same drops to

| Ticks | Ask | Changes in ASK | | | Bid | Changes in BID | | |
|---|---|---|---|---|---|---|---|---|
| | | Buys | $\Delta \neq 0$ | $\Delta > 3c$ | | Sells | $\Delta \neq 0$ | $\Delta > 3c$ |
| Obs | | 6,852 | 3,259 | 1,165 | | 7,358 | 4,052 | 1,910 |
| 0 | 99.5 | 28.2 | 10.7 | 9.7 | 99.5 | 22.7 | 7.2 | 6.1 |
| $\geq$-5 | 12.8 | 0.1 | 0.0 | 0.0 | 17.2 | 0.3 | 0.4 | 0.2 |
| -4 | 4.5 | 0.1 | 0.1 | 0.3 | 4.5 | 0.0 | 0.0 | 0.0 |
| -3 | 5.6 | 0.1 | 0.1 | 0.0 | 5.6 | 0.1 | 0.1 | 0.0 |
| -2 | 8.6 | 0.1 | 0.1 | 0.1 | 7.0 | 0.1 | 0.1 | 0.1 |
| -1 | 22.6 | 1.1 | 0.4 | 0.7 | 17.9 | 0.8 | 0.4 | 0.5 |
| 1 | 14.2 | 19.5 | 20.5 | 0.9 | 13.3 | 15.7 | 14.3 | 0.9 |
| 2 | 7.4 | 13.6 | 13.6 | 0.9 | 7.1 | 11.8 | 11.3 | 0.9 |
| 3 | 5.6 | 12.3 | 11.9 | 1.9 | 5.5 | 10.4 | 10.2 | 0.9 |
| 4 | 4.4 | 9.8 | 9.2 | 15.3 | 4.3 | 9.5 | 9.4 | 12.2 |
| $\leq$5 | 14.3 | 43.4 | 44.2 | 80.0 | 17.5 | 51.3 | 53.9 | 84.4 |

**Table 4.10** Market Impact of an execution on the best price – own side (AAPL 20130730).

28 (23) percent. This percentage falls even further if we only look at executions that sweep the order book ($\Delta \neq 0$), that is after a buy (sell) order that generates an immediate change in the ask (bid). Such 'large' executions are more long-lived in the sense that 10 ms after such a change the probability that the ask (bid) has returned to its pre-order arrival level drops to 11 (7) percent. The columns labelled '$\Delta > 3c$' look at the subset of the executions that sweep the order book, and we also observe a large (greater than three cent) change in the ask (bid) price respectively. The likelihood of returning is smaller than that for all sweep orders but not by much.

The rows of Table 4.10 (except the 'Obs' and '0' rows) reflect the distribution of price movements conditional on different non-zero price changes. The benchmark distributions for bid and ask price movements are symmetric and very similar, something that is not true for the distributions after MOs arrivals. After a buy (sell) order, the distribution of the ask (bid) clearly shifts away from its previous level and is almost never better (closer to the midprice) than before the arrival of the MO 10 ms later. The difference we observe for a sweep order seems to be centred on the probability of returning to the pre-arrival level, but does not seem to have much effect on the distribution of price changes for non-zero changes. However, large price swings do seem to be followed by changes in the distribution of bid/ask price changes, and we see little evidence that these large price movements are reversed within 10 ms.

In Table 4.11 we repeat the analysis but looking at the effect of an order arrival on the other side of the book, that is how the arrival of a buy (sell) MO affects the bid (ask). We keep the signs so that a positive move in Table 4.10 is also a positive move on the other side of the book in Table 4.11. That is, suppose the

| | | Changes in BID | | | | Changes in ASK | | |
|---|---|---|---|---|---|---|---|---|
| Ticks | Bid | Buys | $\Delta \neq 0$ | $\Delta > 3c$ | Ask | Sells | $\Delta \neq 0$ | $\Delta > 3c$ |
| Obs | | 6852 | 3259 | 1165 | | 7358 | 4052 | 1910 |
| 0 | 99.5 | 81.6 | 78.7 | 75.4 | 99.5 | 82.2 | 80.3 | 79.1 |
| $\geq$-5 | 17.5 | 6.9 | 6.5 | 7.7 | 14.3 | 6.0 | 6.0 | 8.3 |
| -4 | 4.3 | 1.8 | 1.7 | 2.8 | 4.4 | 1.7 | 1.8 | 2.0 |
| -3 | 5.5 | 1.8 | 1.9 | 3.1 | 5.6 | 2.1 | 2.1 | 2.8 |
| -2 | 7.1 | 3.2 | 3.9 | 3.1 | 7.4 | 3.4 | 3.3 | 4.3 |
| -1 | 13.3 | 8.7 | 8.2 | 9.1 | 14.2 | 8.0 | 8.1 | 8.5 |
| 1 | 17.9 | 29.5 | 28.7 | 34.8 | 22.6 | 29.2 | 27.7 | 28.8 |
| 2 | 7.0 | 10.8 | 11.8 | 11.8 | 8.6 | 11.6 | 12.9 | 13.5 |
| 3 | 5.6 | 7.0 | 6.3 | 7.3 | 5.6 | 6.5 | 6.4 | 6.0 |
| 4 | 4.5 | 6.1 | 5.0 | 2.4 | 4.5 | 6.8 | 5.6 | 4.5 |
| $\leq$5 | 17.2 | 24.2 | 25.9 | 17.8 | 12.8 | 24.7 | 26.1 | 21.3 |

**Table 4.11** Market Impact of an execution on the best price – other side (AAPL 20130730).

ask price is \$453.02 and the bid is \$452.96. After a buy order, a +1 cent change in the ask is an increase from \$453.02 to \$453.03 (Table 4.10), and a +1 cent move in the bid is an increase from \$452.96 to \$452.97 (Table 4.11). Whereas after a sell order, a +1 cent change in the bid results in a decrease from \$452.96 to \$452.95 (Table 4.10), and a +1 cent move in the ask results is a change from \$453.02 to \$453.01 (Table 4.11).

With this convention, we see that the effect of an arrival on one side of the LOB is followed by a similar but weaker effect on the other. The probability of the price remaining/returning to the pre-arrival level drops from 99.5 to 82 for both the bid and the ask after a buy and a sell order arrive, respectively. This probability is slightly smaller for (intermarket) **sweep orders**. We also see a shift in the distribution of non-zero price changes that (weakly) follows that of the changes on the other side of the book. So we see how the arrival of a buy order is followed by a shift in the (non-zero) bid price changes away from the midprice, so the conditional probability of a 1 cent move away from the pre-arrival bid price goes from 17.9 to 29.5 percent after a buy order, and that of a 1 cent move away from the pre-arrival ask price goes from 22.6 to 29.2 percent after a sell order. The pattern is very similar after a buy (sell) order, a sweep buy (sell) order, or a sweep buy (sell) order with a large price move. Combining this observation with the price moves in Table 4.10, we find evidence that the quoted spread increases after a buy or sell order, and substantially so after a large sweep order.

To conclude our look at the impact of MOs, in Table 4.12 we look at the effect on the changes we observed at the 10 ms horizon, we consider longer (30 ms, 100 ms and 1,000 ms) horizons. Table 4.12 is split horizontally into three sections:

| | Ticks | Changes in ASK | | | | Changes in BID | | | |
|---|---|---|---|---|---|---|---|---|---|
| | | 10 | 30 | 100 | 1,000 | 10 | 30 | 100 | 1,000 |
| Bench | <= −3 | 0.1 | 0.3 | 1.0 | 7.7 | 0.1 | 0.4 | 1.2 | 8.9 |
| | {−1,−2} | 0.2 | 0.4 | 1.2 | 7.7 | 0.1 | 0.3 | 0.7 | 4.0 |
| | 0 | 99.5 | 98.7 | 96.0 | 72.9 | 99.5 | 98.6 | 95.7 | 70.3 |
| | {1,2} | 0.1 | 0.3 | 0.7 | 3.9 | 0.1 | 0.4 | 1.1 | 6.9 |
| | >= 3 | 0.1 | 0.3 | 1.0 | 7.7 | 0.1 | 0.4 | 1.3 | 9.9 |
| Buys | <= −3 | 0.2 | 0.3 | 0.4 | 1.3 | 6.9 | 9.8 | 13.2 | 26.2 |
| | {−1,−2} | 0.8 | 1.0 | 1.4 | 3.1 | 7.4 | 7.8 | 8.3 | 11.3 |
| | 0 | 28.2 | 26.7 | 24.3 | 16.0 | 81.6 | 77.8 | 72.4 | 49.2 |
| | {1,2} | 23.8 | 22.4 | 21.8 | 18.4 | 2.2 | 2.4 | 2.6 | 4.1 |
| | >= 3 | 47.0 | 49.6 | 52.1 | 61.3 | 1.9 | 2.2 | 3.5 | 9.2 |
| Sells | <= −3 | 6.8 | 8.3 | 12.0 | 24.0 | 4.1 | 4.2 | 4.6 | 6.8 |
| | {−1,−2} | 7.3 | 8.3 | 9.3 | 13.3 | 4.1 | 4.0 | 4.3 | 6.2 |
| | 0 | 82.2 | 78.8 | 73.4 | 52.3 | 47.3 | 44.6 | 41.1 | 28.1 |
| | {1,2} | 2.0 | 2.1 | 2.6 | 4.0 | 15.9 | 15.7 | 15.0 | 14.5 |
| | >= 3 | 1.8 | 2.4 | 2.7 | 6.4 | 28.5 | 31.4 | 35.0 | 44.4 |

**Table 4.12** Market Impact of an execution on the midprice over time (AAPL 20130730).

the first ('Bench') is the benchmark table that looks at changes in the bid and ask over the corresponding horizons for all such time intervals; the bottom two sections consider the effects of the arrival of a buy and a sell order respectively on bid and ask prices. For this table we continue to keep the signs matched on the bid and ask sides, but to avoid confusion we keep the sign of changes on the bid (ask) side the same as in the benchmark case, as well as after a buy or a sell order, that is, the interpretation of the sign does not depend on whether it follows a buy or a sell MO, but only on which side of the book we are looking at. So, suppose the ask price is $453.02 and the bid is $452.96. After a buy order, a +1 cent in the ask is a move from $453.02 to $453.03, and a +1 cent move in the bid is a move from $452.96 to $452.95 (a 1 cent move away from the midprice). The same happens after a sell order (and in the benchmark case): a +1 cent in the ask results in an increase from $453.02 to $453.03, and a +1 cent move in the bid is move from $452.96 to $452.95 (one cent away from the midprice). Note also, that all percentiles reflect probabilities (we are not conditioning on non-zero price movements in this table).

The first thing to notice in Table 4.12 is the natural effect of time on all prices: as we expand the horizon, prices tend to move more, and the distributions become more dispersed. We also see that the initial price movements are not followed by quick reversals and that even one second (1,000 ms) after a buy order there is a marked shift of the bid and ask away from its pre-execution level, with worse prices and a hint of a delayed price impact on future executions and wider spreads.

All these results must be interpreted in context, and not causally. As we will now see, MOs do not arrive at random times. They tend to arrive when spreads are narrow, and opportunistically hit orders that are posted closer to the midprice, so it is only natural that we should observe a wider spread after an execution.

### 4.3.5        Walking the LOB and Permanent Price Impact

We have seen that one of the key ingredients in trading algorithms is how the investor's own actions together with the order flow of the other market participants affect the price of the assets she is trading in. In the trading algorithms developed in Part III we show how strategies depend on the market impact of trades. For example in Chapter 6 we show how to trade large positions when the investor's own trades walk the LOB, in addition to adversely affecting the midprice by exerting upward (downward) pressure in the drift of the midprice if the investor is buying (selling). In Chapter 7 we study the problem of an agent wishing to liquidate a large position when the order flow from other traders in the market also impacts the midprice. In this case, if the agent's execution programme is going with or against net order flow, the strategy adapts to maximise the revenues from liquidating the position.

Here we want to empirically assess the parameter values for the different effects a trade can have on prices: the permanent and the temporary price impact. We look at these impacts for five stocks using data from NASDAQ and for the year 2013. A first approach is to estimate these separately. We first estimate the permanent price impact by looking at the impact of order flow on the change in price over five-minute intervals. Let $\Delta S_n = S_{n\tau} - S_{(n-1)\tau}$ be the change in the midprice during the time interval $[(n-1)\tau, n\tau]$ where $\tau = 5\,\mathrm{min}$. Let $\mu_n$ be the net order flow defined as the difference between the volumes of buy and sell MOs during the same time interval. We then estimate the permanent price impact as the parameter $b$ in the following robust linear regression:

$$\Delta S_n = b\,\mu_n + \varepsilon_n, \tag{4.2}$$

where $\varepsilon_n$ is the error term (assumed normal). The model (4.2) is estimated every day, using Winsorised data, excluding the upper and lower 0.5% tails. The first row of Table 4.13 shows the average value of the estimated parameters for permanent price impact and the second row shows its standard deviation.

In the third and fourth rows of the table we show the parameter estimate for temporary impact and its standard deviation respectively. To estimate this parameter, which we denote by $k$, we assume that temporary price impact is linear in the volume traded. Specifically, the difference between the execution price that the investor receives and the best quote is $k\,Q$, where $Q$ is the total volume traded. To perform the estimation, we take a snapshot of the LOB each second, determine the price per share $S_t^{exec}(Q_i)$ for various volumes $\{Q_1, Q_2, \ldots, Q_N\}$ (by walking the LOB), compute the difference between the execution price per share and the best quote at that time, and perform a linear regression. That is we regress,

$$S_{i,t}^{exec,bid} = S_t^{bid} - k^{bid} Q_i + \varepsilon_{i,t}^{bid}, \qquad S_{i,t}^{exec,ask} = S_t^{ask} + k^{ask} Q_i + \varepsilon_{i,t}^{ask},$$

where $\varepsilon_{i,t}$ represent the estimation error of the $i^{th}$ volume for the $t^{th}$ timestamp.

The slope argument of the linear regression $\hat{k}$ is an estimate of the temporary price impact per share at that time. We do this for every second of every trading day and in the table we report the mean and standard deviation of these daily estimates (for the buy side) when we exclude the first and last half-hour of the trading day and Winsorise the data. Moreover, the fifth row shows the mean of the daily ratio $\widehat{b/k}$, and the sixth row shows its standard deviation. We observe that FARO shows the smallest ratio of 1.02 and SMH shows the largest at 7.43 – at the end of this section we discuss this ratio in more detail.

|  | FARO | SMH | NTAP | ORCL | INTC |
|---|---|---|---|---|---|
| $\hat{b}$ | $1.41 \times 10^{-4}$ | $5.45 \times 10^{-6}$ | $5.93 \times 10^{-6}$ | $1.82 \times 10^{-6}$ | $6.15 \times 10^{-7}$ |
|  | $(9.61 \times 10^{-5})$ | $(4.20 \times 10^{-6})$ | $(2.31 \times 10^{-6})$ | $(7.19 \times 10^{-7})$ | $(2.16 \times 10^{-7})$ |
| $\hat{k}$ | $1.86 \times 10^{-4}$ | $8.49 \times 10^{-7}$ | $3.09 \times 10^{-6}$ | $8.23 \times 10^{-7}$ | $2.50 \times 10^{-7}$ |
|  | $(2.56 \times 10^{-4})$ | $(8.22 \times 10^{-7})$ | $(1.75 \times 10^{-6})$ | $(3.78 \times 10^{-7})$ | $(1.25 \times 10^{-7})$ |
| $\widehat{b/k}$ | 1.02 | 7.43 | 2.04 | 2.28 | 2.55 |
|  | (0.83) | (6.24) | (0.77) | (0.74) | (0.70) |

**Table 4.13** Permanent and temporary price impact parameters for NASDAQ stocks for 2013. Below each parameter estimate we show its standard deviation.

Moreover, to showcase the variability of the permanent price impact parameter, the first panel of Figure 4.10 depicts the estimate of $b$ for each day of 2013 – the dashed line shows the average $\hat{b}$. The second panel in the figure shows a histogram of the five-minute net order flow using all the data in 2013. Finally, the last panel shows the expected net order flow (with error bars) conditional on a given price change being observed.[3] As already shown by the regression results there is a positive relationship between net order flow and price changes. The figure shows further details of this relationship to support the finding that when net order flow is positive (negative), that is more (less) buy than sell MOs, the midprice tends to increase (decrease). Moreover, we see that assuming a linear relationship between price changes and net order flow is plausible for a wide range of midprice changes. Only in the two extremes, very high or very low price changes, does the relationship fails to be linear, but we note that there are fewer

---

[3] For the year 2013, 99% of the 5 minute price changes for INTC were within the range [-0.1,0.1].

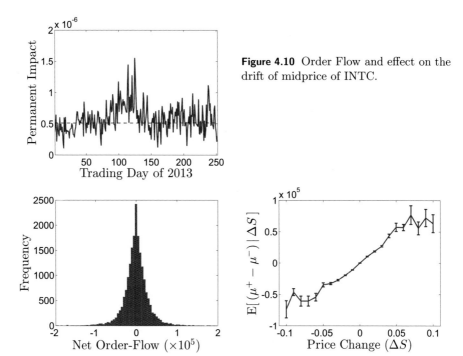

Figure 4.10 Order Flow and effect on the drift of midprice of INTC.

observations in the tails as shown by the histogram and this is also shown by the confidence intervals around the estimates.

Figure 4.11 explores the temporary price impact for INTC. The top panel shows a snapshot of the LOB for INTC on Nov 1, 2013 at 11am. The bottom left panel captures the empirical temporary price impact curve generated by hypothetical MOs of various quantities as they walk through the buy side of the LOB. Each curve represents the curve at every second from 11:00 to 11:01. We also include a linear regression with intercept set to the half-spread (the dash line) which would correspond to the model used to estimate the parameter $k$ above. Notice that the impact function fluctuates within the minute, and with it the impact that trades of different size could have. The linear regression provides an approximation of the temporary impact during that one minute.

The third picture in the Figure shows how the slope of this linear impact model fluctuates throughout the entire day. We see that the largest impact tends to occur in the morning, then this impact flattens and stays flat throughout the day, and towards the end of the day it lessens. Such a pattern is seen in a number of assets and is consistent with the reduction in spreads and increases in depth we have documented earlier.

The analysis above looks at temporary and permanent effects separately but their joint dynamics is a relevant quantity in execution algorithms. Liquidation and acquisition strategies take into account the trade-off between costs that stem from walking the book and the permanent impact. In particular, when both types

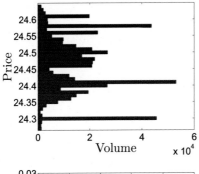

**Figure 4.11** An illustration of how the temporary impact may be estimated from snapshots of the LOB using INTC on Nov 1, 2013. The first panel is at 11:00am, the second from 11:00am to 11:01am and the third picture contains the entire day.

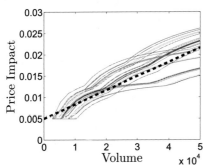

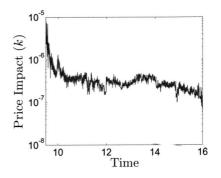

of impact are linear in the rates of trading, this trade-off is in part captured by the ratio $b/k$ (see for example Section 7.3 in the context of a liquidation algorithm and Chapter 9 for strategies that track volume such as POV and VWAP). In the left panel of Figure 4.12 we show a scatter plot of the daily pair $(b, k)$ for INTC which shows a clear positive relationship between temporary and permanent impact. This is consistent with the theoretical relationship between price impact and depth, so that days with little depth will be associated with high price impact, both permanent and temporary, while a deep market will be associated with lower price impacts. Finally, in the right panel of the figure we see the histogram for the ratio $b/k$ which ranges between 1 and 4 and is symmetric around approximately 2.5.

## 4.4      Messages and Cancellation Activity

An important feature in the way exchanges operate is the ability to cancel LOs which have not been filled. Traders who provide liquidity must be able to change their views on the market and therefore cancel their LOs or reposition them in the light of new information. Later, when we look at algorithms that provide liquidity to the market (see for instance Chapters 7, 8 and 10) we will see that these rely on the ability to reposition LOs in the LOB. For example, when we develop market making algorithms that require low latency, the agent is

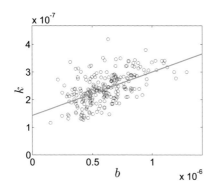 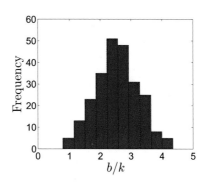

**Figure 4.12** Price Impact INTC using daily observations for 2013.

constantly cancelling LOs to reposition them as new information arrives and the agent's view on short-lived deviations in the midprice are taken into account.

Here we employ our detailed ITCH data to measure trading activity by the number of messages recorded by the exchange, where a 'message' is a line of data in the ITCH dataset, as we saw at the beginning of the chapter. The total number of messages is slightly greater than twice the number of posted orders, as most posted orders are either cancelled or executed in full.

| Asset | Mean | StdDev | P01 | Q1 | Median | Q3 | P99 |
|-------|------|--------|-----|-----|--------|-----|-----|
| ISNS | 1,711 | 6,078 | 173 | 450 | 760 | 1,745 | 8,943 |
| FARO | 24,038 | 10,871 | 8,524 | 16,277 | 22,347 | 29,232 | 71,445 |
| MENT | 59,661 | 21,755 | 23,157 | 43,477 | 53,972 | 72,639 | 131,715 |
| AAPL | 531,728 | 166,652 | 280,242 | 417,576 | 500,680 | 614,437 | 1,067,248 |

**Table 4.14** Daily Number of Messages (in 000s).

Table 4.14 contains the descriptive statistics for daily messages for our four assets (in thousands; P01 and P99 refer to the 1st and 99th percentile, respectively). We can see how, as with trading activity, the number of messages for each asset is different by orders of magnitude (except that for of MENT which is about 2.5 times that for FARO). In order to adjust for trading activity, a usual procedure is to normalise by the number of trades (as we do in Table 4.15) or by trading volume. The results in Table 4.15 suggest an interesting phenomenon: more frequently traded assets 'require' fewer messages per trade than less frequently traded ones.

The reverse of this phenomenon is captured in Table 4.16, where we look at the percentage of cancellations (out of all messages: posts, cancels and executions). Only for AAPL do we see less than 45% of orders being cancelled most of the time.

| Asset | Mean | StdDev | P01 | Q1 | Median | Q3 | P99 |
|-------|------|--------|------|------|--------|-------|--------|
| ISNS | 226.7 | 749.1 | 10.8 | 44.4 | 80.7 | 159.1 | 2885.1 |
| FARO | 88.2 | 55.8 | 20.7 | 57.8 | 79.4 | 106.1 | 223.5 |
| MENT | 70.0 | 21.8 | 29.8 | 54.2 | 66.5 | 83.2 | 134.2 |
| AAPL | 22.6 | 4.9 | 12.6 | 19.3 | 22.3 | 25.3 | 39.4 |

**Table 4.15** Messages per Number of Trades.

| Asset | Mean | StdDev | P01 | Q1 | Median | Q3 | P99 |
|-------|------|--------|------|------|--------|------|------|
| ISNS | 45.8 | 3.2 | 36.3 | 44.2 | 46.4 | 48.3 | 49.9 |
| FARO | 48.1 | 1.0 | 44.4 | 47.6 | 48.3 | 48.7 | 49.5 |
| MENT | 47.2 | 1.0 | 44.1 | 46.7 | 47.4 | 48.0 | 48.9 |
| AAPL | 43.1 | 1.9 | 37.8 | 41.8 | 43.3 | 44.3 | 47.1 |

**Table 4.16** Cancellations as percentage of Messages.

For intraday trading, it is very important to understand the posting and cancellation dynamics, especially around the bid and ask. Table 4.17 looks at the orders posted by their distance to the midprice ('Distance to Mid', $k$) for AAPL on July 30th, 2013 and to illustrate the contents of the table we use Figure 4.13. The second column ('Posts') counts the number of messages posted $k$ ticks (cents) from the midprice. In Figure 4.13 we display this visually using a hypothetical midprice of $101.05, and split the quantities evenly between the bid and the ask. Thus, for example, the total quantity posted two ticks from the midprice (3,053 units) is displayed as 1,527 units posted at $101.07 and 1,527 units posted at $101.03 (the total length of the bars). The third column ('% Exe') looks at the percentage of those posted messages that were executed, that is, the posted order was crossed with an incoming MO. In Figure 4.13 this is illustrated by using a lighter colour for the orders that were executed (and a darker one for those cancelled). Thus, for example, of the 1,527 units posted at $101.07, 64.7 percent (988 units) were executed. Finally, the fourth column ('Exe') describes the percentage of the total number of executed orders that were executed at that level. So, these 988 orders executed at $101.07 plus the 988 executed at $101.03, represent 9.2% of the total number of orders executed that day.

If we consider that the (one-minute time-average) quoted spread for that day is 10.3 cents on average (Q1: 8.5, median: 10.2, Q3: 11.7), most of the time the distance to the midprice (half of the quoted spread) is between 4 and 6 cents. Using the information in Table 4.17 we compute that 26% of orders executed were initially posted at between 1 and 3 cents from the midprice, but only 32% are posted between 4 and 6 and the remaining 42% of orders executed had been originally posted relatively far from the midprice (7+ cents away).

If, on the other hand, we look at the distance from the midprice at the time

| Distance | At Post | | | At Exit | | |
| to Midprice | Posts | % Exec | Exec | Posts | % Exec | Exec |
|---|---|---|---|---|---|---|
| < 2 | 905 | 78.6 | 3.3 | 988 | 88.4 | 4.1 |
| 2 | 3,053 | 64.7 | 9.2 | 3,508 | 76.4 | 12.5 |
| 3 | 5,193 | 55.4 | 13.5 | 6,236 | 67.5 | 19.7 |
| 4 | 5,617 | 44.4 | 11.7 | 6,448 | 51.5 | 15.6 |
| 5 | 6,374 | 34.9 | 10.4 | 7,557 | 45.5 | 16.1 |
| 6 | 7,626 | 27.6 | 9.8 | 7,586 | 29.9 | 10.6 |
| 7 | 7,996 | 20.2 | 7.6 | 7,624 | 20.4 | 7.3 |
| 8 | 7,826 | 15.9 | 5.8 | 8,062 | 14.2 | 5.4 |
| 9 | 7,675 | 12.3 | 4.4 | 7,946 | 7.5 | 2.8 |
| 10 | 7,967 | 8.6 | 3.2 | 7,487 | 6.1 | 2.1 |
| > 10 | 195,415 | 2.3 | 21 | 192,205 | 0.4 | 3.8 |

**Table 4.17** Messages by distance to midprice at post and at exit (AAPL 2013-07-30).

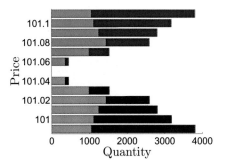

**Figure 4.13** Illustration of orders posted and executed as described in Table 4.17.

the trade was *executed*, not posted, in Table 4.17, we find that the distance to the midprice is (naturally) shorter, and we can compute that 36% of orders were executed at prices between 1 and 3 cents from the midprice, and 42% at prices between 4 and 6, so that only 22 % of executions were relatively far from the midprice (7+ cents away). This illustrates the point made earlier (when comparing the volatility of the effective and the quoted spreads) that executions tend to occur more often when the spread is narrower, and hence the effective spread will naturally be less volatile than the quoted spread.

In Figure 4.14 we display the survivor function, $S(x)$, (one minus the CDF: $S(x) = Pr(X > x) = 1 - F(x)$) of total executions, as the distance from the price at which the original LO was posted increases. This represents an approximation to the 'fill probability' – the probability that a posted order is executed. The thick blue line describes the distribution in Table 4.17. We have also included the same distribution separating executions on the bid and ask side, and it is interesting that the distribution for bid-(ask-)side executions lies systematically below (above) the one for all executions. This indicates that market buy orders

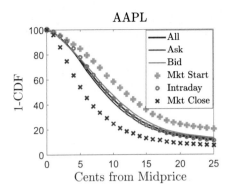

**Figure 4.14** Survivor function for executions as a function of distance from midprice.

tended to occur much closer to the midprice than market sell orders on this particular day, which had an overall positive order flow for AAPL shares and a slight price increase from market open to market close.

In Figure 4.14, we have also included total executions separated by the time of day: the first half hour after the market opens (Mkt Start), the last half hour before the market closes (Mkt Close), and the time in between (Intraday). We observe that Mkt Close tends to be below that of Intraday, implying that during the last half hour of trading, executions tend to be close to the midprice, which is consistent with the pattern of the quoted spread in Figure 4.6. But the difference does not seem to be very large and may be statistically insignificant.

What happens at the market open does look very different, as the distribution is above and quite far away from that for Intraday. It appears that the wider spreads we observed in Figure 4.6 and the uncertainty from Figure 4.7 combine to generate executions for orders posted quite far from the midprice.

Figure 4.15 looks at the same data from a different angle. In it we consider (in logs) the proportion of orders posted a certain distance from the midprice, that were eventually executed. Interpreting this proportion as a probability, the figure displays the natural decreasing relationship between the distance from the midprice and the probability of the order being executed. We have drawn these curves for: all executions, aggressive buys and sells, and executions by time of day: around the market open, the market close, and the rest of the day. All of them are very similar with only one exception: that for the first half hour of the trading day (Mkt Start). What we observe (looking at the underlying data) is that, at Mkt Start, an unusually high proportion of trades which were posted six cents from the midprice were later executed, and this generates the shift in the CDF we observe in Figure 4.15. Looking at the quoted spreads during that time, we find that the mean was 15.2 cents on average (Q1: 12.5, median: 14.2, Q3: 19.0), which suggests that as early morning uncertainty over the 'true market price' was reduced, the quoted spread was slow to react and a relatively large number of executions occurred – and this happened when the quoted spread had fallen to around 12 cents.

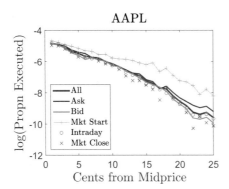

AAPL

**Figure 4.15** Log of the proportion of posted orders that are executed as a function of distance from midprice.

## 4.5 Hidden Orders

When discussing market quality earlier (Section 4.3), and spreads in particular, we saw that one of the reasons why the quoted spread is generally greater than the effective spread is the presence of posted orders that are not visible to market participants, but that will match with incoming MOs ahead of existing visible ones (at a price at or better than the current bid/ask). These are **hidden orders**.

| Asset | Mean | StdDev | P01 | Q1 | Median | Q3 | P99 |
|-------|------|--------|-----|----|--------|-----|-----|
| ISNS  | 4    | 59     | 0   | 0  | 0      | 0   | 100 |
| FARO  | 31   | 154    | 0   | 0  | 0      | 0   | 600 |
| MENT  | 117  | 568    | 0   | 0  | 0      | 0   | 2,150 |
| AAPL  | 3,849 | 5,905 | 0   | 1,052 | 2,220 | 4,504 | 26,547 |
| ISNS  | 1.2  | 10.7   | 0.0 | 0.0 | 0.0   | 0.0 | 99.6 |
| FARO  | 9.9  | 27.1   | 0.0 | 0.0 | 0.0   | 0.0 | 100.0 |
| MENT  | 9.4  | 24.1   | 0.0 | 0.0 | 0.0   | 0.0 | 100.0 |
| AAPL  | 44.6 | 16.9   | 0.0 | 33.5 | 44.9 | 56.0 | 83.7 |

**Table 4.18** Execution against hidden orders (volume ($Q$) and percentage).

Table 4.18 is split into two panels. The top panel of the table describes the quantity executed against hidden orders in NASDAQ per minute, for each minute of 2013. As we can see, for the less traded assets, ISNS, FARO and MENT, there is little trading taking place against hidden orders (less than 25 percent of the time), though when it happens it can be quite significant. But for AAPL, the case is quite different. We find trading against hidden orders more than 75 percent of the time, and for a substantial amount of shares (more than 1,000 units per minute). Note that these large quantities are not indicative of large trades, but rather of quite frequent ones: the distribution of the average size of an MO executed against a hidden order (per minute) has a mean of 127, with Q1 equal to 94 and Q3 to 148 shares per trade.

The bottom panel of Table 4.18 considers the same variable, the quantity of shares executed against hidden orders, but rather than in absolute numbers, as a proportion of the total number of shares executed (in that minute). For ISNS, FARO and MENT, executions are relatively infrequent, and when they occur against hidden orders they tend to be isolated trades. In those cases, the hidden order is a large proportion, if not one hundred percent, of all shares traded during that minute. For AAPL, execution against hidden orders is a common phenomenon and half the time they represent between 33 and 56 percent of all trades. An agent posting visible offers for AAPL at the bid and ask (during 2013) found her offers trumped by more aggressive hidden ones relatively often.

## 4.6    Bibliography and Selected Readings

Boehmer, Fong & Wu (2012), Chaboud, Hjalmarsson, Vega & Chiquoine (2009), Moallemi & Saglam (2013), SEC (2010), CFTC & SEC (2010), Hendershott, Jones & Menkveld (2011), Biais, Bisiere & Spatt (2010), Biais et al. (2005), Hagströmer & Norden (2013), Andersen & Bondarenko (2014), Pascual & Veredas (2009), Hirschey (2013), Martínez, Nieto, Rubio & Tapia (2005), Hasbrouck (2013), Hasbrouck & Saar (2013), Cartea & Meyer-Brandis (2010), Cartea (2013), Brogaard, Hendershott & Riordan (2014), Menkveld (2013), Riordan & Storkenmaier (2012), Hendershott & Riordan (2013), Foucault, Kadan & Kandel (2013), Moro, Vicente, Moyano, Gerig, Farmer, Vaglica, Lillo & Mantegna (2009), Gerig (2012), Hall & Hautsch (2007), Gould, Porter, Williams, McDonald, Fenn & Howison (2013).

# Part II

## Mathematical Tools

# Introduction to Part II

In this part of the book, we develop the mathematical tools for the analysis of trading algorithms: stochastic optimal control and stopping. It is written so that readers without previous exposure to these techniques equip themselves with the necessary tools to understand the mathematical models behind some algorithmic trading strategies. Readers are expected to have basic knowledge of continuous-time finance; however, the approach taken here is a pragmatic one and we do not delve into subtle mathematical issues. Focus is instead placed on the mechanics which allow for immediate application to algorithmic trading problems employed at low and high frequencies. To provide readers who need a refresher, and to keep the book self contained, we include in Appendix A a concise review of the main tools and results of Stochastic Calculus required to study stochastic control and stopping problems.

# 5 Stochastic Optimal Control and Stopping

## 5.1 Introduction

Stochastic control problems arise in many facets of financial modelling. The classical example is the optimal investment problem introduced and solved in continuous-time by Merton (1971). Of course there is a multitude of other applications, such as optimal dividend setting, optimal entry and exit problems, utility indifference valuation and so on. In general, the all-encompassing goal of stochastic control problems is to maximise (or minimise) some expected profit (cost) function by tuning a strategy which itself affects the dynamics of the underlying stochastic system, and to find the strategy which attains the maximum (minimum). For example, in the simplest form of the Merton problem, the agent is trying to maximise expected utility of future wealth by trading a risky asset and a risk-free bank account. The agent's actions affect her wealth, but at the same time the uncertain dynamics in the traded asset modulate the agent's wealth in a stochastic manner. The resulting optimal strategies are tied to the dynamics of the asset and perhaps also to the agent's wealth. It is a surprising fact that, in many cases, the optimal strategies turn out to be Markov in the underlying state variables, even if the agent is considering non-Markovian controls (which may depend on the entire history of the system).

One tool keeps coming to the forefront when solving stochastic control problems: the **dynamic programming principle** (DPP) and the related non-linear partial differential equation (PDE) known as the **Hamilton–Jacobi–Bellman** (HJB) equation – also called the **dynamic programming equation** (DPE). The DPP allows a stochastic control problem to be solved from the terminal date backwards and the HJB equation / DPE can be viewed as its infinitesimal version.

Here, the subtle mathematical issues are not addressed and focus is instead placed on the mechanics which allow for immediate application to algorithmic trading problems employed at low and high frequencies. The interested reader is referred to the many excellent texts which focus on the theoretical aspects of stochastic control for a thorough treatment of the subject: Yong & Zhou (1999), Fleming & Soner (2006), Øksendal & Sulem (2007), Pham (2010), and Touzi (2013).

## 5.2     Examples of Control Problems in Finance

This section provides a few examples of financially motivated stochastic control problems. The first example is a classical one in finance and pertains to optimal investment over long time horizons. The second is one of the first algorithmic trading control problems and pertains to the optimal liquidation of assets. The third refers to optimal placement of orders in a limit order book (LOB). All of these are essentially toy models and the last two will encompass the focus of many of the future chapters.

### 5.2.1     The Merton Problem

As a first example let us consider the classical portfolio optimisation problem of Merton (1971), in which the agent seeks to maximise expected (discounted) wealth by trading in a risky asset and the risk-free bank account (see Merton (1992) for many more examples and generalisations). Specifically, at time $t$, she places $\pi_t$ dollars of her total wealth $X_t$ in the risky asset $S_t$ and seeks to obtain the so-called **value function**

$$H(S,x) = \sup_{\pi \in \mathcal{A}_{0,T}} \mathbb{E}_{S,x}\left[U\left(X_T^\pi\right)\right], \tag{5.1}$$

which depends on the current wealth $x$ and asset price $S$, and the resulting **optimal trading strategy** $\pi$, where,

$$dS_t = (\mu - r)\,S_t\,dt + \sigma\,S_t\,dW_t, \qquad\qquad S_0 = S, \qquad \text{risky asset,} \quad (5.2a)$$
$$dX_t^\pi = (\pi_t\,(\mu - r) + r\,X_t^\pi)\,dt + \pi_t\,\sigma\,dW_t, \quad X_0^\pi = x, \quad \text{agent's wealth.} \quad (5.2b)$$

In the above, $\mu$ represents the (expected) continuously compounded rate of growth of the traded asset, $r$ is the continuously compounded rate of return of the risk-free bank account,

- $W = (W_t)_{\{0 \le t \le T\}}$ is a Brownian motion,
- $S = (S_t)_{\{0 \le t \le T\}}$ is the discounted price process of a traded asset,
- $\pi = (\pi_t)_{\{0 \le t \le T\}}$ is a self-financing trading strategy corresponding to having $\pi_t$ invested in the risky asset at time $t$ (with the remaining funds in the risk-free bank account),
- $X^\pi = (X_t^\pi)_{\{0 \le t \le T\}}$ is the agent's discounted wealth process given that she follows the self-financing strategy $\pi$,
- $U(x)$ is the agent's utility function (e.g., power $x^\gamma$, exponential $-e^{-\gamma x}$, and HARA $\frac{1}{\gamma}(x - x_0)^\gamma$),
- $\mathcal{A}_{t,T}$ is a set of strategies, called the **admissible set**, corresponding to all $\mathcal{F}$-predictable self-financing strategies that have $\int_t^T \pi_s^2\,ds < +\infty$. This constraint excludes doubling strategies and allows strong solutions to (5.2b) to exist.

In this classical example, the agent's trading decisions affect only her wealth process, but not the dynamics of the asset which she is trading. On long time scales, and if the agent's strategy does not change "too quickly", this is a reasonable assumption. However, if an agent is attempting to acquire (or sell) a large number of shares in a short period of time, her actions most certainly affect the dynamics of the price itself – in addition to her wealth process. This issue is ignored in the Merton problem, but is at the heart of research into algorithmic trading and specifically optimal execution problems which we introduce next.

## 5.2.2     The Optimal Liquidation Problem

As mentioned above, imagine that an agent has a large number of shares $\mathfrak{N}$ of an asset whose price is $S_t$. Furthermore, suppose her fundamental analysis on the asset shows that it is no longer a valuable investment to hold. She therefore wishes to liquidate these shares by the end of the day, say at time $T$. The fact that the market does not have infinite liquidity (to absorb a large sell order) at the best available price implies that the agent will obtain poor prices if she attempts to liquidate all units immediately. Instead, she should spread this out over time, and solve a stochastic control problem to address the issue. She may also have a certain sense of urgency to get rid of these shares, represented by penalising holding inventories different from zero throughout the strategy. If $\nu_t$ denotes the rate at which the agent sells her shares at time $t$, then the agent seeks the value function

$$H(x, S, q) = \sup_{\nu \in \mathcal{A}_{0,T}} \mathbb{E}\left[ X_T^\nu + Q_T^\nu(S_T^\nu - \alpha Q_T^\nu) - \phi \int_0^T (Q_s^\nu)^2 \, ds \right] \qquad (5.3)$$

and the resulting optimal liquidation trading strategy $\nu^*$, where,

$$dQ_t^\nu = -\nu_t \, dt, \qquad\qquad Q_0^\nu = q, \qquad\qquad \text{agent's inventory,} \qquad (5.4a)$$
$$dS_t^\nu = -g(\nu_t) \, dt + \sigma \, dW_t, \qquad S_0^\nu = S, \qquad \text{fundamental asset price,} \qquad (5.4b)$$
$$\hat{S}_t^\nu = S_t^\nu - h(\nu_t), \qquad\qquad \hat{S}_0^\nu = S, \qquad\qquad \text{execution price,} \qquad (5.4c)$$
$$dX_t^\nu = \nu_t \, \hat{S}_t^\nu \, dt, \qquad\qquad X_0^\nu = x, \qquad\qquad \text{agent's cash.} \qquad (5.4d)$$

In the above,

- $\nu = (\nu_t)_{\{0 \le t \le T\}}$ is the (positive) rate at which the agent trades (liquidation rate) and is what the agent can control,
- $Q^\nu = (Q_t^\nu)_{\{0 \le t \le T\}}$ is the agent's inventory,
- $W = (W_t)_{\{0 \le t \le T\}}$ is a Brownian motion,
- $S^\nu = (S_t^\nu)_{\{0 \le t \le T\}}$ is the fundamental price process,
- $g : \mathbb{R}_+ \to \mathbb{R}_+$ denotes the permanent (negative) impact that the agent's trading action has on the fundamental price,
- $\hat{S}^\nu = (\hat{S}_t^\nu)_{\{0 \le t \le T\}}$ corresponds to the execution price process at which the agent can sell the asset,

- $h : \mathbb{R}_+ \to \mathbb{R}_+$ denotes the temporary (negative) impact that the agent's trading action has on the price they can execute the trade at,
- $X^\nu = (X_t^\nu)_{\{0 \le t \le T\}}$ is the agent's cash process,
- $\mathcal{A}_{t,T}$ is the admissible set of strategies: $\mathcal{F}$-predictable non-negative bounded strategies. This constraint excludes repurchasing of shares and keeps the liquidation rate finite.

### 5.2.3 Optimal Limit Order Placement

In the optimal liquidation problem above, the agent is assumed to post market orders spread through time to liquidate her shares. Such a strategy is intuitively sub-optimal since she will consistently be crossing the spread and potentially walking the book in order to sell her shares. Since she may also place limit orders, she can at least save the cost of crossing the spread, and perhaps even achieve better performance by posting deeper in the limit order book LOB – at a depth of $\delta_t$ relative to the midprice $S_t$. The risk in doing so is that she may not execute her shares. Conditional on a market sell order arriving, the probability that it lifts the agent's posted offer at a price of $S_t + \delta_t$ can be modelled as a function of $\delta_t$ which we call the fill probability and denote by $P(\delta_t)$. The agent therefore can pose a control problem to decide how deep she must post in the LOB to optimise the value of liquidating her shares, subject to crossing the spread at the end of the trading horizon.

In this case, the agent's value function is given by

$$H(x, S, q) = \sup_{\delta \in \mathcal{A}_{[0,T]}} \mathbb{E}\left[ X_T^\delta + Q_T^\delta \left(S_T^\delta - \alpha\, Q_T^\delta\right) - \phi \int_0^T \left(Q_s^\delta\right)^2 ds \right],$$

where

| | | | |
|---|---|---|---|
| $M_t$ | | market sell orders, | (5.5a) |
| $S_t = S_0 + \sigma\, W_t,$ | | asset midprice, | (5.5b) |
| $dX_t^\delta = (S_t + \delta_t)\,(-dQ_t^\delta),$ | $X_0^\delta = x,$ | agent's cash, | (5.5c) |
| $dQ_t^\delta = -\mathbb{1}_{\left\{U_{M_t^{-1}+1} > P(\delta_t)\right\}} dM_t,$ | $Q_0^\delta = q,$ | agent's inventory. | (5.5d) |

and where $U_1, U_2, \dots$ are i.i.d. uniform random variables.

## 5.3 Control for Diffusion Processes

In this section, the control problems are of the general form

$$H(\boldsymbol{x}) = \sup_{\boldsymbol{u} \in \mathcal{A}_{0,T}} \mathbb{E}\left[ G(\boldsymbol{X}_T^{\boldsymbol{u}}) + \int_0^T F(s, \boldsymbol{X}_s^{\boldsymbol{u}}, \boldsymbol{u}_s)\, ds \right], \qquad (5.6)$$

where $\boldsymbol{u} = (\boldsymbol{u}_t)_{0 \leq t \leq T}$ is the vector (dim $p$) valued control process, $\boldsymbol{X}^u = (\boldsymbol{X}_t)_{0 \leq t \leq T}$ is the vector (dim $n$) valued controlled process assumed (in this section) to be an Itô diffusion satisfying

$$dX_t^u = \boldsymbol{\mu}(t, \boldsymbol{X}_t^u, \boldsymbol{u}_t)\, dt + \boldsymbol{\sigma}(t, \boldsymbol{X}_t^u, \boldsymbol{u}_t)\, d\boldsymbol{W}_t, \qquad \boldsymbol{X}_0^u = \boldsymbol{x}, \qquad (5.7)$$

where $(\boldsymbol{W}_t)_{0 \leq t \leq T}$ is a vector of independent Brownian motions, $\mathcal{A}$ is a set (called the **admissible set**) of $\mathcal{F}$-predictable processes such that (5.7) admits a strong solution (and may contain other constraints such as the process being bounded), $G : \mathbb{R}^n \mapsto \mathbb{R}$ is a terminal reward and $F : \mathbb{R}_+ \times \mathbb{R}^{n+p} \mapsto \mathbb{R}$ is a running penalty/reward. The running penalty/reward may, in general, be dependent on time $t$, the current position of the controlled process $\boldsymbol{X}_t^u$, and the control itself $\boldsymbol{u}_t$, while the terminal reward depends solely on the terminal value of the controlled process. For simplicity, the functions $G$ and $F$ are assumed uniformly bounded and the vector of drifts $\boldsymbol{\mu}_t$ and volatilities $\boldsymbol{\sigma}_t$ are, as usual, Lipschitz continuous. The integrability assumption on the controls, drift and volatility are necessary to ensure that the steps outlined below can be made rigorous. The predictability assumption on the controls is necessary since otherwise the agent may be able to peek into the future to optimise her strategy, and strategies which do peek into the future cannot be implemented in the real world.

The **value function** (5.6) has the interpretation that the agent wishes to maximise the total of terminal reward function $G$ and running reward/penalty by acting in an optimal manner. Her actions $u$ affect the dynamics of the underlying system in some generic way given by (5.7). Thus, her past actions affect the future dynamics and she must therefore adapt and tune her actions to account for this feedback effect.

For an arbitrary admissible control $u$, define the so-called **performance criteria** $H^u(\boldsymbol{x})$ by

$$H^u(\boldsymbol{x}) = \mathbb{E}\left[ G\left(\boldsymbol{X}_T^u\right) + \int_0^T F(s, \boldsymbol{X}_s^u, \boldsymbol{u}_s)\, ds \right]. \qquad (5.8)$$

The agent therefore seeks to maximise this performance criteria, and naturally

$$H(\boldsymbol{x}) = \sup_{\boldsymbol{u} \in \mathcal{A}_{0,T}} H^u(\boldsymbol{x}). \qquad (5.9)$$

As mentioned in the introduction, rather than optimising $H^u(\boldsymbol{x})$ directly, it is more convenient (and powerful) to introduce a time-indexed collection of optimisation problems on which a dynamic programming principle (DPP) can be derived. The DPP in infinitesimal form leads to a DPE – the Hamilton–Jacobi–Bellman (HJB) equation – which is a non-linear PDE whose solution is a tentative solution to the original problem. If a classical solution[1] to the DPE exists, it is possible to prove, through a verification argument, that it is in fact the solution to the original control. The next three subsections take on this programme.

---

[1] Here, a classical solution means that the solution is once differentiable in time and twice in all (diffusive) state variables, so that the infinitesimal generator can be applied to it.

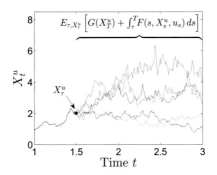

**Figure 5.1** The DPP allows the value function to be written as an expectation of the future value function. The key idea is to flow the dynamics of the controlled process from $t$ to $\tau$ and then rewrite the remaining expectation as the future performance criteria.

### 5.3.1  The Dynamic Programming Principle

The usual trick[2] to solving stochastic (and deterministic!) control problems is to embed the original problem into a larger class of problems indexed by time $t \in [0, T]$ but equal to the original problem at $t = 0$. To this end, first define (with a slight abuse of notation)

$$H(t, \boldsymbol{x}) := \sup_{\boldsymbol{u} \in \mathcal{A}_{t,T}} H^{\boldsymbol{u}}(t, \boldsymbol{x}), \quad \text{and} \tag{5.10a}$$

$$H^{\boldsymbol{u}}(t, \boldsymbol{x}) := \mathbb{E}_{t,\boldsymbol{x}} \left[ G(\boldsymbol{X}_T^{\boldsymbol{u}}) + \int_t^T F(s, \boldsymbol{X}_s^{\boldsymbol{u}}, \boldsymbol{u}_s) \, ds \right], \tag{5.10b}$$

where the notation $\mathbb{E}_{t,\boldsymbol{x}}[\cdot]$ represents expectation conditional on $\boldsymbol{X}_t^{\boldsymbol{u}} = \boldsymbol{x}$. These two objects are the time indexed analog of the original control problem and the performance criteria. In particular, $H(0, \boldsymbol{x})$ coincides with the original control problem (5.6) and $H^{\boldsymbol{u}}(0, \boldsymbol{x})$ with the performance criteria (5.8).

Next take an arbitrary admissible strategy $\boldsymbol{u}$ and imagine flowing the $\boldsymbol{X}$ process forward in time from $t$ to an arbitrary stopping time $\tau \leq T$. Then, conditional on $\boldsymbol{X}_\tau^{\boldsymbol{u}}$, the contribution of the running reward/penalty from $\tau$ to $T$ and the terminal reward can be viewed as the performance criteria starting from the new value of $\boldsymbol{X}_\tau^{\boldsymbol{u}}$ (see Figure 5.1). This allows the value function to be written in terms of the expectation of its future value at $\tau$ plus the reward between now and $\tau$.

More precisely, by iterated expectations the time-indexed performance criteria

---

[2]  There is another approach to controlling both deterministic and stochastic systems which makes use of Pontryagin's maximum principle. See Yong & Zhou (1999) for a wonderful exposition of its use for stochastic systems and the connection between it and the DPP formulation.

becomes

$$H^{\boldsymbol{u}}(t, \boldsymbol{x}) = \mathbb{E}_{t,\boldsymbol{x}} \left[ G(\boldsymbol{X}_T^{\boldsymbol{u}}) + \int_{\tau}^{T} F(s, \boldsymbol{X}_s^{\boldsymbol{u}}, \boldsymbol{u}_s)\, ds + \int_{t}^{\tau} F(s, \boldsymbol{X}_s^{\boldsymbol{u}}, \boldsymbol{u}_s)\, ds \right]$$

$$= \mathbb{E}_{t,\boldsymbol{x}} \left[ \mathbb{E}_{\tau, \boldsymbol{X}_\tau^{\boldsymbol{u}}} \left[ G(\boldsymbol{X}_T^{\boldsymbol{u}}) + \int_{\tau}^{T} F(s, \boldsymbol{X}_s^{\boldsymbol{u}}, \boldsymbol{u}_s)\, ds \right] + \int_{t}^{\tau} F(s, \boldsymbol{X}_s^{\boldsymbol{u}}, \boldsymbol{u}_s)\, ds \right]$$

$$= \mathbb{E}_{t,\boldsymbol{x}} \left[ H^{\boldsymbol{u}}(\tau, \boldsymbol{X}_\tau^{\boldsymbol{u}}) + \int_{t}^{\tau} F(s, \boldsymbol{X}_s^{\boldsymbol{u}}, \boldsymbol{u}_s)\, ds \right]. \tag{5.11}$$

Now, $H(t, \boldsymbol{x}) \geq H^{\boldsymbol{u}}(t, \boldsymbol{x})$ for an arbitrary admissible control $\boldsymbol{u}$ (with equality holding if $\boldsymbol{u}$ is the optimal control $\boldsymbol{u}^*$ – assuming that $\boldsymbol{u}^* \in \mathcal{A}_{t,T}$, i.e. the supremum is attained by an admissible strategy[3]) and an arbitrary $\boldsymbol{x}$. Hence, on the right-hand side of (5.11), the performance criteria $H^{\boldsymbol{u}}(\tau, \boldsymbol{X}_\tau^{\boldsymbol{u}})$ at the stopping time $\tau$ is bounded above by the value function $H(\tau, \boldsymbol{X}_\tau^{\boldsymbol{u}})$. The equality can then be replaced by an inequality with the value function (and not the performance criteria) showing up under the expectation:

$$H^{\boldsymbol{u}}(t, \boldsymbol{x}) \leq \mathbb{E}_{t,\boldsymbol{x}} \left[ H(\tau, \boldsymbol{X}_\tau^{\boldsymbol{u}}) + \int_{t}^{\tau} F(s, \boldsymbol{X}_s^{\boldsymbol{u}}, \boldsymbol{u}_s)\, ds \right]$$

$$\leq \sup_{u \in \mathcal{A}} \mathbb{E}_{t,\boldsymbol{x}} \left[ H(\tau, \boldsymbol{X}_\tau^{\boldsymbol{u}}) + \int_{t}^{\tau} F(s, \boldsymbol{X}_s^{\boldsymbol{u}}, \boldsymbol{u}_s)\, ds \right].$$

Note that on the right-hand side of the above, the arbitrary control $\boldsymbol{u}$ only acts over the interval $[t, \tau]$ and the optimal one is implicitly incorporated in the value function $H(\tau, \boldsymbol{X}_\tau^{\boldsymbol{u}})$ but starting at the point to which the arbitrary control $\boldsymbol{u}$ caused the process $\boldsymbol{X}$ to flow, namely $\boldsymbol{X}_\tau^{\boldsymbol{u}}$.

Taking supremum over admissible strategies on the left-hand side, so that the left-hand side also reduces to the value function, we have that

$$\boxed{H(t, \boldsymbol{x}) \leq \sup_{u \in \mathcal{A}} \mathbb{E}_{t,\boldsymbol{x}} \left[ H(\tau, \boldsymbol{X}_\tau^{\boldsymbol{u}}) + \int_{t}^{\tau} F(s, \boldsymbol{X}_s^{\boldsymbol{u}}, \boldsymbol{u}_s)\, ds \right].} \tag{5.12}$$

This provides us with a first inequality.

Next, we aim to show that the inequality above can be reversed. Take an arbitrary admissible control $\boldsymbol{u} \in \mathcal{A}$ and consider what is known as an $\boldsymbol{\varepsilon}$-**optimal control** denoted by $\boldsymbol{v}^\varepsilon \in \mathcal{A}$ and defined as a control which is better than $H(t, \boldsymbol{x}) - \varepsilon$, but of course not as good as $H(t, x)$, i.e. a control such that

$$H(t, \boldsymbol{x}) \geq H^{\boldsymbol{v}^\varepsilon}(t, \boldsymbol{x}) \geq H(t, \boldsymbol{x}) - \varepsilon. \tag{5.13}$$

Such a control exists, assuming that the value function is continuous in the space of controls. Consider next the modification of the $\varepsilon$-optimal control

$$\tilde{\boldsymbol{v}}^\varepsilon = \boldsymbol{u}_t \, \mathbb{1}_{t \leq \tau} + \boldsymbol{v}^\varepsilon \, \mathbb{1}_{t > \tau}, \tag{5.14}$$

---

[3] It may be the case that the supremum is obtained by a limiting sequence of admissible strategies for which the limiting strategy is in fact not admissible.

i.e. the modification is $\varepsilon$-optimal after the stopping time $\tau$, but potentially sub-optimal on the interval $[t, \tau]$. Then we have that

$$H(t, \boldsymbol{x}) \geq H^{\tilde{\boldsymbol{v}}^\varepsilon}(t, \boldsymbol{x})$$

$$= \mathbb{E}_{t,\boldsymbol{x}} \left[ H^{\tilde{\boldsymbol{v}}^\varepsilon}(\tau, \boldsymbol{X}_\tau^{\tilde{\boldsymbol{v}}^\varepsilon}) + \int_t^\tau F(s, \boldsymbol{X}_s^{\tilde{\boldsymbol{v}}^\varepsilon}, \tilde{\boldsymbol{v}}^\varepsilon{}_s)\, ds \right], \qquad \text{(from (5.11))},$$

$$= \mathbb{E}_{t,\boldsymbol{x}} \left[ H^{\tilde{\boldsymbol{v}}}(\tau, \boldsymbol{X}_\tau^{\boldsymbol{u}}) + \int_t^\tau F(s, \boldsymbol{X}_s^{\boldsymbol{u}}, \boldsymbol{u}_s)\, ds \right], \qquad \text{(using (5.14))},$$

$$\geq \mathbb{E}_{t,\boldsymbol{x}} \left[ H(\tau, \boldsymbol{X}_\tau^{\boldsymbol{u}}) + \int_t^\tau F(s, \boldsymbol{X}_s^{\boldsymbol{u}}, \boldsymbol{u}_s)\, ds \right] - \varepsilon, \qquad \text{(by (5.13))}.$$

Taking the limit as $\varepsilon \searrow 0$, we have

$$H(t, \boldsymbol{x}) \geq \mathbb{E}_{t,\boldsymbol{x}} \left[ H(\tau, \boldsymbol{X}_\tau^{\boldsymbol{u}}) + \int_t^\tau F(s, \boldsymbol{X}_s^{\boldsymbol{u}}, \boldsymbol{u}_s)\, ds \right].$$

Moreover, since the above holds true for every $\boldsymbol{u} \in \mathcal{A}$ we have that

$$H(t, \boldsymbol{x}) \geq \sup_{\boldsymbol{u} \in \mathcal{A}} \mathbb{E}_{t,\boldsymbol{x}} \left[ H(\tau, \boldsymbol{X}_\tau^{\boldsymbol{u}}) + \int_t^\tau F(s, \boldsymbol{X}_s^{\boldsymbol{u}}, \boldsymbol{u}_s)\, ds \right]. \qquad (5.15)$$

The upper bound (5.12) and lower bound (5.15) form the dynamic programming inequalities. Putting them together, we obtain the theorem below.

THEOREM 5.1 **Dynamic Programming Principle for Diffusions.** *The value function (5.10) satisfies the DPP*

$$H(t, \boldsymbol{x}) = \sup_{\boldsymbol{u} \in \mathcal{A}} \mathbb{E}_{t,\boldsymbol{x}} \left[ H(\tau, \boldsymbol{X}_\tau^{\boldsymbol{u}}) + \int_t^\tau F(s, \boldsymbol{X}_s^{\boldsymbol{u}}, \boldsymbol{u}_s)\, ds \right], \qquad (5.16)$$

*for all $(t, \boldsymbol{x}) \in [0, T] \times \mathbb{R}^n$ and all stopping times $\tau \leq T$.*

This equation is really a sequence of equations that tie the value function to its future expected value, plus the running reward/penalty. Since it is a sequence of equations, an even more powerful equation can be found by looking at its infinitesimal version – the so-called DPE.

## 5.3.2 Dynamic Programming Equation / Hamilton–Jacobi–Bellman Equation

The DPE is an infinitesimal version of the dynamic programming principle (DPP) (5.16). There are two key ideas involved:

(i) Setting the stopping time $\tau$ in the DPP to be the minimum between (a) the time it takes the process $\boldsymbol{X}_t^{\boldsymbol{u}}$ to exit a ball of size $\epsilon$ around its starting point, and (b) a fixed (small) time $h$ – all while keeping it bounded by $T$. This can be viewed as in Figure 5.2 and can be stated precisely as

$$\tau = T \wedge \inf \left\{ s > t : (s - t, |\boldsymbol{X}_s^{\boldsymbol{u}} - \boldsymbol{x}|) \notin [0, h) \times [0, \epsilon) \right\}.$$

Notice that as $h \searrow 0$, $\tau \searrow t$, a.s. and that $\tau = t + h$ whenever $h$ is

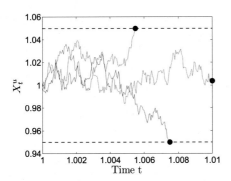

**Figure 5.2** The DPE is an infinitesimal version of the DPP where the stopping time $\tau$ is the first exit time from a ball of size $\epsilon$ or a small time $h$, whichever occurs first. This sample plot of three paths, and the corresponding value $X_\tau^u$ and stopping time $\tau$ indicated by the black circles for $\epsilon = 0.05$ and $h = 0.01$.

sufficiently small – since as the time span $h$ shrinks, it is less and less likely that $\boldsymbol{X}$ will exit the ball first.

(ii) Writing the value function (for an arbitrary admissible control $\boldsymbol{u}$) at the stopping time $\tau$ in terms of the value function at $t$ using Itô's lemma. Specifically, assuming enough regularity of the value function, we can write

$$
\begin{aligned}
H(\tau, &\boldsymbol{X}_\tau^{\boldsymbol{u}}) \\
&= H(t, \boldsymbol{x}) + \int_t^\tau (\partial_t + \mathcal{L}_s^{\boldsymbol{u}}) H(s, \boldsymbol{X}_s^{\boldsymbol{u}}) ds + \int_t^\tau \boldsymbol{D}_x H(s, \boldsymbol{X}_s^{\boldsymbol{u}})' \boldsymbol{\sigma}_s^{\boldsymbol{u}} d\boldsymbol{W}_s,
\end{aligned} \tag{5.17}
$$

where $\boldsymbol{\sigma}_t^{\boldsymbol{u}} := \sigma(t, \boldsymbol{X}_t^{\boldsymbol{u}}, \boldsymbol{u}_t)$ for compactness, $\mathcal{L}_t^{\boldsymbol{u}}$ represents the infinitesimal generator of $\boldsymbol{X}_t^{\boldsymbol{u}}$, and $\boldsymbol{D}_x H(\cdot)$ denotes the vector of partial derivatives with components $[\boldsymbol{D}_x H(\cdot)]_i = \partial_{x^i} H(\cdot)$. For example, in the one-dimensional case,

$$
\begin{aligned}
\mathcal{L}_t^u &= \mu_t^u \, \partial_x + \tfrac{1}{2}(\sigma_t^u)^2 \, \partial_{xx} \\
&= \mu(t, x, u) \, \partial_x + \tfrac{1}{2}\sigma^2(t, x, u) \, \partial_{xx} \ .
\end{aligned}
$$

As before, we derive the DPE in two stages by obtaining two inequalities. First, taking $\boldsymbol{v} \in \mathcal{A}$ to be constant over the interval $[t, \tau]$, applying the lower bound (5.15), and substituting (5.17) into the right-hand side implies that

$$
\begin{aligned}
H(t, \boldsymbol{x}) &\geq \sup_{\boldsymbol{u} \in \mathcal{A}} \mathbb{E}_{t, \boldsymbol{x}} \left[ H(\tau, \boldsymbol{X}_\tau^{\boldsymbol{u}}) + \int_t^\tau F(s, \boldsymbol{X}_s^{\boldsymbol{u}}, \boldsymbol{u}_s) \, ds \right] \\
&\geq \mathbb{E}_{t, \boldsymbol{x}} \left[ H(\tau, \boldsymbol{X}_\tau^{\boldsymbol{v}}) + \int_t^\tau F(s, \boldsymbol{X}_s^{\boldsymbol{v}}, \boldsymbol{v}) \, ds \right] \\
&= \mathbb{E}_{t, \boldsymbol{x}} \left[ H(t, \boldsymbol{x}) + \int_t^\tau (\partial_t + \mathcal{L}_s^{\boldsymbol{v}}) \, H(s, \boldsymbol{X}_s^{\boldsymbol{v}}) \, ds \right. \\
&\qquad \left. + \int_t^\tau \boldsymbol{D}_x H(s, X_s^{\boldsymbol{v}})' \boldsymbol{\sigma}_s^{\boldsymbol{v}} \, dW_s + \int_t^\tau F(s, \boldsymbol{X}_s^{\boldsymbol{v}}, \boldsymbol{v}) \, ds \right] \ .
\end{aligned}
$$

The integrand in the stochastic integral above, i.e. $\boldsymbol{D}_x H(s, \boldsymbol{X}_s^{\boldsymbol{v}})' \boldsymbol{\sigma}_s^{\boldsymbol{v}}$, is bounded on the interval $[t, \tau]$ since we have ensured that $|\boldsymbol{X}_t^{\boldsymbol{v}} - \boldsymbol{x}| \leq \epsilon$ on the interval. Hence, this stochastic integral is the increment of a martingale and we can be assured that its expectation is zero. Therefore,

$$H(t, \boldsymbol{x}) \geq \mathbb{E}_{t, \boldsymbol{x}} \left[ H(t, \boldsymbol{x}) + \int_t^\tau \left\{ (\partial_t + \mathcal{L}_s^{\boldsymbol{v}}) H(s, \boldsymbol{X}_s^{\boldsymbol{v}}) + F(s, \boldsymbol{X}_s^{\boldsymbol{v}}, \boldsymbol{v}) \right\} ds \right],$$

and recall that $\tau = t + h$.

Moving the $H(t, \boldsymbol{x})$ on the left-hand side over to the right-hand side, dividing by $h$ and taking the limit as $h \searrow 0$ yields

$$0 \geq \lim_{h \downarrow 0} \mathbb{E}_{t, \boldsymbol{x}} \left[ \frac{1}{h} \int_t^\tau \left\{ (\partial_t + \mathcal{L}_s^{\boldsymbol{v}}) H(s, \boldsymbol{X}_s^{\boldsymbol{v}}) + F(s, \boldsymbol{X}_s^{\boldsymbol{v}}, \boldsymbol{v}) \right\} ds \right]$$

$$= (\partial_t + \mathcal{L}_t^{\boldsymbol{v}}) H(t, \boldsymbol{x}) + F(t, \boldsymbol{x}, \boldsymbol{v}).$$

The second line follows from:

(i) as $h \searrow 0$, $\tau = t + h$ a.s. since the process will not hit the barrier of $\epsilon$ in extremely short periods of time,

(ii) the condition that $|\boldsymbol{X}_\tau^{\boldsymbol{u}} - \boldsymbol{x}| \leq \epsilon$, which implies that if the process does hit the barrier it is bounded,

(iii) the Mean-Value Theorem allows us to write $\lim_{h \downarrow 0} \frac{1}{h} \int_t^{t+h} w_s \, ds = w_t$, and

(iv) the process starts at $\boldsymbol{X}_t^{\boldsymbol{v}} = \boldsymbol{x}$.

Since the above inequality holds for arbitrary $\boldsymbol{v} \in \mathcal{A}$, it follows that

$$\partial_t H(t, \boldsymbol{x}) + \sup_{\boldsymbol{u} \in \mathcal{A}} \left( \mathcal{L}_t^{\boldsymbol{u}} H(t, \boldsymbol{x}) + F(t, \boldsymbol{x}, \boldsymbol{u}) \right) \leq 0. \tag{5.18}$$

Next, we show that the inequality is indeed an equality. To show this, suppose that $\boldsymbol{u}^*$ is an optimal control, then from (5.16), we have

$$H(t, \boldsymbol{x}) = \mathbb{E}_{t, \boldsymbol{x}} \left[ H(\tau, \boldsymbol{X}_\tau^{\boldsymbol{u}^*}) + \int_t^\tau F(s, \boldsymbol{X}_s^{\boldsymbol{u}^*}, \boldsymbol{u}^*) \, ds \right].$$

As above, by applying Itô's lemma to write $H(\tau, \boldsymbol{X}_\tau^{\boldsymbol{u}^*})$ in terms of $H(t, \boldsymbol{x})$ plus the integral of its increments, taking expectations, and then the limit as $h \searrow 0$, we find that

$$\partial_t H(t, \boldsymbol{x}) + \mathcal{L}_t^{\boldsymbol{u}^*} H(t, \boldsymbol{x}) + F(t, \boldsymbol{x}, \boldsymbol{u}^*) = 0.$$

Combined with (5.18), we finally arrive at the DPE (also known in this context as the **Hamilton–Jacobi–Bellman equation**)

$$\boxed{\begin{aligned} \partial_t H(t, \boldsymbol{x}) + \sup_{\boldsymbol{u} \in \mathcal{A}} \left( \mathcal{L}_t^{\boldsymbol{u}} H(t, \boldsymbol{x}) + F(t, \boldsymbol{x}, \boldsymbol{u}) \right) &= 0, \\ H(T, \boldsymbol{x}) &= G(\boldsymbol{x}). \end{aligned}} \tag{5.19}$$

The terminal condition above follows from the definition of the value function in (5.10) from which we see that the running reward/penalty drops out and $G(\boldsymbol{X}_T^{\boldsymbol{u}})$ is $\mathcal{F}_T$ measurable.

Notice that the optimisation of the control in (5.19) is only over its value at time $t$, rather than over the whole path of the control. Hence, it appears that the optimal control can be obtained pointwise. Treating the value function as known, the optimal control can often be found in **feedback control** form in terms of the value function itself. Substituting the feedback control back into (5.19) results in non-linear PDEs. In fact the function

$$\mathfrak{H}(t,\, x,\, \boldsymbol{D}_x H,\, \boldsymbol{D}_x^2 H) = \sup_{u \in \mathcal{A}} \left( \mathcal{L}_t^u H(t,x) + F(t,x,u) \right)$$

is called the **Hamiltonian** of the associated stochastic control problem. Here $\boldsymbol{D}_x H$ and $\boldsymbol{D}_x^2 H$ represent the collection of first and second order derivatives of the value function $H$ respectively (recall that in general $X$ is vector-valued) and in particular

$$[\boldsymbol{D}H]_k = \partial_{x^k} H, \qquad \text{and} \qquad [\boldsymbol{D}_x^2 H]_{jk} = \partial_{x^j x^k} H.$$

These objects appear in the infinitesimal generator.

### Example: The Merton Problem

Consider now the Merton optimisation problem described in Section 5.2.1. The optimisation problem is given in (5.1) and has the associated time dependent performance criteria

$$H^\pi(t, x, S) = \mathbb{E}_{t,x,S} \left[ U\left(X_T^\pi\right) \right], \tag{5.20}$$

where $X^\pi$ (investor wealth) and $S$ (risky asset price) satisfy the SDEs (5.2) with $\pi$ representing the dollar value of wealth invested in the risky asset $S$. From (5.2b), the infinitesimal generator of the pair of processes $(X_t^\pi, S_t)_{0 \le t \le T}$ is then

$$\mathcal{L}_t^\pi = (r\,x + (\mu - r)\,\pi)\,\partial_x + \tfrac{1}{2}\sigma^2\,\pi\,\partial_{xx} + (\mu - r)\,S\,\partial_S + \tfrac{1}{2}\sigma^2\,S^2\,\partial_{SS} + \sigma\,\pi\,\partial_{xS}.$$

According to (5.19), the value function $H(t, x, S) = \sup_{\pi \in \mathcal{A}_{[t,T]}} H^\pi(t, x, S)$ should satisfy the equation

$$0 = \left( \partial_t + r\,x\,\partial_x + \tfrac{1}{2}\sigma^2\,S^2\,\partial_{SS} \right) H$$
$$+ \sup_\pi \left\{ \pi \left( (\mu - r)\,\partial_x + \sigma\,\partial_{xS} \right) H + \tfrac{1}{2}\sigma^2\,\pi^2\,\partial_{xx} H \right\},$$

subject to the terminal condition $H(T, x, S) = U(x)$. Note that the argument of the sup is quadratic in $\pi$ and as long as $\partial_{xx} H(t, x, S) < 0$, the sup attains a maximum. By completing the squares we have

$$\pi \left( (\mu - r)\,\partial_x + \sigma \partial_{xS} \right) H + \tfrac{1}{2}\sigma^2\,\pi^2\,\partial_{xx} H$$
$$= \tfrac{1}{2}\sigma^2\,\partial_{xx} H \left( (\pi - \pi^*)^2 - (\pi^*)^2 \right),$$

where

$$\boxed{\pi^* = -\frac{(\mu - r)\,\partial_x H + \sigma\,\partial_{xS} H}{\sigma^2\,\partial_{xx} H}}$$

is the optimal control in feedback form – i.e. it is the optimal control given the known value function $H(t, x, S)$. Substituting this optimum back into the DPE yields the non-linear PDE for the value function

$$0 = \left(\partial_t + r\,x\,\partial_x + \tfrac{1}{2}\,\sigma^2\,S^2\,\partial_{SS}\right) H - \frac{\left((\mu - r)\,\partial_x H + \sigma\,\partial_{xS} H\right)^2}{2\,\sigma^2\,\partial_{xx} H}\,.$$

This simplifies somewhat by observing that the terminal condition $H(t, x, S) = U(x)$ is independent of $S$. Hence, it suggests the ansatz $H(t, x, S) = h(t, x)$ in which case we obtain a simpler, but still non-linear, equation for $h(t, x)$

$$0 = \left(\partial_t + r\,x\,\partial_x\right) h(t, x) - \frac{\lambda}{2\sigma}\,\frac{(\partial_x h(t, x))^2}{\partial_{xx} h(t, x)}\,,$$

with terminal condition $h(T, x) = U(x)$ and where $\lambda = \frac{\mu - r}{\sigma}$ is the market price of risk, also referred to as the Sharpe ratio. Moreover, the optimal control simplifies to

$$\pi^* = -\frac{\lambda}{\sigma}\left(\frac{\partial_x h}{\partial_{xx} h}\right)\,.$$

The explicit solution of the non-linear PDE depends on the precise form of the utility function $U(x)$. Two classic examples are

(i) exponential utility

$$U(x) = -e^{-\gamma x}\,, \qquad \gamma > 0\,,$$

which is defined for all $x \in \mathbb{R}$, and

(ii) hyperbolic absolute risk aversion (HARA) utility

$$U(x) = \tfrac{1-\gamma}{\gamma}\left(a + \tfrac{b}{1-\gamma}\,x\right)^\gamma\,, \qquad \gamma > 1,\, b > 0,\, x \in \left(-(1-\gamma)\tfrac{a}{b}, +\infty\right)\,.$$

Admissible wealth has a lower bound in this case and the investor is infinitely averse to dropping below this level.

For exponential utility, the value function admits an affine solution and we can write an ansatz

$$h(t, x) = -\alpha(t)\,e^{-\gamma\,x\,\beta(t)}\,,$$

where $\alpha(t)$ and $\beta(t)$ are yet to be determined functions of time alone. From the terminal condition $h(T, X) = -e^{-\gamma x}$ we have that $\alpha(T) = \beta(T) = 1$ and upon substituting this back into the non-linear PDE above, we find that

$$\left(\partial_t \alpha - \tfrac{\lambda}{2\sigma}\,\alpha\right) - \gamma\left(\partial_t \beta + r\,\beta\right)\alpha\,x = 0\,.$$

Since this must hold for every $x$ and $t$, the terms in braces must individually vanish. These two equations $\partial_t \alpha - \tfrac{\lambda}{2\sigma}\,\alpha = 0$ and $\partial_t \beta + r\,\beta = 0$, together with the terminal conditions, are easily solved to find

$$\alpha(t) = e^{-\tfrac{\lambda}{2\sigma}(T-t)}\,, \qquad \beta(t) = e^{r(T-t)}\,.$$

Upon back substitution, we find that the optimal amount to invest in the risky asset is a deterministic function of time

$$\pi^*(t) = \frac{\lambda}{\gamma\,\sigma}\,e^{-r(T-t)}\,.$$

As risk-aversion increases, the investor puts less into the risky asset. The fact that the amount invested is independent of wealth results from the agent's **absolute risk aversion**, defined as $-U''(x)/U'(x) = \gamma$, being a constant since she uses exponential utility. For HARA utilities, e.g., neither the absolute risk aversion nor the **relative risk aversion**, defined as $R(x) = -x\,U''(x)/U'(x)$, are constants. In the HARA case, the agent's risky investment is a non-trivial function of both wealth and time.

### 5.3.3    Verification

The derivation of the DPE (5.19) in the previous section provides a necessary condition for the value function. A pertinent question, however, is whether or not a solution of the DPE does indeed provide the solution to the original control problem. The main workhorse for showing this is indeed the case, when classical solutions to the associated DPE exist, is the so-called **verification theorem**. We state the result below and refer the reader to many of the excellent texts on optimal control for its proof, see e.g., Yong & Zhou (1999), Fleming & Soner (2006), Øksendal & Sulem (2007), and Pham (2010).

THEOREM 5.2    **Verification Theorem.** *Let $\psi \in C^{1,2}([0,T] \times \mathbb{R}^n)$ and satisfies, for all $\boldsymbol{u} \in \mathcal{A}$,*

$$\partial_t \psi(t,\boldsymbol{x}) + (\mathcal{L}_t^{\boldsymbol{u}} \psi(t,\boldsymbol{x}) + F(t,\boldsymbol{x},\boldsymbol{u})) \leq 0, \quad \forall\, (t,\boldsymbol{x}) \in [0,T] \times \mathbb{R}^n,$$
$$G(\boldsymbol{x}) - \psi(T,\boldsymbol{x}) \leq 0.$$

*Then*

$$\psi(t,\boldsymbol{x}) \geq H^{\boldsymbol{u}}(t,\boldsymbol{x}), \quad \forall\, (t,\boldsymbol{x}) \in [0,T] \times \mathbb{R}^n,$$

*for all Markov controls $\boldsymbol{u} \in \mathcal{A}$.*

*Moreover, if for every $(t,\boldsymbol{x}) \in [0,T] \times \mathbb{R}^n$, there exists measurable $\boldsymbol{u}^*(t,\boldsymbol{x})$ such that*

$$0 = \partial_t \psi(t,\boldsymbol{x}) + \left( \mathcal{L}_t^{\boldsymbol{u}^*(t,\boldsymbol{x})} \psi(t,\boldsymbol{x}) + F(t,\boldsymbol{x},\boldsymbol{u}^*(t,\boldsymbol{x})) \right)$$
$$= \partial_t \psi(t,\boldsymbol{x}) + \sup_{\boldsymbol{u} \in \mathcal{A}} \left( \mathcal{L}_t^{\boldsymbol{u}} \psi(t,\boldsymbol{x}) + F(t,\boldsymbol{x},\boldsymbol{u}) \right), \quad \forall\, (t,\boldsymbol{x}) \in [0,T] \times \mathbb{R}^n,$$

*with $\psi(T,\boldsymbol{x}) = G(\boldsymbol{x})$, and the SDE*

$$d\boldsymbol{X}_s^* = \mu(t, \boldsymbol{X}_s^*, \boldsymbol{u}^*(t,\boldsymbol{X}_s^*))\, dt + \sigma(t, \boldsymbol{X}_s^*, \boldsymbol{u}^*(t,\boldsymbol{X}_s^*))d\boldsymbol{W}_s, \quad \boldsymbol{X}_t^* = \boldsymbol{x},$$

*admits a unique solution and $\{\boldsymbol{u}^*(s, \boldsymbol{X}_s^*)\}_{t \leq s \leq T} \in \mathcal{A}$, then*

$$H(t,\boldsymbol{x}) = \psi(t,\boldsymbol{x}), \quad \forall\, (t,\boldsymbol{x}) \in [0,T] \times \mathbb{R}^n,$$

*and $\boldsymbol{u}^*$ is an optimal Markov control.*

The theorem states that if we can find a solution to the DPE and demonstrate that it is a classical solution, i.e. once differentiable in time and twice differentiable in the state variables, and the resulting control is admissible, then the solution is indeed the value function we seek and the resulting control is the optimal one – at least an optimal Markov control. Another key result is that, under some more technical assumptions, the optimal control is Markov, even if we search over general $\mathcal{F}$-predictable controls. See, e.g., Theorem 11.2.3 in Øksendal (2010).

## 5.4    Control for Counting Processes

In the previous section, diffusion processes were the driving sources of uncertainty in the control problem. In many circumstances, and in particular problems related to algorithmic and high-frequency trading, counting processes will be used to drive uncertainty. There are many features that can be incorporated into the analysis, but the general approach remains the same, and as such only the case of a single counting process with controlled intensity will be investigated. This amounts to treating doubly stochastic Poisson processes, or Cox processes, which are counting processes with intensity that itself is a stochastic process and in this case at least partially controlled.

Consider the situation in which the agent can control the frequency of the jumps in a counting process $N$ and does so to maximise some target. In this case, the control problem is of the general form

$$H(n) = \sup_{u \in \mathcal{A}_{0,T}} \mathbb{E}\left[ G(N_T^u) + \int_0^T F(s, N_s^u, u_s)\, ds \right], \qquad (5.21)$$

where $u = (u_t)_{0 \leq t \leq T}$ is the control process, $(N_t^u)_{0 \leq t \leq T}$ is a controlled doubly stochastic Poisson process (starting at $N_{0-} = n$) with intensity $\lambda_t^u = \lambda(t, N_{t-}^u, u_t)$ so that $(\widehat{N}_t^u)_{0 \leq t \leq T}$, where

$$\widehat{N}_t^u = N_t - \int_0^t \lambda_s^u\, ds$$

is a martingale, $\mathcal{A}$ is a set of $\mathcal{F}$-predictable processes such that $\widehat{N}$ is a true martingale, $G : \mathbb{R} \mapsto \mathbb{R}$ is a terminal reward and $F : \mathbb{R}_+ \times \mathbb{R}^2 \mapsto \mathbb{R}$ is a running reward/penalty. As before, the functions $G$ and $F$ are assumed uniformly bounded.

For an arbitrary admissible control $u$, the **performance criteria** $H^u(n)$ is given by

$$H^u(n) = \mathbb{E}\left[ G\left(N_T^u\right) + \int_0^T F(s, N_s^u)\, ds \right], \qquad (5.22)$$

and the agent seeks to maximise this performance criteria, i.e.

$$H(n) = \sup_{u \in \mathcal{A}_{0,T}} H^u(n). \qquad (5.23)$$

In the next two subsections we provide a concise derivation of the DPP and the associated DPE.

### 5.4.1     The Dynamic Programming Principle

As before, the original problem is embedded into a larger class of problems indexed by time $t \in [0, T]$ by first defining

$$H(t, n) := \sup_{u \in \mathcal{A}_{t,T}} H^u(t, n), \quad \text{and} \tag{5.24a}$$

$$H^u(t, n) := \mathbb{E}_{t,n} \left[ G(N_T^u) + \int_t^T F(s, N_s^u, u_s) \, ds \right], \tag{5.24b}$$

where the notation $\mathbb{E}_{t,n}[\cdot]$ represents expectation conditional on $N_{t-} = n$. Next, take an arbitrary admissible strategy $u$ and flow the $N$ process forward in time from $t$ to an arbitrary stopping time $\tau \leq T$. Then, conditional on $N_\tau^u$, the contribution of the running reward/penalty from $\tau$ to $T$ and the terminal reward can be viewed as the performance criteria starting from the new value of $N_\tau^u$. This allows the value function to be written in terms of the expectation of its future value at $\tau$ plus the reward between now and $\tau$.

More precisely, by iterated expectations the time-indexed performance criteria becomes

$$H^u(t, n) = \mathbb{E}_{t,n} \left[ G(N_T^u) + \int_\tau^T F(s, N_s^u, u_s) \, ds + \int_t^\tau F(s, N_s^u, u_s) \, ds \right]$$

$$= \mathbb{E}_{t,n} \left[ \mathbb{E}_{\tau, N_\tau^u} \left[ G(N_T^u) + \int_\tau^T F(s, N_s^u, u_s) \, ds \right] + \int_t^\tau F(s, N_s^u, u_s) \, ds \right]$$

$$= \mathbb{E}_{t,n} \left[ H^u(\tau, N_\tau^u) + \int_t^\tau F(s, N_s^u, u_s) \, ds \right]. \tag{5.25}$$

Now, $H(t, n) \geq H^u(t, n)$ for an arbitrary admissible control $u$ (with equality holding if $u$ is the optimal control $u^*$) and an arbitrary $n$. Hence, on the right-hand side of (5.25) the performance criteria $H^u(\tau, N_\tau^u)$ is bounded above by the value function $H(\tau, N_\tau^u)$ and the equality can be replaced by an inequality with the value function (and not the performance criteria) showing up under the expectation:

$$H^u(t, n) \leq \mathbb{E}_{t,n} \left[ H(\tau, N_\tau^u) + \int_t^\tau F(s, N_s^u, u_s) \, ds \right]$$

$$\leq \sup_{u \in \mathcal{A}} \mathbb{E}_{t,n} \left[ H(\tau, N_\tau^u) + \int_t^\tau F(s, N_s^u, u_s) \, ds \right]. \tag{5.26}$$

Note that on the right-hand side of the first line of (5.26), the arbitrary control $u$ only acts over the interval $[t, \tau]$ and the optimal control is implicitly incorporated in the value function $H(\tau, N_\tau^u)$, but with the state variable $N$ beginning at the point where the arbitrary control $u$ caused $N$ to flow, namely $N_\tau^u$. Next, taking the best control on the right-hand side must also dominate the left-hand side, since the arbitrary one does. Then, the right-hand side is no longer dependent on the arbitrary control. Finally, taking a supremum over all admissible controls on the left-hand side allows us to replace the left-hand side with the value function and provides us with the first inequality

$$H(t,n) \leq \sup_{u \in \mathcal{A}} \mathbb{E}_{t,n} \left[ H(\tau, N_\tau^u) + \int_t^\tau F(s, N_s^u, u_s) \, ds \right]. \tag{5.27}$$

To obtain the reverse inequality, take an $\epsilon$-optimal control denoted by $v^\epsilon \in \mathcal{A}$ such that

$$H(t,x) \geq H^{v^\epsilon}(t,x) \geq H(t,x) - \epsilon. \tag{5.28}$$

Such a control exists if the value function is continuous in the space of controls. Consider its modification up to time $\tau$:

$$\tilde{v}^\epsilon = u_t \mathbb{1}_{t \leq \tau} + v^\epsilon \mathbb{1}_{t > \tau}, \tag{5.29}$$

where $u \in \mathcal{A}$ is an arbitrary admissible control. Then we have that

$$H(t,n) \geq H^{\tilde{v}^\epsilon}(t,n)$$

$$= \mathbb{E}_{t,n} \left[ H^{\tilde{v}^\epsilon}(\tau, N_\tau^{\tilde{v}^\epsilon}) + \int_t^\tau F(s, N_s^{\tilde{v}^\epsilon}, \tilde{v}_s^\epsilon) \, ds \right], \qquad \text{(from (5.25))},$$

$$= \mathbb{E}_{t,n} \left[ H^{\tilde{v}^\epsilon}(\tau, N_\tau^u) + \int_t^\tau F(s, N_s^u, u_s) \, ds \right], \qquad \text{(using (5.29))},$$

$$\geq \mathbb{E}_{t,n} \left[ H(\tau, N_\tau^u) + \int_t^\tau F(s, N_s^u, u_s) \, ds \right] - \varepsilon, \qquad \text{(by (5.28))}.$$

Since the above holds for every $u$ and every $\epsilon$, it holds for the sup, and we take the limit as $\varepsilon \searrow 0$ to find the inequality

$$H(t,n) \geq \sup_{u \in \mathcal{A}} \mathbb{E}_{t,n} \left[ H(\tau, N_\tau^u) + \int_t^\tau F(s, N_s^u, u_s) \, ds \right]. \tag{5.30}$$

Putting (5.30) together with (5.27), we obtain the theorem below.

THEOREM 5.3 *Dynamic Programming Principle for Counting Processes. The value function (5.23) satisfies the DPP*

$$\boxed{H(t,n) = \sup_{u \in \mathcal{A}} \mathbb{E}_{t,n} \left[ H(\tau, N_\tau^u) + \int_t^\tau F(s, N_s^u, u_s) \, ds \right],} \tag{5.31}$$

*for all $(t,n) \in [0,T] \times \mathbb{Z}_+$ and all stopping times $\tau \leq T$.*

## 5.4.2 Dynamic Programming Equation / Hamilton–Jacobi–Bellman Equation

This section develops the DPE satisfied by the value function, which we obtain by looking at the DPP over infinitesimal amounts of time. Analogous to the diffusion case, set the stopping time $\tau$ in the DPP to be the minimum of (i) the time it takes the process $N_t^u$ to exit a ball of size $\epsilon$ around its starting point, and (ii) a fixed (small) time $h$, all while keeping it bounded by $T$.

$$\tau = T \wedge \inf\{s > t : (s - t, |N_s^u - n|) \notin [0,h) \times [0,\epsilon)\}.$$

Next, write the value function (for an arbitrary admissible control $u$) at the stopping time $\tau$ in terms of the value function at $t$ using Itô's lemma. Specifically,

$$H(\tau, N_\tau^u) = H(t, n) + \int_t^\tau (\partial_t + \mathcal{L}_s^u) H(s, N_s^u) \, ds$$

$$+ \int_t^\tau [H(s, N_{s-}^u + 1) - H(s, N_{s-}^u)] \, d\widehat{N}_s^u \,,$$

(5.32)

where $\mathcal{L}_t^u$ represents the infinitesimal generator of $N_t^u$ and acts on functions $h : \mathbb{R}_+ \times \mathbb{Z}_+ \mapsto \mathbb{R}$ as follows:

$$\mathcal{L}_t^u h(t, n) = \lambda(t, n, u) \left[ h(t, n+1) - h(t, n) \right] .$$

Taking $v \in \mathcal{A}$ to be constant over the interval $[t, \tau]$, applying the lower bound (5.30), and substituting (5.32) into the right-hand side implies that

$$H(t, n) \geq \sup_{u \in \mathcal{A}} \mathbb{E}_{t,n} \left[ H(\tau, N_\tau^u) + \int_t^\tau F(s, N_s^u, u_s) \, ds \right]$$

$$\geq \mathbb{E}_{t,n} \left[ H(\tau, N_\tau^v) + \int_t^\tau F(s, N_s^v, v) \, ds \right]$$

$$= \mathbb{E}_{t,n} \left[ H(t, n) + \int_t^\tau (\partial_t + \mathcal{L}_s^v) H(s, N_s^v) \, ds \right.$$

$$\left. + \int_t^\tau (H(s, N_{s-}^v + 1) - H(s, N_{s-}^v)) \, d\widehat{N}_s^v + \int_t^\tau F(s, N_s^v, v) \, ds \right]$$

$$= \mathbb{E}_{t,n} \left[ H(t, n) + \int_t^\tau ((\partial_t + \mathcal{L}_s^v) H(s, N_s^v) + F(s, N_s^v, v)) \, ds \right] .$$

The third equality follows because the stochastic integral with respect to $\widehat{N}_t$ has zero mean since $\widehat{N}_t$ is a martingale. The martingale property is guaranteed because we have $|N_t^v - n| \leq \epsilon$, and so $\mathcal{L}_t^v H(t, N_t^v)$ is bounded. Subtracting $H(t, n)$, dividing by $h$, and taking the limit as $h \searrow 0$ implies that

$$0 \geq \lim_{h \downarrow 0} \mathbb{E}_{t,n} \left[ \frac{1}{h} \int_t^\tau ((\partial_t + \mathcal{L}_s^v) H(s, N_s^v) + F(s, N_s^v, v)) \, ds \right]$$

$$= (\partial_t + \mathcal{L}_t^v) H(t, n) + F(t, n, v) .$$

(5.33)

In the above, the control is constant, and equals $v$, on the interval $[t, \tau]$. The second line follows, as before, from:

(i) as $h \searrow 0$, $\tau = t + h$ a.s. since the process will not hit the barrier of $\epsilon$ in extremely short periods of time,

(ii) the condition that $|N_\tau^v - n| \leq \epsilon$, so that if the process does hit the barrier, it is bounded,

(iii) the Fundamental Theorem of Calculus: $\lim_{h \downarrow 0} \frac{1}{h} \int_t^{t+h} w_s \, ds = w_t$, and

(iv) $N_{t-}^u = n$.

Since the above inequality holds for arbitrary $v \in \mathcal{A}$, it follows that

$$\partial_t H(t, n) + \sup_{u \in \mathcal{A}} \left( \mathcal{L}_t^u H(t, n) + F(t, n, u) \right) \leq 0 .$$

(5.34)

Next, we show the inequality is indeed an equality. For this purpose, suppose that $u^* \in \mathcal{A}$ is an optimal control, then from the DPP (5.31) we have

$$H(t,n) = \mathbb{E}_{t,n}\left[ H(\tau, N_\tau^{u^*}) + \int_t^\tau F(s, N_s^{u^*}, u_s^*)\, ds \right].$$

Repeating the steps from above (i.e. applying Itô's lemma, taking expectations, and the limit as $h \searrow 0$) we find that

$$\partial_t H(t,n) + \mathcal{L}_t^{u^*} H(t,n) + F(t,n,u^*) = 0.$$

Combining this result with (5.34), we arrive at the DPE (also known in this context as the **Hamilton–Jacobi–Bellman equation**)

$$\begin{cases} \partial_t H(t,n) + \sup\limits_{u \in \mathcal{A}_t} \left( \mathcal{L}_t^u H(t,n) + F(t,n,u) \right) = 0, \\ \qquad\qquad\qquad\qquad\qquad\qquad\quad H(T,n) = G(n). \end{cases} \tag{5.35}$$

The terminal condition follows from the observation that the integral in the optimisation problem (5.24) vanishes as $t \nearrow T$. Recall that the generator $\mathcal{L}_t^u$ for the controlled process acts as follows:

$$\mathcal{L}_t^u H(t,n) = \lambda(t,n,u)\left[ H(t,n+1) - H(t,n) \right].$$

As in the diffusion case, the optimisation of the control in (5.35) is only over its value at time $t$, rather than over the whole path of the control. Hence, it appears that the optimal control can be obtained pointwise. Treating the value function as known, the optimal control can often be found in **feedback control** form in terms of the value function itself. Substituting the feedback control back into (5.35) typically results in non-linear PDEs. In this Poisson case, the supremum in (5.35) has a very simple form since

$$\sup_{u \in \mathcal{A}_t} \left( \mathcal{L}_t^u H(t,n) + F(t,n,u) \right)$$
$$= \sup_{u \in \mathcal{A}_t} \left\{ \lambda(s,n,u)\left[ H(t,n+1) - H(t,n) \right] + F(t,n,u) \right\},$$

and hence, if $F = 0$, the optimal choice of the control is to make $\lambda(s,n,u)$ as large as possible if $H(t,n+1) - H(t,n) > 0$ and as small as possible if $H(t,n+1) - H(t,n) < 0$. Such controls are called **bang-bang controls** because the quantity being controlled (in this case the intensity) reaches its extremal points. To make the problem more interesting we either need to (i) include a running reward/penalty (i.e. $F \neq 0$) or, more interestingly (and relevant), (ii) introduce another stochastic process which is driven by the counting process, and therefore controlled indirectly, and have this stochastic process affect the agent's terminal and running rewards.

## Using the Poisson Process to Drive a Secondary Controlled Process

One such way to do this, which will also appear in later chapters as a result of optimisation problems in algorithmic trading, is to let $(X_t^u)_{0 \le t \le T}$ denote a controlled process satisfying the SDE

$$dX_t^u = \mu(t, X_t^u, N_t^u, u_t)\, dt + \sigma(t, X_{t-}^u, N_{t-}^u, u_t)\, dN_t^u. \tag{5.36}$$

In this manner, the counting process $N^u$ acts as the source of jumps in $X^u$, and the control $u$ may modulate the size of those jumps as well as the drift of the $X$ process – in addition to the arrival rate of the jumps themselves.

In this more general context, the performance criteria can depend on both $N$ and $X$. Specifically, let

$$H(t, x, n) := \sup_{u \in \mathcal{A}} H^u(t, x, n), \quad \text{and} \tag{5.37a}$$

$$H^u(t, x, n) := \mathbb{E}_{t,x,n}\left[ G(X_T^u, N_T^u) + \int_t^T F(s, X_s^u, N_s^u, u_s)\, ds \right], \tag{5.37b}$$

where $\mathbb{E}_{t,x,n}[\cdot]$ denotes expectation conditional on $N_{t-} = n$ and $X_{t-} = x$. Following the same arguments as in the previous sections, the DPP for this problem can be written as

$$H(t, x, n) = \sup_{u \in \mathcal{A}} \mathbb{E}_{t,x,n}\left[ H\left(\tau, X_\tau^u, N_\tau^u\right) + \int_t^\tau F\left(s, X_s^u, N_s^u, u_s\right) ds \right], \tag{5.38}$$

and a DPE can be derived to find

$$\begin{cases} \partial_t H(t, x, n) + \sup_{u \in \mathcal{A}} \left( \mathcal{L}_t^u H(t, x, n) + F(t, x, n) \right) = 0, \\[2mm] \hspace{4cm} H(T, x, n) = G(x, n). \end{cases} \tag{5.39}$$

Here, the infinitesimal generator $\mathcal{L}_t^u$ acts as follows:

$$\begin{aligned} \mathcal{L}_t^u H(t, x, n) = {}& \mu(t, x, n, u)\, \partial_x H(t, x, n) \\ & + \lambda(t, x, n, u)\left[ H(t, x + \sigma(t, x, n, u),\, n+1) - H(t, x, n) \right]. \end{aligned}$$

Notice that the control appears in both the intensity factor and in the difference operator, and there is a partial derivative with respect to the state variable $x$. This represents the trade-off between increasing/decreasing intensity through the control and what that change does to the process $X_t^u$, as well as what the control does to the drift.

## Example: Maximising Expected Wealth using Round-Trip Trades

Here, we provide an example of an agent who uses a market order (MO) to purchase one share at the best offer and then seeks to unwind her position by posting a limit order (LO) at the midprice plus the depth $u$ which she controls. She repeats this operation over and over again until a future date $T$. Her cost from acquiring the share is $S_t + \Delta/2$, where $\Delta$ is the spread between the best bid and best ask and is assumed constant, and since $S_t$ is the midprice, the best ask is resting in the LOB at $S_t + \Delta/2$. The revenue from selling (if her LO is lifted by an MO) is $S_t + u_t$. Therefore, the wealth that is accrued to the agent from this round-trip trade is $u_t - \Delta/2$.

There is, however, no guarantee that the sell LO will be filled, and in this case the agent's wealth $X$ satisfies (5.36) with $\mu = 0$ and $\sigma_t^u = \left( u_t - \frac{\Delta}{2} \right)$ so that

$$dX_t^u = \left( u_t - \frac{\Delta}{2} \right) dN_t^u .$$

Here $N_t^u$ counts the number of round-trip trades that the agent has completed up until time $t$. The agent controls the intensity of this counting process because the larger she chooses $u$, the lower is the probability of her LO being filled.

There are many ways in which we can model the probability of the LO being filled given how deep in the LOB the agent posts the order. We need two ingredients. First, we need to assume an arrival rate for the buy MOs that are sent by other market participants. Here we assume, for simplicity, that this rate is a constant $\Lambda > 0$. The other ingredient is the probability of the LO being filled conditional on the MO arriving. A popular choice in the literature is to assume that when posted $u \geq 0$ away from the midprice, the probability of being filled, given that an MO arrives, is $P(u) = e^{-\kappa u_t}$, and another is $P(u) = (1 + \kappa u_t)^{-\gamma}$, where $\kappa$ and $\gamma$ are positive constants. Putting these ingredients together gives us two choices to model the fill probability: $\lambda_t^u = e^{-\kappa u_t} \Lambda$ and $\lambda_t^u = (1 + \kappa u_t)^{-\gamma} \Lambda$.

To solve the agent's optimisation problem we use the performance criteria and value function as in (5.37) with $F = 0$ and $G(x, n) = x$. If we assume that the fill rate is $\lambda_t^u = e^{-\kappa u_t} \Lambda$, the DPE becomes

$$\partial_t H + \sup_{u \geq 0} \Lambda e^{-\kappa u} \left( H\left(t, x + \left( u - \tfrac{1}{2}\Delta \right), n + 1 \right) - H(t, x, n) \right) = 0 ,$$

subject to $H(T, x, n) = x$. Since there is no explicit dependence on $n$ itself, we can assume $H(t, x, n) = h(t, x)$ so the value function depends solely on wealth and time. Furthermore, due to the linear nature of the problem, we can further write $h(t, x) = x + g(t)$ for some deterministic function $g(t)$ with terminal condition $g(T) = 0$. Hence, the above reduces to

$$\partial_t g + \sup_{u \geq 0} \Lambda e^{-\kappa u} \left( u - \tfrac{1}{2}\Delta \right) = 0 . \tag{5.40}$$

This shows that the optimal control is independent of $t$, $x$ and $n$. In particular,

$$u^* = \operatorname*{argmax}_{u \geq 0} \left\{ e^{-\kappa u} \left( u - \tfrac{1}{2}\Delta \right) \right\} ,$$

and it is straightforward to show that

$$u^* = \tfrac{1}{2}\Delta + \frac{1}{\kappa} .$$

It is not difficult to check that this is indeed a maximum and not a minimum, and upon substituting the feedback control (which in this case is just a constant) back into (5.40), we find that $g$ satisfies

$$\partial_t g + \frac{\Lambda}{\kappa} e^{-\kappa \left( \frac{1}{2}\Delta + \frac{1}{\kappa} \right)} = 0 .$$

The value function is therefore given by the very compact expression

$$H(t,x,n) = x + \frac{\Lambda}{\kappa}\, e^{-\kappa\left(\frac{1}{2}\Delta+\frac{1}{\kappa}\right)}\,(T-t)\,.$$

The optimal posting strategy (5.4.2) has a simple interpretation. The agent must recover the half-spread cost she incurred when using an MO to acquire the asset, and this is given by $\frac{1}{2}\Delta$ in the optimal posting (5.4.2). In addition she posts further away from the midprice by an amount which maximises how much deeper her posting can rest in the book, given the probability of being filled, and this is the term $\frac{1}{\kappa}$.

The strategy derived here is optimal, but naive because it is a result of our simplifying assumptions in the way we model the state variables, and the simple performance criteria employed by the agent. For instance, this strategy does not make any adjustments to the optimal posting based on important quantities and costs such as: accumulated inventory, wealth, remaining time to trade, and adverse selection costs. The economic principles that underpin the link between these quantities and costs to the optimal postings were discussed in Chapter 2. In the latter parts of this book we incorporate these issues in the trading strategies when developing algorithms where the agent maximises profits executing round-trip trades in a more realistic and general setting than the one developed here.

## 5.4.3  Combined Diffusion and Jumps

As already hinted above, there are many situations in which the agent is exposed to more than one source of uncertainty. Typically, an agent is faced with control problems where both diffusive and jump uncertainty appear, and she may be able to control all or only parts of the system. Such scenarios will appear in several of the algorithmic trading problems that arise in the sections ahead and here we simply state the main results for a fairly general class of models.

First, let $\boldsymbol{N^u} = (\boldsymbol{N_t^u})_{0 \le t \le T}$ denote a collection of counting processes (of dim $p$) with controlled intensities $\boldsymbol{\lambda^u} = (\boldsymbol{\lambda_t^u})_{0 \le t \le T}$, and let $\boldsymbol{u} = (\boldsymbol{u_t})_{0 \le t \le T}$ denote the control processes (of dim $m$). Furthermore, let $\boldsymbol{W} = (\boldsymbol{W_t})_{0 \le t \le T}$ denote a collection of independent Brownian motions (of dim $m$). Next, let $\boldsymbol{X^u} = (\boldsymbol{X_t^u})_{0 \le t \le T}$ denote the controlled processes (of dim $m$) which will appear directly in the agent's performance criteria. If one (or more) of the counting processes $N_t^i$ should appear in the performance criteria or in the dependence in $\mu_t^u$, $\sigma_t^u$ and/or $\gamma_t^u$, shown below, then take $X_t^j = N_t^i$ for some $j$, i.e. include it through one of the components in $X$. We next assume that the controlled $\boldsymbol{X^u}$ processes satisfy the SDEs

$$d\boldsymbol{X_t^u} = \boldsymbol{\mu_t^u}\, dt + \boldsymbol{\sigma_t^u}\, d\boldsymbol{W_t} + \boldsymbol{\gamma_t^u}\, d\boldsymbol{N_t^u}\,. \tag{5.41a}$$

With a slight abuse of notation we write the controlled drift, volatility and jump size as

$$\boldsymbol{\mu}_t^u := \boldsymbol{\mu}(t, \boldsymbol{X}_t^u, \boldsymbol{u}_t), \qquad (m \times 1 \text{ vector}), \qquad (5.41b)$$

$$\boldsymbol{\sigma}_t^u := \boldsymbol{\sigma}(t, \boldsymbol{X}_t^u, \boldsymbol{u}_t), \qquad (m \times m \text{ matrix}), \qquad (5.41c)$$

$$\boldsymbol{\gamma}_t^u := \boldsymbol{\gamma}(t, \boldsymbol{X}_t^u, \boldsymbol{u}_t), \qquad (m \times p \text{ matrix}). \qquad (5.41d)$$

Furthermore, we assume that the controlled intensity takes the form

$$\boldsymbol{\lambda}_t^u := \boldsymbol{\lambda}(t, \boldsymbol{X}_t^u, \boldsymbol{u}_t) . \qquad (5.41e)$$

The modelling approach above implies that the agent can control, in general, the drift, volatility, jump size and jump arrivals. Precisely how much control she has depends on the specific form of the various functions appearing in (5.41b), (5.41c), (5.41d) and (5.41e).

The agent's performance criteria is given by

$$H^u(t, \boldsymbol{x}) = \mathbb{E}_{t, \boldsymbol{x}} \left[ G(\boldsymbol{X}_T^u) + \int_t^T F(s, \boldsymbol{X}_s^u, \boldsymbol{u}_s) \, ds \right], \qquad (5.42a)$$

and the value function is then given by the usual expression

$$H(t, \boldsymbol{x}) = \sup_{\boldsymbol{u} \in \mathcal{A}_{[t,T]}} H^u(t, \boldsymbol{x}). \qquad (5.42b)$$

Repeating the analysis along similar lines to the previous sections, we arrive at the DPP for the combined problem.

THEOREM 5.4 ***Dynamic Programming Principle for Jump-Diffusions.**
The value function (5.42) satisfies the DPP*

$$\boxed{H(t, \boldsymbol{x}) = \sup_{\boldsymbol{u} \in \mathcal{A}_{[t,T]}} \mathbb{E}_{t, \boldsymbol{x}} \left[ H(\tau, \boldsymbol{X}_\tau^u) + \int_t^\tau F(s, \boldsymbol{X}_s^u, \boldsymbol{u}_s) \, ds \right],} \qquad (5.43)$$

*for all $(t, \boldsymbol{x}) \in [0, T] \times \mathbb{R}^m$ and all stopping times $\tau \leq T$.*

Moreover, following similar arguments as before, one can develop a DPE

$$\boxed{\begin{cases} \partial_t H(t, \boldsymbol{x}) + \sup_{\boldsymbol{u} \in \mathcal{A}_t} (\mathcal{L}_t^u H(t, \boldsymbol{x}) + F(t, \boldsymbol{x}, \boldsymbol{u})) = 0, \\ \\ \hspace{5cm} H(T, \boldsymbol{x}) = G(\boldsymbol{x}), \end{cases}} \qquad (5.44)$$

where the infinitesimal generator $\mathcal{L}_t^u$ acts as follows:

$$\mathcal{L}_t^u H(t, \boldsymbol{x}) = \boldsymbol{\mu}(t, \boldsymbol{x}, \boldsymbol{u}) \cdot \boldsymbol{D}_x H(t, \boldsymbol{x}) + \tfrac{1}{2} \boldsymbol{\sigma}(t, \boldsymbol{x}, \boldsymbol{u}) \boldsymbol{\sigma}(t, \boldsymbol{x}, \boldsymbol{u})' \boldsymbol{D}_x^2 H(t, \boldsymbol{x})$$

$$+ \sum_{j=1}^p \lambda_j(t, \boldsymbol{x}, \boldsymbol{u}) \left[ H(t, \boldsymbol{x} + \boldsymbol{\gamma}_{\cdot j}(t, \boldsymbol{x}, \boldsymbol{u})) - H(t, \boldsymbol{x}) \right],$$

where $\boldsymbol{\gamma}_{\cdot j}$ denotes the vector corresponding to the $j^{th}$ column of $\boldsymbol{\gamma}$, $\boldsymbol{D}_x H$ represents the vector of partial derivatives with respect to $x$, and $\boldsymbol{D}_x^2 H$ represents the matrix of (mixed) second order partial derivatives with respect to $x$.

The various terms in the generator can be easily interpreted: in the first line, the first term represents the (controlled) drift of $\boldsymbol{X}$, and the second term represents the (controlled) volatility of $\boldsymbol{X}$; in the second line, the intensity for each counting process is shown separately, and each term in the sum has the (controlled) rate of arrival of that jump component through $\lambda_j$, the difference terms that appear are due to the jump arriving and causing that component of $\boldsymbol{N}$ to increment, and simultaneously cause (potentially all components of) $\boldsymbol{X}$ to jump by $\boldsymbol{\gamma}_{\cdot j}$.

## 5.5      Optimal Stopping

In many circumstances the agent wishes to find the best time at which to enter or exit a given strategy. The classical finance example of this is the 'American put option' in which an agent who owns the option has the right, at any point in time up to, and including, the maturity date $T$, to exercise the option by receiving the amount of cash $K$ in exchange for the underlying asset. The net cash value of this transaction at the exercise date $\tau$ is $(K - S_\tau)$. Naturally, the agent will not exercise when $S_\tau > K$, hence the effective payoff is $(K - S_\tau)_+$ where $(\cdot)_+ = \max(\cdot, 0)$. This simple observation provides only a trivial part of the strategy: to determine the full strategy, the agent seeks the stopping time $\tau$ which maximises the discounted value of the payoff, i.e. she searches for the stopping time which attains the supremum (if possible)

$$\sup_{\tau \in \mathcal{T}} \mathbb{E}\left[ e^{-r\tau} \left( K - S_\tau \right)_+ \right] ,$$

where $\mathcal{T}$ are the $\mathcal{F}$-stopping times bounded by $T$. This is just one of many such problems, and in general, problems which seek optimal stopping times are called **optimal stopping problems**. Similar to optimal control problems, optimal stopping problems admit a DPP and have an infinitesimal version, i.e. a DPE. In this section we provide a concise outline of the DPP and DPE for optimal stopping problems.

Rather than first developing the diffusion case, then the jump, and then the jump-diffusion, here we begin immediately with a fairly general jump-diffusion model. For this purpose, let $\boldsymbol{X} = (\boldsymbol{X}_t)_{0 \le t \le T}$ denote a vector-valued process of dim $m$ satisfying the SDE

$$d\boldsymbol{X}_t = \boldsymbol{\mu}(t, \boldsymbol{X}_t) \, dt + \boldsymbol{\sigma}(t, \boldsymbol{X}_t) \, d\boldsymbol{W}_t + \boldsymbol{\gamma}(t, \boldsymbol{X}_t) \, d\boldsymbol{N}_t ,$$

where $\boldsymbol{\mu} : [0, T] \times \mathbb{R}^m \mapsto \mathbb{R}^m$, $\boldsymbol{\sigma} : [0, T] \times \mathbb{R}^m \mapsto \mathbb{R}^m \times \mathbb{R}^m$, $\boldsymbol{\gamma} : [0, T] \times \mathbb{R}^m \mapsto \mathbb{R}^m \times \mathbb{R}^m$, $\boldsymbol{W}$ is an $m$-dimensional Brownian motion with independent components, and $\boldsymbol{N}$ is an $m$-dimensional counting process with intensities $\boldsymbol{\lambda}(t, \boldsymbol{X}_t)$. The filtration $\mathcal{F}$ is the natural one generated by $\boldsymbol{X}$, and the generator of the process $\mathcal{L}_t$ acts on twice differentiable functions as follows:

$$\mathcal{L}_t h(t, \boldsymbol{x}) = \boldsymbol{\mu}(t, \boldsymbol{x}) \cdot \boldsymbol{D}_x h(t, \boldsymbol{x}) + \tfrac{1}{2} \boldsymbol{Tr} \boldsymbol{\sigma}(t, \boldsymbol{x}) \boldsymbol{\sigma}(t, \boldsymbol{x})' \boldsymbol{D}_x^2 h(t, \boldsymbol{x})$$

$$+ \sum_{j=1}^{p} \lambda_j(t, \boldsymbol{x}) \left[ h(t, \boldsymbol{x} + \boldsymbol{\gamma}_{\cdot j}(t, \boldsymbol{x})) - h(t, \boldsymbol{x}) \right] .$$

Recall that $\boldsymbol{D}_x h$ represents the vector of partial derivatives with respect to $x$, and $\boldsymbol{D}_x^2 h$ represents the matrix of (mixed) second order partial derivatives with respect to $x$.

The agent then has a performance criterion, for each $\mathcal{F}$-stopping time $\tau \in \mathcal{T}_{[t,T]}$, given by

$$H^\tau(t, \boldsymbol{x}) = \mathbb{E}_{t,\boldsymbol{x}} \left[ G(\boldsymbol{X}_\tau) \right] , \tag{5.45a}$$

where $G(\boldsymbol{X}_\tau)$ is the reward upon exercise, and she seeks to find the value function

$$H(t, \boldsymbol{x}) = \sup_{\tau \in \mathcal{T}_{[t,T]}} H^\tau(t, \boldsymbol{x}) , \tag{5.45b}$$

and the stopping time $\tau$ which attains the supremum if it exists. At first glance, it appears that we have omitted the running reward/penalty $\int_t^\tau F(s, \boldsymbol{X}_s) \, ds$ that we included when studying optimal control. Such terms can, however, be cast into the above form by choosing one of the components of $\boldsymbol{X}$, say $X^1$, to satisfy

$$dX_t^1 = F(t, X_t^2, \ldots, X_t^m) \, dt ,$$

and writing $G(\boldsymbol{x}) = x^1 + \tilde{G}(x^2, \ldots, x^m)$. It is therefore no loss of generality to consider only terminal rewards. In the stochastic control case, we kept the explicit running reward/penalty since it traditionally appears there. Now it will prove more convenient to absorb the running reward/penalty in $G$. Similarly, we can incorporate discounting by introducing a second state process which equals the discount factor up to that point in time, and modify $G$ accordingly.

From the posing of the stopping problem in (5.45), intuitively, we see that the agent is attempting to decide between stopping 'now' and receiving the reward $G$, or continuing to hold off in hopes of receiving a larger reward in the future. However, on closer examination it seems clear that she should continue to wait at the point $(t, \boldsymbol{x}) \in [0, T] \times \mathbb{R}^m$ as long as the value function has not attained a value of $G(\boldsymbol{x})$ there. This motivates the definition of the **stopping region**, $\mathcal{S}$, which we define as

$$\mathcal{S} = \left\{ (t, \boldsymbol{x}) \in [0, T] \times \mathbb{R}^m : H(t, \boldsymbol{x}) = G(\boldsymbol{x}) \right\} .$$

It should be evident that whenever $(t, \boldsymbol{x}) \in \mathcal{S}$, i.e. if the state of the system lies in $\mathcal{S}$, it is optimal to stop immediately – since the agent cannot improve beyond $G$ by the definition of the value function in (5.45). The complement of this region, $\mathcal{S}^c$, is known as the **continuation region**. In this region, the agent can still improve her value and therefore continues to wait. Both regions can be generically visualised as in Figure 5.3.

The difficulties in optimal stopping problems arise because the region $\mathcal{S}$ (or

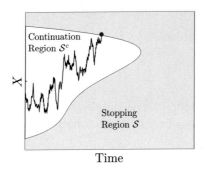

**Figure 5.3** A generic depiction of the continuation and stopping regions for the optimal stopping problem (5.45). A sample path is also shown and the red dot corresponds to the optimal time to stop.

equivalently its boundary $\partial\mathcal{S}$) must be solved simultaneously with the value function itself. Hence, the DPEs which arise in this context are free-boundary problems, also called obstacle problems, or variational inequalities. Such PDEs are more difficult to solve, and even very simple examples typically do not admit explicit solutions – take for example the American put option when the underlying asset is a geometric Brownian motion. Nonetheless, a non-linear PDE is often enough to characterise the solution and many numerical schemes exist for solving them.

### 5.5.1    The Dynamic Programming Principle

Reasoning along similar lines as before, we can show that the following DPP applies.

THEOREM 5.5    ***Dynamic Programming Principle for Stopping Problems.** The value function* (5.45) *satisfies the DPP*

$$H(t,\boldsymbol{x}) = \sup_{\tau\in\mathcal{T}_{[t,T]}} \mathbb{E}_{t,\boldsymbol{x}}\left[ G(\boldsymbol{X}_\tau)\,\mathbb{1}_{\tau<\theta} + H(\theta,\boldsymbol{X}_\theta)\,\mathbb{1}_{\tau\geq\theta}\right], \qquad (5.46)$$

*for all* $(t,\boldsymbol{x}) \in [0,T] \times \mathbb{R}^m$ *and all stopping times* $\theta \leq T$.

The intuition of this DPP is that if the optimal stopping time $\tau^*$ occurs prior to $\theta$, then the agent's value function equals the reward at $\tau^*$. If, however, the agent has not stopped by $\theta$, then at $\theta$ she receives the value function evaluated at the current state of the system.

### 5.5.2    Dynamic Programming Equation

As in the optimal control problems studied earlier, the DPP for optimal stopping can be recast in its infinitesimal version – a much more useful form for computation – i.e. it can be used to derive a DPE. This time, we state the result first and then provide the proof. We follow closely the proof in Touzi (2013) as it provides a nice alternative to the usual construction. It also opens the door for the interested reader to move on to studying the **viscosity solution** approach to stochastic optimal control and optimal stopping, which relies heavily on the proof by contradiction approach taken here.

THEOREM 5.6 **Dynamic Programming Equation for Stopping Problems.** *Assume that the value function $H(t, x)$ is once differentiable in $t$ and all second order derivatives in $x$ exist, i.e. $H \in C^{1,2}([0, T], \mathbb{R}^m)$, and that $G : \mathbb{R}^m \mapsto \mathbb{R}$ is continuous. Then $H$ solves the **variational inequality**, also known as an obstacle problem, or free-boundary, problem:*

$$\max \left\{ \partial_t H + \mathcal{L}_t H , \, G - H \right\} = 0, \quad \text{on} \quad \mathcal{D}, \tag{5.47}$$

*where $\mathcal{D} = [0, T] \times \mathbb{R}^m$.*

*Proof* The proof is broken up into two steps by showing (on $\mathcal{D}$) that (i) the left-hand side is first smaller than or equal to zero, and that (ii) by contradiction the left-hand side is also greater than or equal to zero. Therefore, equality must hold.

(i) First we establish that

$$\max \left\{ \partial_t H + \mathcal{L}_t H , \, G - H \right\} \le 0, \quad \text{on} \quad \mathcal{D}.$$

To show this, first note that the constant stopping rule $\tau = t$ is admissible. Therefore, the performance criteria on this constant rule equals $G$ and we must have $H \ge G$ so that $G - H \le 0$.

Next, take any point $(t_0, x_0) \in \mathcal{D}$. We show that the desired inequality holds at this point. Consider the sequence of stopping times indexed by $h > 0$,

$$\theta_h = \inf \left\{ t > t_0 : (t, \|X_t - x_0\|) \notin [t_0, t_0 + h] \times 1 \right\},$$

where $\|\cdot\|$ denotes the Euclidean norm. That is, we take a stopping time equal to the minimum of $t_0 + h$ and the time it takes $X$ to exit a ball of size 1 from its current position. As long as $h < T - t_0$, this is an admissible stopping time. Therefore, by taking $\tau = t_0$, the DPP (5.46) implies that

$$H(t_0, x_0) \ge \mathbb{E}_{t_0, x_0} \left[ H \left( \theta_h, X_{\theta_h} \right) \right].$$

We then expand the term under the expectation using Itô's lemma for jump-diffusions to write

$$
\begin{aligned}
H \left( \theta_h, X_{\theta_h} \right) \\
= H(t_0, x_0) \\
+ \int_{t_0}^{\theta_h} \left\{ \partial_t H(t, X_t) + \mathcal{L}_t H(t, X_t) \right\} dt + \int_{t_0}^{\theta_h} \left( \sigma(t, X_t) D_x H(t, X_t) \right)' dW_t \\
+ \sum_{j=1}^{p} \int_{t_0}^{\theta_h} \left[ H \left( t, X_t + \gamma_{\cdot j}(t, X_t) \right) - H \left( t, X_t \right) \right] d\widehat{N}_t^j,
\end{aligned}
$$

where $\widehat{N}_t = N_t - \int_0^t \lambda(s, X_s)\, ds$ are the compensated versions of the counting process, and $\gamma_{\cdot j}$ denotes the $j^{th}$ column of $\gamma$.

Since the stopping time $\theta_h$ is chosen so that the process $X$ remains bounded by the ball of size 1 around $x_0$ plus the potential of a jump (which we assume bounded), it follows that the stochastic integrals with respect to both the Brownian motions and the compensated counting processes vanish under the expectation. Hence, we have

$$0 \geq \mathbb{E}_{t_0, x_0} \left[ \int_{t_0}^{\theta_h} \left\{ \partial_t H(t, X_t) + \mathcal{L}_t H(t, X_t) \right\} dt \right] .$$

Dividing by $h$ and taking $h \searrow 0$, in which case $\theta_h \searrow h$ a.s. (since $X_t$ will a.s. not hit the edge of the bounding ball), the Mean-Value Theorem implies that

$$\partial_t H(t, x_0) + \mathcal{L}_t H(t_0, x_0) \leq 0 .$$

This completes the first part of the proof.

(ii) We next show by contradiction that

$$\max \left\{ \partial_t H + \mathcal{L}_t H , \ G - H \right\} \geq 0, \quad \text{on} \quad \mathcal{D} .$$

If the above inequality does not hold on $\mathcal{D}$, then there exists a point $(t_0, x_0) \in \mathcal{D}$ such that

$$G(t_0, x_0) - H(t_0, x_0) < 0 \quad \text{and} \quad (\partial_t + \mathcal{L}_t) H(t_0, x_0) < 0 . \tag{5.48}$$

We show that (5.48) contradicts the DPP in (5.46). To this end, introduce a new function $\varphi_\varepsilon$ which approximates the value function near $(t_0, x_0)$ but locally dominates it:

$$\varphi_\varepsilon(t, x) := H(t, x) + \varepsilon \left( ||x - x_0||^4 + |t - t_0|^2 \right) , \quad \forall (t, x) \in \mathcal{D}, \qquad \varepsilon > 0 .$$

Under the assumption (5.48), for $\varepsilon > 0$ but sufficiently small, there is a small neighbourhood around $(t_0, x_0)$ on which the value function is at least $\delta$ larger than the reward $G$ and for which the operator $(\partial_t + \mathcal{L}_t)$ renders the approximation $\varphi_\varepsilon$ negative. More precisely, there exists $h > 0$ and $\delta > 0$ such that

$$H \geq G + \delta \quad \text{and} \quad (\partial_t + \mathcal{L}_t) \varphi_\varepsilon \leq 0 \quad \text{on} \quad \mathcal{D}_h := [t_0, t_0 + h] \times \mathcal{B}_h , \tag{5.49}$$

where $\mathcal{B}_h$ is a ball of size $h$ around $x_0$, i.e. $\mathcal{B}_h = \{ x \in \mathbb{R}^m : ||x - x_0|| \leq h \}$. Also, near $(t_0, x_0)$, $\varphi_\varepsilon$ is locally larger than $H$, hence,

$$-\zeta := \max_{\partial \mathcal{D}_h} (H - \varphi_\varepsilon) < 0 , \tag{5.50}$$

where $\partial \mathcal{D}_h$ represents the boundary of the set $\mathcal{D}_h$.

We now take a stopping time equal to the first time the process exits this ball:

$$\theta := \inf \left\{ t > t_0 : (t, X_t) \notin \mathcal{D}_h \right\} .$$

Take a second stopping rule, this time arbitrary, $\tau \in \mathcal{T}_{[t,T]}$, and let $\psi = \tau \wedge \theta$. Then we have

$$
\begin{aligned}
H(\psi \,,\, \boldsymbol{X}_\psi) - H(t_0, \boldsymbol{x}_0) \\
= (H - \varphi_\epsilon)(\psi \,,\, \boldsymbol{X}_\psi) + (\varphi_\epsilon(\psi \,,\, \boldsymbol{X}_\psi) - \varphi_\epsilon(t_0, \boldsymbol{x}_0)) \,,
\end{aligned}
\tag{5.51}
$$

since $\varphi_\epsilon$ and $H$ coincide at $(t_0, \boldsymbol{x}_0)$. From Itô's lemma, and the fact that $\boldsymbol{X}_\psi$ is bounded due to stopping the first time $\boldsymbol{X}$ exits the ball $\mathcal{B}_h$, we have

$$
\mathbb{E}_{t_0, \boldsymbol{x}_0} \left[ \varphi_\epsilon(\psi \,,\, \boldsymbol{X}_\psi) - \varphi_\epsilon(t_0, \boldsymbol{x}_0) \right] = \mathbb{E}_{t_0, \boldsymbol{x}_0} \left[ \int_{t_0}^\psi (\partial_t + \mathcal{L}_t) \, \varphi_\varepsilon(t, \boldsymbol{X}_t) \, dt \right] \le 0 \,.
$$

The diffusive and jump terms vanish because they are martingales, and the inequality follows from the second inequality in (5.49).

Hence, putting this together with (5.51), we have

$$
\mathbb{E}_{t_0, \boldsymbol{x}_0} \left[ H(\psi \,,\, \boldsymbol{X}_\psi) - H(t_0, \boldsymbol{x}_0) \right] \le \mathbb{E}_{t_0, \boldsymbol{x}_0} \left[ (H - \varphi_\varepsilon)(\psi \,,\, \boldsymbol{X}_\psi) \right] \le -\zeta \, \mathbb{P}(\tau \ge \theta) \,,
$$

where the second inequality follows from (5.50). By rearranging to isolate $H(t_0, \boldsymbol{x}_0)$, we have

$$
\begin{aligned}
H(t_0, \boldsymbol{x}_0) &\ge \zeta \, \mathbb{P}(\tau \ge \theta) + \mathbb{E}_{t_0, \boldsymbol{x}_0} \left[ H(\psi \,,\, \boldsymbol{X}_\psi) \right] \\
&= \zeta \, \mathbb{P}(\tau \ge \theta) + \mathbb{E}_{t_0, \boldsymbol{x}_0} \left[ H(\tau \,,\, \boldsymbol{X}_\tau) \mathbb{1}_{\tau < \theta} + H(\theta \,,\, \boldsymbol{X}_\theta) \mathbb{1}_{\tau \ge \theta} \right] \,.
\end{aligned}
$$

By the first inequality in (5.49), $H \ge G + \delta$ on $\mathcal{D}_h$, so therefore

$$
H(t_0, \boldsymbol{x}_0) \ge \zeta \, \mathbb{P}(\tau \ge \theta) + \mathbb{E}_{t_0, \boldsymbol{x}_0} \left[ (G(\boldsymbol{X}_\tau) + \delta) \, \mathbb{1}_{\tau < \theta} + H(\theta \,,\, \boldsymbol{X}_\theta) \mathbb{1}_{\tau \ge \theta} \right] \,,
$$

where we have replaced $H(\tau \,,\, \boldsymbol{X}_\tau)$ by its lower bound $(G(\boldsymbol{X}_\tau) + \delta)$ on $\{\tau < \theta\}$, since in that event we are still in $\mathcal{D}_h$. Finally, since $\mathbb{E}_{t_0, \boldsymbol{x}_0} \left[ \mathbb{1}_{\tau < \theta} \right] = \mathbb{P}(\tau < \theta)$, we have

$$
H(t_0, \boldsymbol{x}_0) \ge \zeta \, \mathbb{P}(\tau \ge \theta) + \delta \, \mathbb{P}(\tau < \theta) + \mathbb{E}_{t_0, \boldsymbol{x}_0} \left[ G(\boldsymbol{X}_\tau) \mathbb{1}_{\tau < \theta} + H(\theta \,,\, \boldsymbol{X}_\theta) \mathbb{1}_{\tau \ge \theta} \right] \,.
$$

By the arbitrariness of $\tau \in \mathcal{T}_{[t,T]}$, and the fact that the constants added to the expectation above are positive, we arrive at a contradiction to the DPP (5.46) and the proof is complete.

$\square$

The approximating function $\varphi_\varepsilon$ in the second part of the above proof can be seen as an upper-semicontinuous approximation to the value function, and the viscosity solution approach makes heavy use of both upper and lower semi-continuous envelopes of the value function when it is not smooth enough to differentiate. We do not divert into these discussions here, and instead refer the interested reader to the excellent monographs Touzi (2013) as well as Pham (2010) and Fleming & Soner (2006).

It is worth briefly exploring the interpretation of the variational inequality appearing above, repeated here for convenience:

$$
\max \left\{ \partial_t H + \mathcal{L}_t H \,,\, G - H \right\} = 0 \,, \quad \text{on} \quad \mathcal{D} \,.
$$

This really represents two possibilities: either we have

$$\text{(i)} \qquad \partial_t H + \mathcal{L}_t H = 0 \quad \text{and} \quad H < G,$$

or we have

$$\text{(ii)} \qquad H = G \quad \text{and} \quad \partial_t H + \mathcal{L}_t H < 0.$$

The first of these possibilities corresponds to the value function $H$ being lower than the reward $G$, and in this region the value function satisfies a linear PDE. If we introduce the stochastic process $h_t = H(t, \boldsymbol{X}_t)$, then due to the linear PDE having zero right-hand side, $h_t$ is a martingale. The region in which (i) holds is what we identified earlier as the continuation region $\mathcal{S}^c$, since it is suboptimal to stop there. The second of these possibilities corresponds to the value function $H$ equalling the reward $G$, and hence occurs in the stopping region $\mathcal{S}$. Moreover, the linear operator $\partial_t + \mathcal{L}_t$ renders the value function negative. In this region, if we did not pin the value function to the reward, we see that the process $h_t$ would be a supermartingale; however, pinning it to the reward constrains the process to become a martingale. Hence, we see that the stochastic process $h_t$ corresponding to the flow of the value function is in fact a martingale on the entire $\mathcal{D}$.

## 5.6     Combined Stopping and Control

There are many instances in which an agent wishes to solve a **combined optimal stopping and control** problem – i.e. she wishes to solve simultaneously for the optimal timing and optimal strategy in order to maximise a reward. Such problems inherit features from both problem types and we will see that the resulting DPEs are in effect a combination of the HJB equation and variational inequality.

In this case, we adopt the modelling assumptions and notation from subsection 5.4.3 but repeat it here. First, let $\boldsymbol{N}^{\boldsymbol{u}} = (\boldsymbol{N}_t^{\boldsymbol{u}})_{0 \le t \le T}$ denote a collection of counting processes (of dim $p$) with controlled intensities $\boldsymbol{\lambda}^{\boldsymbol{u}} = (\boldsymbol{\lambda}_t^{\boldsymbol{u}})_{0 \le t \le T}$, and let $\boldsymbol{u} = (\boldsymbol{u}_t)_{0 \le t \le T}$ denote the control processes (of dimension $m$). Furthermore, let $\boldsymbol{W} = (\boldsymbol{W}_t)_{0 \le t \le T}$ denote a collection of independent Brownian motions (of dim $m$). Next, let $\boldsymbol{X}^{\boldsymbol{u}} = (\boldsymbol{X}_t^{\boldsymbol{u}})_{0 \le t \le T}$ denote the controlled processes (of dim $m$) which will appear directly in the agent's performance criteria. If one (or more) of the counting processes $N_t^i$ should appear in the performance criteria or in the dependence in $\boldsymbol{\mu}_t^{\boldsymbol{u}}$, $\boldsymbol{\sigma}_t^{\boldsymbol{u}}$ and/or $\boldsymbol{\gamma}_t^{\boldsymbol{u}}$, shown below, then take $X_t^j = N_t^i$ for some $j$. i.e. include it through one of the components in $X$. We next assume that the controlled $\boldsymbol{X}^{\boldsymbol{u}}$ processes satisfy the SDEs

$$d\boldsymbol{X}_t^{\boldsymbol{u}} = \boldsymbol{\mu}_t^{\boldsymbol{u}} \, dt + \boldsymbol{\sigma}_t^{\boldsymbol{u}} \, d\boldsymbol{W}_t + \boldsymbol{\gamma}_t^{\boldsymbol{u}} \, d\boldsymbol{N}_t^{\boldsymbol{u}} \,. \tag{5.52a}$$

With a slight abuse of notation we write the controlled drift, volatility and jump size as

$$\boldsymbol{\mu}_t^{\boldsymbol{u}} := \boldsymbol{\mu}(t, \boldsymbol{X}_t^{\boldsymbol{u}}, \boldsymbol{u}_t), \qquad (m \times 1 \text{ vector}), \qquad (5.52\text{b})$$

$$\boldsymbol{\sigma}_t^{\boldsymbol{u}} := \boldsymbol{\sigma}(t, \boldsymbol{X}_t^{\boldsymbol{u}}, \boldsymbol{u}_t), \qquad (m \times m \text{ matrix}), \qquad (5.52\text{c})$$

$$\boldsymbol{\gamma}_t^{\boldsymbol{u}} := \boldsymbol{\gamma}(t, \boldsymbol{X}_t^{\boldsymbol{u}}, \boldsymbol{u}_t), \qquad (m \times p \text{ matrix}). \qquad (5.52\text{d})$$

Furthermore, we assume that the controlled intensity takes the form

$$\boldsymbol{\lambda}_t^{\boldsymbol{u}} := \boldsymbol{\lambda}(t, \boldsymbol{X}_t^{\boldsymbol{u}}, \boldsymbol{u}_t) . \qquad (5.52\text{e})$$

The modelling approach above implies that the agent can control, in general, the drift, volatility, jump size and jump arrivals. Precisely how much control she has depends on the specific form of the various functions appearing in (5.52b), (5.52c), (5.52d) and (5.52e).

Next, the agent's performance criteria for a given admissible control $\boldsymbol{u}$ and admissible stopping time $\tau$ is given by

$$H^{\boldsymbol{u},\tau}(t, \boldsymbol{x}) = \mathbb{E}_{t,\boldsymbol{x}}\left[ G\left( \boldsymbol{X}_\tau^{\boldsymbol{u}} \right) \right], \qquad (5.53\text{a})$$

and her value function is

$$H(t, \boldsymbol{x}) = \sup_{\tau \in \mathcal{T}_{[t,T]}} \sup_{\boldsymbol{u} \in \mathcal{A}_{[t,T]}} H^{\boldsymbol{u},\tau}(t, \boldsymbol{x}). \qquad (5.53\text{b})$$

In this manner, the agent selects a stopping rule, seeks the best strategy, and then selects the stopping rule which provides the best overall performance. As such, she is aiming to decide whether to stop 'now' and receive the reward at the point in state space to which her strategy took her, or wait and continue to control the process in hopes of receiving a better reward later on.

A DPP still applies in the current setting; however, now we must take care of both optimal stopping and control. Following similar arguments as in the previous sections we arrive at the following DPP.

THEOREM 5.7 *Dynamic Programming Principle for Optimal Stopping and Control. The value function (5.53) satisfies the DPP*

$$\boxed{H(t, \boldsymbol{x}) = \sup_{\tau \in \mathcal{T}_{[t,T]}} \sup_{\boldsymbol{u} \in \mathcal{A}_{[t,T]}} \mathbb{E}_{t,\boldsymbol{x}}\left[ G(X_\tau^{\boldsymbol{u}}) \, \mathbb{1}_{\tau < \theta} + H(\theta, \boldsymbol{X}_\theta^{\boldsymbol{u}}) \, \mathbb{1}_{\tau \geq \theta} \right],} \qquad (5.54)$$

*for all $(t, \boldsymbol{x}) \in [0, T] \times \mathbb{R}^m$ and all stopping times $\theta \leq T$.*

The intuition here is similar to the previous DPPs. If stopping has not yet occurred $(\tau \geq \theta)$, then the agent receives the value function at that point in time, at that point in state space, where the optimal control has driven her to. If stopping has already occurred $(\tau < \theta)$, then the agent has already received the reward, and her value equals the reward at the time it was paid, at the point in state space where the optimal control drove her to.

Next, we state the DPE arising in this class of problems.

THEOREM 5.8   *Dynamic Programming Equation for Stopping and Control Problems.* *Assume that the value function $H(t, \boldsymbol{x})$ is once differentiable in $t$ and all second order derivatives in $\boldsymbol{x}$ exist, i.e. $H \in C^{1,2}([0, T], \mathbb{R}^m)$, and that $G : \mathbb{R}^m \mapsto \mathbb{R}$ is continuous. Then $H$ solves the **quasi-variational inequality (QVI)**,*

$$\max\left\{ \partial_t H + \sup_{\boldsymbol{u} \in \mathcal{A}_t} \mathcal{L}_t^{\boldsymbol{u}} H \, , \, G - H \right\} = 0, \quad on \quad \mathcal{D}, \tag{5.55}$$

*where $\mathcal{D} = [0, T] \times \mathbb{R}^m$.*

This is similar in spirit to the variational inequality (5.47) for the optimal stopping problem, but now contains an optimisation over the control process as well. The optimisation results in the equation becoming non-linear, as in the HJB equation (5.44), and hence the above equation is referred to as a quasi-variational inequality rather than a variational inequality.

We once again have the notion of a stopping region $\mathcal{S}$ and continuation region $\mathcal{S}^c$, where

$$\mathcal{S} = \left\{ (t, \boldsymbol{x}) \in [0, T] \times \mathbb{R}^m \, : \, H(t, \boldsymbol{x}) = G(\boldsymbol{x}) \right\}.$$

The only difference is that in the continuation region $\mathcal{S}^c$, the value function should satisfy an HJB equation

$$\partial_t H + \sup_{\boldsymbol{u} \in \mathcal{A}_t} \mathcal{L}_t^{\boldsymbol{u}} H = 0, \quad on \quad \mathcal{S}^c,$$

rather than a linear PDE as in the stopping problem.

## 5.7     Bibliography and Selected Readings

Merton (1971), Bertsekas & Shreve (1978), Merton (1992), Yong & Zhou (1999), Chang (2004), Fleming & Soner (2006), Øksendal & Sulem (2007), Øksendal (2010), Pham (2010), Touzi (2013).

# Part III

## Algorithmic and High-Frequency Trading

# Introduction to Part III

In this part of the book we delve into the modelling of algorithmic trading strategies. The first two chapters are concerned with optimal execution strategies where the agent must liquidate or acquire a large position over a pre-specified window and trades continuously using only market orders. Chapter 6 covers the classical execution problem when the investor's trades impact the price of the asset and also adjusts the level of urgency with which she desires to execute the programme. In Chapter 7 we develop three execution models where the investor: i) carries out the execution programme as long as the price of the asset does not breach a critical boundary, ii) incorporates order flow in her strategy to take advantage of trends in the midprice which are caused by one-sided pressure in the buy or sell side of the market, and iii) trades in both a lit venue and a dark pool.

In Chapter 8 we assume that the investor's objective is to execute a large position over a trading window, but she employs only limit orders, or uses both limit and market orders. Moreover, we show execution strategies where the investor also tracks a particular schedule as part of the liquidation programme.

Chapter 9 is concerned with execution algorithms that target volume-based schedules. We develop strategies for investors who wish to track the overall volume traded in the market by targeting: Percentage of Volume, Percentage of Cumulative Volume, and Volume Weighted Average Price, also known as VWAP.

The final three chapters cover various topics in algorithmic trading. Chapter 10 shows how market makers choose where to post limit orders in the book. The models that are developed look at how the strategies depend on different factors including the market maker's aversion to inventory risk, adverse selection, and short-term lived trends in the dynamics of the midprice.

Finally, Chapter 11 is devoted to statistical arbitrage and pairs trading, and Chapter 12 shows how information on the volume supplied in the limit order book is employed to improve execution algorithms.

# 6 Optimal Execution with Continuous Trading I

## 6.1 Introduction

A classical problem in finance is how an agent can sell or buy a large amount of shares and yet minimise adverse price movements which are a consequence of her own trades. Here, the term 'large' means that the amount the agent is interested in buying or selling is too big to execute in one trade. One way to think about a trade being too large is to compare it to the size of an average trade or to the volume posted on the limit order book (LOB) at the best bid or best offer. Clearly, if the number of shares that the agent seeks to execute is significantly larger than the average size of a trade, then it is probably not a good idea to try to execute all the shares in one trade.

Investors who regularly come to the market with large orders (orders that are a significant fraction of average daily volume) are institutional traders such as pension funds, hedge funds, mutual funds, and sovereign funds. These investors often delegate their trades to an agency broker (the agent) who acts on their behalf. The agent will slice the **parent order** into smaller parts (sometimes called child orders) and try to execute each one of these **child orders** over a period of time, taking into account the balance between price impact (trade quicker) and price risk (take longer to complete all trades). What we mean by this trade-off is the following: imagine the situation in which the agent is selling shares. If she trades quickly, then her orders will walk through the buy side of the LOB and she will obtain worse prices for her orders. Even if she breaks up each order into small bits (so that each one does not walk the book), and sends them quickly to the market, then other traders will notice an excess of sell orders and reshuffle their quotes inducing again a negative price impact. If on the other hand, she trades slowly, so as to avoid this price impact, then she will be exposed to the uncertainty of what precisely the future prices will be. Hence, she must attempt to balance these two factors.

The time the agent takes to space out and execute the smaller orders is crucial. Short time horizons will lead to faster trading (and hence more price impact) and less price uncertainty, but there are also many reasons why a long trading horizon might not be desirable. For instance, it might be that it is decided to sell a large chunk of shares because the price is convenient but by the time the agent executes all child orders, the share price could have dropped to a less desirable level. Another reason which constrains the time needed to sell all the shares, is that this particular operation is part of a bigger one which is the result of a portfolio rebalance which also requires the purchase of a large number of shares in another firm, and both operations need to be completed over approximately the same time period.

Hence the agent must formulate a model to help her decide how to optimally liquidate or acquire shares, where the aim is to minimise the cost of executing her trade(s) and balance it against price risk. **Execution costs** are measured as the difference between a **benchmark price** and the actual price (measured as the average price per share) at which the trade was completed. Our convention is that when the sign of the execution cost is positive, it means that there is loss of value in the operation because the actual price of the trade was worse than the benchmark price.

The benchmark price represents a perfectly executed price in a market with no frictions. It is customary to use the midprice of the asset at the time the order is given to execute the trade. This benchmark is known as the **arrival price** which is generally taken to be the average of the best bid and best ask, i.e. the midprice. Moreover, when the arrival price is the benchmark, the execution cost is known as the **implementation shortfall** or **slippage**, see Almgren (2010).

## 6.2    The Model

To pose the optimal execution problem we require notation to describe the number of shares the agent is holding (inventory), the dynamics of the midprice, and how the agent's market orders (MOs) affect the midprice.

The key stochastic processes are:

- $\nu = (\nu_t)_{\{0 \leq t \leq T\}}$ is the trading rate, the speed at which the agent is liquidating or acquiring shares (it is also the variable the agent controls in the optimisation problem),
- $Q^\nu = (Q_t^\nu)_{\{0 \leq t \leq T\}}$ is the agent's inventory, which is clearly affected by how fast she trades,
- $S^\nu = (S_t^\nu)_{\{0 \leq t \leq T\}}$ is the midprice process, and is also affected in principle by the speed of her trading,
- $\hat{S}^\nu = (\hat{S}_t^\nu)_{\{0 \leq t \leq T\}}$ corresponds to the price process at which the agent can sell or purchase the asset, i.e. the execution price, by walking the LOB, and
- $X^\nu = (X_t^\nu)_{\{0 \leq t \leq T\}}$ is the agent's cash process resulting from the agent's execution strategy.

Whether liquidating or acquiring, the agent's controlled inventory process is given in terms of her trading rate as follows:

$$dQ_t^\nu = \pm \nu_t \, dt \,, \qquad\qquad Q_0^\nu = q \,, \qquad\qquad (6.1a)$$

while the midprice is assumed to satisfy the SDE

$$dS_t^\nu = \pm g(\nu_t)\, dt + \sigma\, dW_t\,, \qquad\qquad S_0^\nu = S\,, \qquad\qquad (6.1\text{b})$$

where

- $W = (W_t)_{\{0 \le t \le T\}}$ is a standard Brownian motion, and
- $g : \mathbb{R}_+ \to \mathbb{R}_+$ denotes the **permanent price impact** that the agent's trading action has on the midprice.

The execution price satisfies the SDE

$$\hat{S}_t^\nu = S_t^\nu \pm \left(\tfrac{1}{2}\Delta + f(\nu_t)\right)\,, \qquad\qquad \hat{S}_0^\nu = \hat{S}\,, \qquad\qquad (6.1\text{c})$$

where

- $f : \mathbb{R}_+ \to \mathbb{R}_+$ denotes the **temporary price impact** that the agent's trading action has on the price they can execute the trade at, and
- $\Delta \ge 0$ is the bid-ask spread, assumed here to be a constant.

Equations (6.1a, 6.1b, and 6.1c) apply to both liquidation and acquisition problems, where the sign $\pm$ changes depending on whether the problem is that of liquidating $(-)$ or acquiring $(+)$ shares.

In equity markets the **fundamental price** of the asset (also known in the literature as the efficient price or true price of the asset) refers to the share price that reflects fundamental information about the value of the firm and this is impounded in the price of the share. In this chapter we assume that during the optimal execution trading period the fundamental price is the same as the midprice of the asset. Thus, as new information about the actual and expected performance of the firm is revealed to the market, the midprice changes. This is partly captured in the model by the increments of the Brownian motion $W$ in (6.1b).

A key element of the model is how the agent's trades affect the midprice. Here the agent's market orders affect the midprice in two ways: through $f(\nu_t)$ in (6.1c), and through $g(\nu_t)$ in (6.1b). These functions capture two different ways in which the agent's trades affect the midprice.

At any one time, the number of shares displayed and available in the market at the quoted bid/ask prices $S_t^\nu \pm \tfrac{1}{2}\Delta$ is limited. A large MO will walk the book, so that the average price per share obtained will be worse than the current bid/ask price. This is captured in our model, as an order of size $\nu\, dt$ will obtain an execution price per share of $S_t^\nu \pm \left(\tfrac{1}{2}\Delta + f(\nu)\right)$ with $f(\nu) \ge 0$. Note, however, that the impact of the order as captured by $f(\nu_t)$ is limited to the execution price and does not affect the midprice of the asset.

In Figure 6.1 we show a snapshot of the LOB (top left panel) for SMH on Oct 1, 2013 at 11am, together with the price impact per share (top right panel) that an MO of various volumes would face as it walks through the buy side of the LOB. The bottom left panel shows the impact every second from 11:00 to 11:01 as well as the average of those curves over the minute (the dash-dotted line).

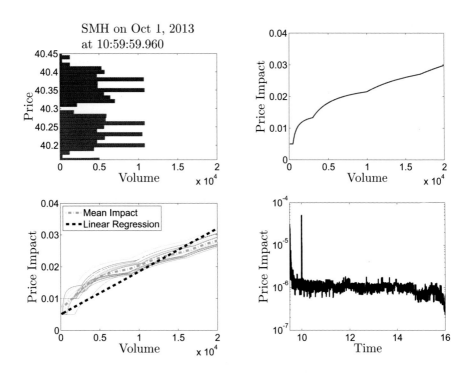

**Figure 6.1** An illustration of how the temporary impact may be estimated from snapshots of the LOB using SMH on Oct 1, 2013. The top two panels are at 11:00am. The bottom left from 11:00am to 11:01am and the bottom right the entire day.

We also include a linear regression (the dashed line) with intercept set to the half-spread; this would correspond to a linear impact function $f(\nu_t)$, which is the simple model we adopt in this chapter, and which is also widely used. Figure 6.1 illustrates that the $f(\nu_t)$ function seems better described by a power law, and the model can be extended to incorporated this. We discuss this extension in Section 6.7.

Notice that the impact function fluctuates within the minute, and with it the impact that trades of different size have. The linear regression provides an approximation of the temporary impact during that one minute. The bottom right panel shows how the slope of this linear impact model fluctuates throughout the entire day. We see that the largest impact tends to occur in the morning, then this impact flattens and stays flat throughout the day, and towards the end of the day it lessens. Such a pattern is seen in a number of assets.

The second way in which the agent's execution can affect the midprice is through $g(\nu_t)$. We refer to this as the permanent impact of the trading rate. If $g(\nu_t) > 0$ in (6.1b) then a trade of size $\nu\, dt > 0$ moves the midprice of the asset upwards. An interpretation of this modelling assumption is that the agent is trading on information that reflects permanent changes in the value of the firm, and market participants adjust their quotes in response to the agent's trades.

Earlier on, in Chapter 4 we discussed linear market impact models and estimated the parameters for various stocks, see in particular subsection 4.3.5.

The model can also be modified to incorporate the situation where the agent's trades exert pressure on the midprice and then the pressure subsides after the agent has completed her execution target. But, as we are focusing on an agent's execution of a single block of shares, this is not relevant as she will never receive any of the benefits of the 'price correction' once she stops trading, and we exclude it from the analysis.

To conclude the description of the model, we turn to the agent's cash process, $X_t^\nu$. This process satisfies the SDE

$$dX_t^v = \hat{S}_t^\nu \, \nu_t \, dt \,, \qquad X_0^v = x \,, \tag{6.2}$$

and the expected revenue from the sale is

$$R^\nu = \mathbb{E}\left[X_T^\nu\right] = \mathbb{E}\left[\int_0^T \hat{S}_t \, \nu_t \, dt\right] \,, \tag{6.3}$$

which is easy to see if we look at the sales proceeds over discrete time-steps. Suppose that the agent must liquidate $Q_0 = \mathfrak{N}$ amount of shares over the time period $[0, T]$. Now, split this trading horizon into equally spaced time intervals $t_0 = 0 < t_1 < t_2 < \cdots < t_N = T$ where $t_n - t_{n-1} = \Delta t$ for $n = 1, 2, \cdots, N$. Next, assuming that over the time interval $[0, t_1)$ the agent sells $Q_0 - Q_{t_1}$ shares at the price $\hat{S}_0$, and over the interval $[t_1, t_2)$ she sells $Q_{t_1} - Q_{t_2}$ at the price $\hat{S}_{t_1}$, and so on, then the total expected revenue from selling shares is

$$R_{\Delta t}^\nu = \mathbb{E}\left[\left(Q_0 - Q_{t_1}\right)\hat{S}_0 + \left(Q_{t_1} - Q_{t_2}\right)\hat{S}_{t_1} + \cdots + \left(Q_{t_{N-1}} - Q_T\right)\hat{S}_{t_{N-1}}\right] \,,$$

and recalling that the speed of trading is given by (6.1a) we observe that as $\Delta t \to 0$ we obtain (6.3).

The rest of this chapter looks at different optimal strategies to trade a block of shares using only MOs, where in each section the setup of the control problem makes different assumptions about how the agent penalises and/or controls inventory, and how her rate of trading affects her execution price as well as the midprice of the asset. We also alternate between share liquidation and share acquisition problems. In Section 6.3 the agent must liquidate a block of shares, and the agent's trades affect her execution price but do not affect the midprice of the asset $(g(\nu) = 0)$. The setup of the problem assumes that the execution strategy is designed so that all shares are liquidated by the terminal date. In Section 6.4 the agent solves for the optimal acquisition rate where any remaining unacquired inventory may be purchased at the terminal date but subject to a penalty (and $g(\nu) = 0$). In Section 6.5 the agent has to liquidate a block of shares, and the agent's actions have both a permanent effect $(g(\nu) \geq 0)$ on the execution price and a temporary effect $(f(\nu) \geq 0)$ on the midprice of the asset. In addition, we incorporate a parameter for the agent's urgency to execute the trade, through a penalisation exposure to inventory throughout the entire life of the strategy.

## 6.3    Liquidation without Penalties only Temporary Impact

We start by discussing how an agent uses only MOs to optimally liquidate $\mathfrak{N}$ shares between $t = 0$ and $T$. We assume that the agent's own trades do not affect the midprice of the asset, thus the stock's midprice is as in (6.1b) with $g(\nu_t) = 0$. On the other hand, the agent's trades have temporary impact on her own execution price because these MOs walk the LOB. We assume that the temporary impact is linear in the speed of trading so $f(\nu_t) = k\,\nu_t$ with $k > 0$ in (6.1c) and recall that the speed of trading $\nu_t$ is what the agent controls. For simplicity, we assume that the bid-ask spread $\Delta = 0$, or equivalently, that $S_t$ represents the best bid price. It is a simple matter to include a non-zero spread and we leave it as an exercise for the interested reader. Finally, we also assume that the agent is adamant that all $\mathfrak{N}$ shares are liquidated by time $T$.

The agent's objective is to choose the rate at which she liquidates $\mathfrak{N}$ shares so that she obtains the maximum amount of revenue from the sale, and her strategy must be such that all shares are liquidated by time $T$, i.e. cannot reach expiry with any inventory left. In other words the agent wishes to find, among all admissible liquidation strategies $\nu$, the one that minimises the execution cost

$$EC^\nu = \mathfrak{N}\,S_0 - \mathbb{E}\left[\int_0^T \hat{S}_t^\nu\,\nu_t\,dt\right],$$

which is equivalent to maximising the expected revenues from the target sale of the $\mathfrak{N}$ shares. Thus the agent's value function is

$$H(t, S, q) = \sup_{\nu \in \mathcal{A}} \mathbb{E}_{t,S,q}\left[\int_t^T (S_u - k\,\nu_u)\,\nu_u\,du\right],$$

where $\mathbb{E}_{t,S,q}[\cdot]$ denotes expectation conditional on $S_t = S$ and $Q_t = q$, and $\mathcal{A}$ is the set of admissible strategies: $\mathcal{F}$-predictable non-negative bounded strategies. This constraint excludes repurchasing of shares and keeps the liquidation rate finite.

To solve this optimal control problem we use the dynamic programming principle (DPP) which suggests that the value function satisfies the dynamic programming equation (DPE)

$$\partial_t H + \tfrac{1}{2}\sigma^2\,\partial_{SS}H + \sup_\nu \left\{(S - k\nu)\,\nu - \nu\,\partial_q H\right\} = 0. \tag{6.4}$$

The agent requires that the optimal strategy liquidates all the inventory by time $T$, thus the value function reflects this by 'penalising' any terminal inventory which is not zero, so we require

$$H(T, S, q) \xrightarrow{t \to T} -\infty, \quad \text{for} \quad q > 0,$$

and

$$H(T, S, 0) \xrightarrow{t \to T} 0.$$

The first order condition applied to DPE (6.4) shows that it attains a supremum at

$$\nu^* = \tfrac{1}{2k}\left(S - \partial_q H\right),\tag{6.5}$$

which is the optimal trading speed in feedback control form. Upon substitution into the DPE we obtain the non-linear partial differential equation

$$\partial_t H + \tfrac{1}{2}\sigma^2 \partial_{SS} H + \tfrac{1}{4k}\left(S - \partial_q H\right)^2 = 0\tag{6.6}$$

for the value function.

To propose an ansatz for the above equation it is helpful to look at the boundary conditions to get an idea of which variables are relevant in the value function. We know that if the strategy reaches the terminal date with a non-zero inventory, the value function must become arbitrarily large and negative – because the optimal strategy must ensure that all shares are liquidated. We propose that the value function be written in terms of the book value of the current inventory (marked-to-market using the midprice as reference) plus the excess value due to optimally liquidating the remaining shares, i.e.

$$H(t, S, q) = q\,S + h(t, q),\tag{6.7}$$

where $h(t, q)$ is still to be determined, though we know that it must blow up as $t$ approaches $T$.

The way the problem is set up, the best the agent can do is achieve the midprice. Hence, the correction $h(t, q)$ to the book value must be negative, and the agent's objective is to minimise this downward adjustment. Substituting this ansatz into the DPE (6.6), we arrive at the following equation for $h(t, q)$:

$$\partial_t h + \tfrac{1}{4k}\left(\partial_q h\right)^2 = 0.$$

Interestingly, the volatility of the asset's midprice drops out of the problem. The reason for this is that the Brownian component is a martingale, and hence on average it contributes zero to the value of liquidating shares.

Focusing on the above non-linear PDE for $h$, we see that writing a separation of variables in the form $h(t, q) = q^2\, h_2(t)$ (note the subscript 2 represents that this function is the coefficient of $q^2$) allows us to factor out $q$ and obtain a simple non-linear ODE for $h_2(t)$:

$$\partial_t h_2 + \tfrac{1}{k} h_2^2 = 0,\tag{6.8}$$

which we solve by integrating between $t$ and $T$ to obtain

$$h_2(t) = \left(\frac{1}{h_2(T)} - \frac{1}{k}(T - t)\right)^{-1}.$$

As discussed above, the optimal strategy must ensure that the terminal inventory is zero and this is equivalent to requiring $h_2(t) \to -\infty$ as $t \to T$. In this way the value function heavily penalises non-zero final inventory. An alternative way to obtain this condition is to calculate the inventory path along the optimal strategy and impose that the terminal inventory be zero. To see this, use the ansatz (6.7) to reduce (6.5) to

$$\nu_t^* = -\tfrac{1}{k}\, h_2(t)\, Q_t^{\nu^*}\,, \tag{6.9}$$

then integrate $dQ_t^{\nu^*} = -\nu_t^*\, dt$ over $[0,t]$ to obtain the inventory profile along the optimal strategy:

$$\int_0^t \frac{dQ_t^{\nu^*}}{Q_t^{\nu^*}} = \int_0^t \frac{h_2(s)}{k}\, ds \quad \Rightarrow \quad Q_t^{\nu^*} = \frac{(T-t) - k/h_2(T)}{T - k/h_2(T)}\, \mathfrak{N}\,.$$

To satisfy the terminal inventory condition $Q_T^{\nu^*} = 0$, and also ensure that the correction $h(t,q)$ to the book value of the outstanding shares that need to be liquidated is negative, we must have

$$h_2(t) \to -\infty \quad \text{as} \quad t \to T\,. \tag{6.10}$$

Returning to solving the optimal problem, we have that

$$h_2(t) = -k\,(T-t)^{-1}\,,$$

so the optimal inventory to hold is

$$\boxed{Q_t^{\nu^*} = \left(1 - \frac{t}{T}\right) \mathfrak{N}\,,} \tag{6.11}$$

and the optimal speed of trading is

$$\boxed{\nu_t^* = \frac{\mathfrak{N}}{T}\,.} \tag{6.12}$$

This final result for the optimal trading speed is quite simple: the shares must be liquidated at a constant rate and this strategy is the same as that of the **time weighted average price** (TWAP).

## 6.4 Optimal Acquisition with Terminal Penalty and Temporary Impact

The problem now is to acquire (not liquidate) $\mathfrak{N}$ shares by time $T$, starting with $Q_0^\nu = 0$. As in the previous section the agent's MOs walk the LOB so her execution price is described by (6.1c) with $f(\nu) = k\nu$, $k > 0$.

Although the agent's objective is to complete the acquisition programme by time $T$, she allows for strategies that fall short of this target, $Q_T^\nu < \mathfrak{N}$, and in this case she must execute a buy MO for the remaining amount and pick up an additional penalty. This terminal inventory penalty is parameterised by $\alpha > 0$, which includes the cost of walking the book at $T$ and any other additional penalties that the agent must incur for the execution of the trade at the terminal date.

Thus, the agent's expected costs from strategy $\nu_t$ is

$$EC^\nu = \mathbb{E}\left[ \int_t^T \hat{S}_u^\nu \, \nu_u \, du + \underbrace{(\mathfrak{N} - Q_T^\nu) \, S_T}_{\text{Terminal execution at mid}} + \underbrace{\alpha \, (\mathfrak{N} - Q_T^\nu)^2}_{\text{Terminal Penalty}} \right]. \qquad (6.13)$$

Compared to the expected costs in the previous section we have two additional terms. In the liquidation problem of the previous section, the agent seeks a strategy that ensures all shares are liquidated by $T$ and the expected costs arise exclusively from continuous trading. Now, the agent can reach $T$ short of her target, but this generates the additional terms that incorporate that sale plus the penalty to purchase the remaining shares at the terminal date.

To simplify notation, we introduce a new stochastic process $Y = (Y_t)_{0 \le t \le T}$ to denote the shares remaining to be purchased between $t$ and the end of the trading horizon $T$:

$$Y_t^\nu = \mathfrak{N} - Q_t^\nu, \qquad \text{so that} \qquad dY_t^\nu = -\nu_t \, dt,$$

and write the value function as

$$H(t, S, y) = \inf_{\nu \in \mathcal{A}} \mathbb{E}_{t,S,y} \left[ \int_t^T \hat{S}_u^\nu \, \nu_u \, du + Y_T^\nu \, S_T + \alpha \, (Y_T^\nu)^2 \right],$$

where it is clear that the strategy seeks to minimise the cash paid to acquire the shares.

Applying the DPP, we expect that the value function should satisfy the DPE

$$0 = \partial_t H + \tfrac{1}{2}\sigma^2 \partial_{SS} H + \inf_\nu \left\{ (S + k\nu) \nu - \nu \partial_y H \right\}, \qquad (6.14)$$

with terminal condition $H(T, S, y) = y \, S + \alpha \, y^2$. Solving for the first order conditions, the optimal speed of trading in feedback form is given by

$$\nu^* = \tfrac{1}{2k} \left( \partial_y H - S \right), \qquad (6.15)$$

and upon substitution into the DPE above, we obtain

$$\partial_t H + \tfrac{1}{2}\sigma^2 \partial_{SS} H - \tfrac{1}{4k} \left( \partial_y H - S \right)^2 = 0.$$

To solve this DPE, we can write the value function in terms of the book value of the assets remaining to be acquired and the excess value function from optimally acquiring these shares. From looking at the terminal condition, and the way $y$ enters into the DPE, we hypothesise that the excess value function can be written in terms of a quadratic function in $y$. The corresponding ansatz is

$$H(t, S, y) = y \, S + h_0(t) + h_1(t) \, y + h_2(t) \, y^2, \qquad (6.16)$$

where $h_2(t)$, $h_1(t)$, $h_0(t)$ are, yet to be determined, deterministic functions of time (note that the subscripts on the functions indicate the power of $y$ which multiplies them in the full ansatz). Recalling that the value function at the terminal date $T$ is $H(T, S, y) = y \, S + \alpha \, y^2$, then

$$h_2(T) = \alpha \quad \text{and} \quad h_1(T) = h_0(T) = 0.$$

Moreover, upon substituting the ansatz into the above non-linear PDE we find that

$$0 = \left\{ \partial_t h_2 - \tfrac{1}{k} h_2^2 \right\} y^2 + \left\{ \partial_t h_1 - \tfrac{1}{2k} h_2 h_1 \right\} y + \left\{ \partial_t h_0 - \tfrac{1}{4k} h_1^2 \right\} .$$

Since this equation must be valid for each $y$, each term in braces must individually vanish. This provides us with three equations for the three functions $h_0$, $h_1$ and $h_2$. Due to the terminal condition $h_1(T) = 0$, we see that the solution we get for $h_1$ (by setting the second term in braces to zero) is $h_1(t) = 0$. Similarly, due to the terminal condition $h_0(T) = 0$, we see that the solution we get for $h_0$ (by setting the third term in braces to zero, and knowing that $h_1(t) = 0$) is $h_0(t) = 0$. Indeed we could have begun with the ansatz $H(t, S, y) = y S + h_2(t) y^2$ and have ended up with the same equation for $h_2$. The final equation (obtained by setting the first term in braces to zero) allows us to obtain $h_2(t)$ and in this case, since $h_2(T) = \alpha$, we obtain the non-trivial solution

$$h_2(t) = \left( \tfrac{1}{k} (T - t) + \tfrac{1}{\alpha} \right)^{-1} .$$

Putting this together with the ansatz for the value function we find that the optimal trading speed is

$$\nu_t^* = \left( (T - t) + \tfrac{k}{\alpha} \right)^{-1} Y_t^{\nu^*} . \tag{6.17}$$

Here we see that as the terminal penalty parameter $\alpha \to \infty$ the acquisition rate converges to that of TWAP. Similarly, the smaller the value of $\alpha$, all else being equal, the slower the acquisition rate will be. Furthermore, in the limiting case $\alpha \to 0$, the optimal strategy is not to purchase any shares until the terminal date is reached, at which point all $\mathfrak{N}$ shares are purchased. In this limiting case, there are no costs of walking the book at date $T$, so it is optimal to purchase all the inventory at the end. In general, however, we expect that $\alpha \gg k$.

As before, we can solve for the optimal inventory path explicitly by integrating $dY_t^{\nu^*} = -\nu_t^* \, dt$ over $[0, t]$, i.e. by solving

$$dY_t^{\nu^*} = - \left( (T - t) + \tfrac{k}{\alpha} \right)^{-1} Y_t^{\nu^*} \, dt$$

for $Y_t^{\nu^*}$. Recalling that $Y_t^{\nu} = \mathfrak{N} - Q_t^{\nu}$, it is straightforward to obtain the optimal inventory path as

$$\boxed{Q_t^{\nu^*} = \frac{t}{T + \tfrac{k}{\alpha}} \, \mathfrak{N}.} \tag{6.18}$$

From this equation we can see that for any finite $\alpha > 0$ and finite $k > 0$, it is always optimal to leave some shares to be executed at the terminal date, and the fraction of shares left to execute at the end decreases with the relative price impact at the terminal date, $k/\alpha$.

To obtain the optimal speed of acquisition, we substitute for $Q_t^{\nu^*}$ into the expression for $\nu_t^*$, so that

$$\nu_t^* = \frac{\mathfrak{N}}{T + \frac{k}{\alpha}} . \tag{6.19}$$

Comparing this with the result from the previous section (see (6.11) and (6.12)), we see that the agent acquires at a constant, but slower rate than that of an agent who heavily penalises (i.e. $\alpha \to \infty$) paths which do not complete the execution fully. Moreover, the agent trades at a constant speed and this speed is the same as that of an agent who must execute everything by the end of the period, but who has a terminal date $T'$ that is further into the future, $T' = T + \frac{k}{\alpha}$.

## 6.5     Liquidation with Permanent Price Impact

In this section we switch from acquisition back to liquidation. The agent continues to use only MOs to liquidate a total of $\mathfrak{N}$ shares, but now her trades have both a temporary and a permanent price impact. The midprice dynamics are given by (6.1b) with drift $g(\nu_t) > 0$, which enters the equation with negative sign because the agent's sell trades exert a permanent downward pressure, and the execution price by (6.1c) with $f(\nu_t) > 0$, which enters the equation with a negative sign because the sell trades have an adverse temporary impact. Here we assume that if the agent's strategy reaches the terminal date $T$ with inventory left, then she must execute an MO to reach $\mathfrak{N}$ for a total revenue of $Q_T^\nu (S_T^\nu - \alpha Q_T^\nu)$, where $\alpha \geq 0$ is the terminal liquidation penalty parameter. The agent's objective is to minimise the execution cost

$$EC^\nu = \mathfrak{N} S_0 - \mathbb{E}\Big[ \underbrace{X_T^\nu}_{\text{Terminal Cash}} + Q_T^\nu ( \underbrace{S_T^\nu}_{\text{Midprice}} - \underbrace{\alpha Q_T^\nu}_{\text{Penalty per Share}} )\Big],$$

where the process corresponding to the investor's wealth $X_t^\nu$ is as in (6.2). Here we have switched from writing out the cash process explicitly in terms of the integrated execution costs, to including the cash process directly. This way the cash process becomes a state variable. Naturally, we could in principle keep using the integrated costs representation; however, it is sometimes easier to motivate the choice of ansatz for the forthcoming problems when value functions are written in terms of $X$ as a state variable.

In this section, we also introduce another element into the model: a running inventory penalty of the form $\phi \int_t^T (Q_u^\nu)^2$ with $\phi \geq 0$. This running inventory penalty is not (and should not be considered) a financial cost to the agent's strategy. The parameter $\phi$ allows us to incorporate the agent's urgency for executing the trade. The higher the value of $\phi$, the quicker the optimal strategy liquidates the shares, as it increases the penalty for the late liquidation of shares and incentivises strategies that front load the liquidation of inventory. Cartea, Donnelly & Jaimungal (2013) show that the running inventory penalty is equivalent to introducing ambiguity aversion on the part of the agent, where the ambiguity is over the midprice which, in their model, may have a non-zero stochastic drift.

Then, the agent's performance criterion is

$$H^\nu(t, x, S, q) = \mathbb{E}_{t,x,S,q} \Big[ \underbrace{X_T^\nu}_{\text{Terminal Cash}} + \underbrace{Q_T^\nu (S_T^\nu - \alpha Q_T^\nu)}_{\text{Terminal Execution}} - \underbrace{\phi \int_t^T (Q_u^\nu)^2 \, du}_{\text{Inventory Penalty}} \Big],$$

(6.20)

and the value function

$$H(t, x, S, q) = \sup_{\nu \in \mathcal{A}} H^\nu(t, x, S, q).$$

The DPP implies that the value function should satisfy the HJB equation

$$0 = \left(\partial_t + \tfrac{1}{2}\sigma^2 \partial_{SS}\right) H - \phi q^2$$
$$+ \sup_\nu \left\{ \left(\nu\left(S - f(\nu)\right) \partial_x - g(\nu) \partial_S - \nu \partial_q\right) H \right\},$$

(6.21)

subject to the terminal condition $H(T, x, S, q) = x + Sq - \alpha q^2$.

We use the simplifying assumption that permanent and temporary price impact functions are linear in the speed of trading, i.e. $f(\nu) = k\nu$ and $g(\nu) = b\nu$ for finite constants $k \geq 0$ and $b \geq 0$. The first order condition allows us to obtain the optimal speed of trading in feedback control form as

$$\nu^* = \tfrac{1}{2k} \frac{(S\partial_x - b\partial_S - \partial_q)H}{\partial_x H}.$$

(6.22)

Upon substituting the optimal feedback control into the DPE, it reduces to

$$0 = \left(\partial_t + \tfrac{1}{2}\sigma^2 \partial_{SS}\right) H - \phi q^2 + \tfrac{1}{4k} \frac{[(S\partial_x - b\partial_S - \partial_q)H]^2}{\partial_x H}.$$

By inspecting the terminal condition $H(T, x, S, q) = x + Sq - \alpha q^2$, it suggests the ansatz

$$H(t, x, S, q) = x + Sq + h(t, S, q),$$

(6.23)

where $h$, with terminal condition $h(T, S, q) = -\alpha q^2$, is yet to be determined. The first term of the ansatz is the accumulated cash of the strategy, the second is the marked-to-market book value (at midprice) of the remaining inventory, and $h$ is the extra value stemming from optimally liquidating the rest of the shares.

Using this ansatz in the equation above and simplifying, we find the following non-linear PDE for $h$:

$$0 = \left(\partial_t + \tfrac{1}{2}\sigma^2 \partial_{SS}\right) h - \phi q^2 + \tfrac{1}{4k} \left[b\left(q + \partial_S h\right) + \partial_q h\right]^2.$$

Since the above PDE contains no explicit dependence on $S$ and the terminal condition is independent of $S$, it follows that $\partial_S h(t, S, q) = 0$, and we can write $h(t, S, q) = h(t, q)$ (with a slight abuse of notation). The equation then simplifies even further to

$$0 = \partial_t h(t, q) - \phi q^2 + \tfrac{1}{4k} \left[bq + \partial_q h(t, q)\right]^2.$$

Furthermore, the optimal control in feedback form from (6.22) takes on the much more compact form

$$\nu^* = -\frac{1}{2k}\left(\partial_q h(t,q) + b\,q\right).$$

$$(6.24)$$

In this form, it appears that the solution admits a separation of variables $h(t,q) = h_2(t)\,q^2$ where $h_2(t)$ satisfies the non-linear ODE (recall that the subscript 2 represents that this function is the coefficient of $q^2$)

$$0 = \partial_t h_2 - \phi + \tfrac{1}{k}\left[h_2 + \tfrac{1}{2}b\right]^2,$$

$$(6.25)$$

subject to the terminal condition $h_2(T) = -\alpha$. This ODE is of Riccati type and can be integrated exactly. First, let $h_2(t) = -\tfrac{1}{2}b + \chi(t)$, then re-arranging the ODE we obtain

$$\frac{\partial_t \chi}{k\phi - \chi^2} = \frac{1}{k},$$

subject to $\chi(T) = \tfrac{1}{2}b - \alpha$. Next, integrating both sides of the above over $[t,T]$ yields

$$\log\frac{\sqrt{k\,\phi} + \chi(T)}{\sqrt{k\,\phi} - \chi(T)} - \log\frac{\sqrt{k\,\phi} + \chi(t)}{\sqrt{k\,\phi} - \chi(t)} = 2\gamma\,(T-t),$$

so that

$$\chi(t) = \sqrt{k\,\phi}\,\frac{1 + \zeta\,e^{2\gamma\,(T-t)}}{1 - \zeta\,e^{2\gamma(T-t)}},$$

where

$$\gamma = \sqrt{\frac{\phi}{k}} \quad\text{and}\quad \zeta = \frac{\alpha - \tfrac{1}{2}b + \sqrt{k\,\phi}}{\alpha - \tfrac{1}{2}b - \sqrt{k\,\phi}}.$$

$$(6.26)$$

At this point the solution of the DPE is fully determined and the optimal speed of trading can now be explicitly shown in terms of the state variables rather than in feedback form. Specifically, from (6.24), the optimal speed to trade at is

$$\boxed{\nu_t^* = \gamma\,\frac{\zeta\,e^{\gamma\,(T-t)} + e^{-\gamma\,(T-t)}}{\zeta\,e^{\gamma(T-t)} - e^{-\gamma\,(T-t)}}\,Q_t^{\nu^*}.}$$

$$(6.27)$$

Interestingly, the optimal speed to trade is still proportional to the investor's current inventory level, as we found in the previous simpler models, but now the proportionality factor depends non-linearly on time.

From this expression, it is also possible to obtain the agent's inventory $Q_t^{\nu^*}$ that results from following this strategy. Recall that the agent's inventory satisfies $dQ_t^\nu = -\nu_t\,dt$, hence

$$dQ_t^{\nu^*} = \frac{\chi(t)}{k}\,Q_t^{\nu^*}\,dt \quad\text{so that}\quad Q_t^{\nu^*} = \mathfrak{N}\exp\left\{\int_0^t \frac{\chi(s)}{k}\,ds\right\}.$$

To obtain the inventory along the optimal strategy we first solve the integral

$$
\int_0^t \frac{\chi(s)}{k}\,ds = \frac{1}{k}\int_0^t \sqrt{k\phi}\,\frac{1+\zeta e^{2\gamma(T-s)}}{1-\zeta e^{2\gamma(T-s)}}\,ds
$$

$$
= \gamma \int_0^t \frac{e^{-2\gamma(T-s)}}{e^{-2\gamma(T-s)}-\zeta}\,ds + \gamma \int_0^t \frac{\zeta e^{2\gamma(T-s)}}{1-\zeta e^{2\gamma(T-s)}}\,ds
$$

$$
= \log\left(e^{-\gamma(T-s)}-\zeta e^{\gamma(T-s)}\right)\Big|_0^t \tag{6.28}
$$

$$
= \log\frac{\zeta e^{\gamma(T-t)}-e^{-\gamma(T-t)}}{\zeta e^{\gamma T}-e^{-\gamma T}}\,, \tag{6.29}
$$

hence

$$
Q_t^{\nu^*} = \frac{\zeta e^{\gamma(T-t)}-e^{-\gamma(T-t)}}{\zeta e^{\gamma T}-e^{-\gamma T}}\,\mathfrak{N}\,. \tag{6.30}
$$

Substituting this expression into (6.27) allows us to write the optimal speed to trade as a simple deterministic function of time

$$
\nu_t^* = \gamma\,\frac{\zeta e^{\gamma(T-t)}+e^{-\gamma(T-t)}}{\zeta e^{\gamma T}-e^{-\gamma T}}\,\mathfrak{N}\,.
$$

In the limit in which the quadratic liquidation penalty goes to infinity, i.e. as $\alpha \to +\infty$, we get $\zeta \to 1$. Then, the optimal inventory to hold and the optimal speed to trade simplify to

$$
Q_t^{\nu^*} \xrightarrow[\alpha\to+\infty]{} \frac{\sinh\left(\gamma(T-t)\right)}{\sinh\left(\gamma T\right)}\,\mathfrak{N}\,,
$$

and

$$
\nu_t^* \xrightarrow[\alpha\to+\infty]{} \gamma\,\frac{\cosh\left(\gamma(T-t)\right)}{\sinh\left(\gamma T\right)}\,\mathfrak{N}\,.
$$

Both of these expressions are independent of $b$. For other values of $\alpha$ the relationship between $\alpha$ and the permanent price impact parameter $b$ is more complex and we look at it after considering some numerical examples.

Figure 6.2 contains plots of the inventory level under the optimal strategy for two levels of the liquidation penalty $\alpha$ and several levels of the running penalty $\phi$. Note that with no running penalty, $\phi = 0$, the strategies are straight lines and in particular, with $\alpha \to \infty$ the strategy is equivalent to a TWAP strategy. As the running penalty $\phi$ increases, the trading curves become more convex and the optimal strategy aims to sell more assets sooner. This is an intuitive result since $\phi$ represents the agent's urgency to liquidate the position, and therefore as it increases she initially liquidates more quickly. Naturally, as the liquidation penalty increases, the terminal inventory is pushed to zero.

As an exercise, one can check that in the limit in which the running penalty vanishes, $\phi \to 0$, the analog of the result from the previous section is recovered,

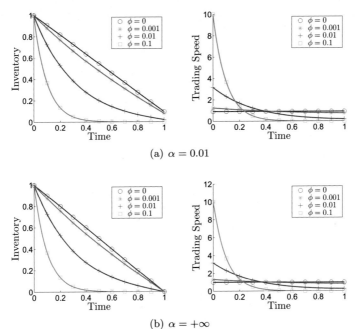

(a) $\alpha = 0.01$

(b) $\alpha = +\infty$

**Figure 6.2** The investor's inventory along the optimal path for various levels of the running penalty $\phi$. The remaining model parameters are $k = 10^{-3}$, $b = 10^{-3}$.

i.e.

$$Q_t^{\nu^*} \xrightarrow[\phi \to 0]{} \left( 1 - \frac{t}{T + \frac{k}{\alpha}} \right) \mathfrak{N} .$$

## Equivalence Between Permanent Price Impact and Terminal Liquidation Penalty

In the previous section we solved the general case when the agent's trades have temporary impact on the execution price and permanent impact on the midprice. We assumed that these two impacts were linear in the speed of trading, $f(\nu) = k\nu$ and $g(\nu) = b\nu$ for constants $k \geq 0$ and $b \geq 0$. One typically observes that $b \ll k$ and we also assume that the liquidation penalty parameter $\alpha \gg k$. In this section we discuss the relationship between the liquidation penalty parameter $\alpha$ and the permanent price impact parameter $b$ – the discussion for acquisition problems is very similar.

The basis for the analysis comes from observing that in the optimal speed of trading, as described in (6.27), the permanent impact and the liquidation penalty always appear in the form $\alpha - \frac{1}{2}b$, see (6.26). This implies that in the current model, where the permanent impact is linear in the speed of trading and the liquidation of terminal inventory is quadratic, $\alpha Q_T^2$, one could define a single parameter $c = \alpha - \frac{1}{2}b$ (so that $c = \chi(T)$) to describe how both the permanent impact and the liquidation penalty affect the optimal speed of trading. Obviously, we cannot do this for other variables in the model, such as for the cash obtained from liquidating shares. The impact of the permanent price impact parameter on this variable is quite distinct from that of the liquidation penalty.

To see this, we consider how the proceeds from selling the $\mathfrak{N}$ shares are affected by the permanent impact that the agent's trades have on the midprice. First, we calculate the agent's terminal cash when she follows an arbitrary strategy $\nu_t$. Recall that the agent's cash position satisfies the SDE

$$dX_t^\nu = (S_t^\nu - k\,\nu_t)\,\nu_t\,dt\,,$$

where

$$dS_t^\nu = -b\,\nu\,dt + \sigma\,dW_t\,,$$

and, for simplicity, assume that $X_0 = 0$, $k = 0$, and $S_0 = 0$. Then, the revenue from liquidating her shares, including the liquidation of the terminal inventory, is

$$
\begin{aligned}
R^\nu &= \int_0^T S_t^\nu\,\nu_t\,dt + Q_T^\nu\,(S_T^\nu - \alpha Q_T^\nu) \\
&= \int_0^T \left\{ -b\int_0^t \nu_u\,du + \sigma\,W_t \right\}\nu_t\,dt + Q_T^\nu\,(S_T^\nu - \alpha Q_T^\nu) \\
&= \int_0^T \left\{ -b\,(\mathfrak{N} - Q_t^\nu) + \sigma\,W_t \right\}\nu_t\,dt + Q_T^\nu\,(S_T^\nu - \alpha Q_T^\nu) \\
&= \int_0^T \left\{ -b\,(\mathfrak{N} - Q_t^\nu) + \sigma W_t \right\}(-dQ_t^\nu) + Q_T^\nu\,(S_T^\nu - \alpha Q_T^\nu) \\
&= -b\int_0^T (\mathfrak{N} - Q_t^\nu)\,d(\mathfrak{N} - Q_t^\nu) - \sigma\int_0^T W_t\,dQ_t^\nu + Q_T^\nu\,(S_T^\nu - \alpha Q_T^\nu) \\
&= -\tfrac{1}{2}\,b\,(\mathfrak{N} - Q_T^\nu)^2 + Q_T^\nu\,(S_T^\nu - \alpha Q_T^\nu) - \sigma\int_0^T W_t\,dQ_t^\nu\,.
\end{aligned}
$$

Having expressed $R^\nu$ in this way, we see that both $\alpha$ and $b$ appear together with $(Q_T^\nu)^2$ and both act to penalise inventories different from zero. Nevertheless, if we isolate the terms in $R^\nu$ that are affected by $\alpha$ and $b$ we obtain

$$R^\nu = -\tfrac{1}{2}\,b\,(\mathfrak{N}^2 - 2\mathfrak{N}Q_T^\nu) - \left(\tfrac{1}{2}\,b + \alpha\right)(Q_T^\nu)^2 + Q_T^\nu S_T^\nu - \sigma\int_0^T W_t\,dQ_t^\nu\,.$$

It is now clear that not only do $\alpha$ and $b$ affect the revenue process in a very different way than they do the speed of trading, but also that the effect of the parameter of the permanent price impact cannot be absorbed into the liquidation penalty.

Indeed, $b$ shows up explicitly in the value function separately from $\alpha$. First note that $\alpha$ and $b$ do appear in $\chi(t)$ together in the form $c = \alpha - \tfrac{1}{2}b$ (through $\zeta$). But, $b$ appears separately through the relationship of $h_2(t) = \chi(t) - \tfrac{1}{2}\,b$. Since $\chi(t)$ is what determines the optimal trading strategy, we see that $b$ can be absorbed into $\alpha$ for the purpose of the trading strategy. But this effect does not extend to the revenue process. We can see this most clearly when the agent follows the optimal strategy in the limiting case where $\alpha \to \infty$. In this limiting case, the agent will complete the trade by the terminal date, hence $Q_T^{\nu^*} = 0$, and any terminal penalty would be applied to a terminal quantity equal to zero.

Nevertheless, the impact of the agent's trades on the midprice will be strictly positive: a loss of $\frac{1}{2} b \mathfrak{N}^2$.

## 6.6     Execution with Exponential Utility Maximiser

In the previous sections, the agent was viewed as a risk-neutral one in the sense that she is maximising her expected terminal wealth (from optimally trading and liquidating any remaining shares at maturity). With the exception of Section 6.3, the agent is not strictly risk-neutral because she is also penalising holding inventory –which is a form of risk aversion. In this section, we demonstrate that if the agent is risk-averse with exponential utility then she acts in the same manner as the risk-neutral, but inventory averse, agent studied in the previous sections.

Let us consider the agent who sets preferences based on expected utility of terminal wealth with exponential utility: $u(x) = -e^{-\gamma x}$. Her performance criteria is

$$H^\nu(t, x, S, q) = \mathbb{E}_{t,x,S,q}\left[ - \exp\left\{ - \gamma\left(X_T^\nu + Q_T^\nu\left(S_T^\nu - \alpha\, Q_T^\nu\right)\right)\right\}\right],$$

and her value function is

$$H(t, x, S, q) = \sup_{\nu \in \mathcal{A}} H^\nu(t, x, S, q),$$

where $S^\nu$, $Q^\nu$ and $X^\nu$ satisfy, as usual, the equations in (6.1) and (6.2). The agent's terminal wealth has two components: the cash that she has accumulated from trading through $X_T^\nu$, and the value she receives from liquidating any remaining assets at the end of the trading horizon through $Q_T^\nu(S_T^\nu - \alpha\,Q_T^\nu)$ – which accounts, as before, for the impact of making a lump trade.

Applying the DPP we expect that $H$ satisfies the DPE

$$0 = \left(\partial_t + \tfrac{1}{2}\sigma^2\,\partial_{SS}\right) H + \sup_\nu \left\{\left(\nu\left(S - k\,\nu\right)\partial_x - b\,\nu\,\partial_S - \nu\,\partial_q\right) H\right\}, \qquad (6.31a)$$

with terminal condition

$$H(T, x, S, q) = - \exp\{-\gamma\left(x + q\left(S - \alpha\,q\right)\right)\}. \qquad (6.31b)$$

The exponential terminal condition suggests that we use the ansatz

$$H(t, x, S, q) = - \exp\{-\gamma\left(x + q\,S + h(t, q)\right)\}, \qquad (6.32)$$

and upon substitution into (6.31), we find that $h$ satisfies the non-linear PDE

$$0 = -\gamma\,h\,\partial_t h + \tfrac{1}{2}\sigma^2\,\gamma^2\,q^2\,h + \sup_\nu \left\{-\gamma\,\nu\left(S - k\,\nu\right) + \gamma\,q\,b\,\nu + \gamma\,\nu\left(S + \partial_q h\right)\right\} h,$$

subject to the terminal condition $h(T, q) = -\alpha\,q^2$. Since we expect that $h$ is negative, due to the terminal condition, we can factor out the common $-\gamma\,h$ terms and obtain the simpler non-linear PDE

$$0 = \partial_t h - \tfrac{1}{2}\sigma^2\,\gamma\,q^2 + \sup_\nu \left\{-k\,\nu^2 - \left(q\,b + \partial_q h\right)\nu\right\}. \qquad (6.33)$$

It is straightforward to obtain the optimal control $\nu^*$ in feedback form as

$$\nu^* = -\frac{1}{2\,k}(q\,b + \partial_q h)\,, \tag{6.34}$$

and upon substitution into (6.33), we further find that $h$ solves

$$0 = \partial_t h - \tfrac{1}{2}\sigma^2\,\gamma\,q^2 + \frac{1}{2\,k}\,(q\,b + \partial_q h)^2\,.$$

A further observation is that if we consider $h$ to be quadratic in $q$, then all the terms in this non-linear equation are quadratic in $q$, and so is the terminal condition. Hence, we expect that $h(t,q) = h_2(t)\,q^2$ for some deterministic function $h_2(t)$ with terminal condition $h_2(T) = -\alpha$ since $h(T,q) = -\alpha\,q^2$. Inserting this second ansatz, and factoring out $q^2$, we find that $h_2(t)$ satisfies the non-linear ODE

$$0 = \partial_t h_2 - \tfrac{1}{2}\sigma^2\,\gamma + \frac{1}{k}\,(h_2 + \tfrac{1}{2}\,b)^2\,. \tag{6.35}$$

Comparing (6.35) to (6.25), we see that the two ODEs coincide whenever $\phi = \frac{1}{2}\gamma\sigma^2$, and since the terminal conditions are identical, the solutions to the two PDEs are identical. Hence, using the same steps that show how to solve (6.25), we find that in the case of an agent with exponential utility preferences, we have

$$h_2(t) = \sqrt{k\,\gamma\,\sigma^2}\,\frac{1 + \zeta\,e^{2\,\xi\,(T-t)}}{1 - \zeta\,e^{2\,\xi\,(T-t)}} - \frac{1}{2}\,b\,,$$

where the constants

$$\xi = \sqrt{\frac{\gamma\,\sigma^2}{2\,k}}\,, \quad \text{and} \quad \zeta = \frac{\alpha - \tfrac{1}{2}b + \sqrt{\tfrac{1}{2}k\,\gamma\,\sigma^2}}{\alpha - \tfrac{1}{2}b - \sqrt{\tfrac{1}{2}k\,\gamma\,\sigma^2}}\,.$$

Recalling that $h(t,q) = q^2\,h_2(t)$ and substituting in the above solution into (6.34), we find that the optimal speed to trade is

$$\boxed{\nu_t^* = \xi\,\frac{\zeta\,e^{\xi\,(T-t)} + e^{-\xi\,(T-t)}}{\zeta\,e^{\xi\,(T-t)} - e^{-\xi\,(T-t)}}\,Q_t^{\nu^*}\,.} \tag{6.36}$$

This strategy is identical in form to the one for the risk-neutral agent who is inventory averse appearing in (6.27). Furthermore, the value functions for the two problems (the exponential utility maximiser and the risk-neutral with inventory aversion) can be mapped to one another. From (6.32), we have

$$H^{exp-util}(t,x,q,S) = -\exp\left\{-\gamma(x + q\,S + q^2\,h_2(t))\right\}\,,$$

where the superscript $exp-util$ emphasises that this is for the exponential utility maximiser. Similarly, from (6.23), we have that

$$H^{inv-aver}(t,x,q,S) = x + q\,S + q^2\,h_2(t)\,,$$

where the superscript $inv-aver$ emphasises that this is for the inventory averse agent. Since the $h_2$ functions coincide when $\phi = \frac{1}{2}\gamma\sigma^2$, we can write the value functions in terms of one another as follows:

$$
\begin{aligned}
&\sup_{\nu\in\mathcal{A}} \mathbb{E}\left[ -\exp\left\{ -\gamma\left( X_T^\nu + Q_T^\nu \left( S_T^\nu - \alpha\, Q_T^\nu \right) \right) \right\} \right] \\
&\quad = -\exp\left\{ -\gamma \sup_{\nu\in\mathcal{A}} \mathbb{E}\left[ X_T^\nu + Q_T^\nu \left( S_T^\nu - \alpha\, Q_T^\nu \right) - \tfrac{\gamma\sigma^2}{2} \int_0^T (Q_u^\nu)^2\, du \right] \right\}.
\end{aligned}
$$

In later sections, we see how the agent with exponential utility can be mapped back to a risk-neutral, but inventory averse agent in several different settings. For example, in Section 8.3 we study the mapping when the agent uses LOs to liquidate, in Section 9.5 we see how an agent who aims to target percentage of volume incorporates utility, and in Section 10.3 we investigate how risk-aversion modifies the behaviour of a market marker.

## 6.7    Non-Linear Temporary Price Impact

In the previous sections we assumed the price impact function $f(\nu)$, see (6.1c), to be linear in the speed of trading. From Figure 6.1, which shows a snapshot of the LOB and how an order of various volumes walks the book, we see that a linear model is a good approximation, but some research has shown that a power law with power less than one fits the data better. Others also argue that, given the extremely low predictive accuracy of market impact models (typically $< 5\%\ R^2$), the cost of increased complexity arising from moving away from a linear model would outweigh any gains from better describing market impact. Nonetheless, it is worthwhile investigating how the problem is modified in the case of non-linear price impact.

To focus on the effects of non-linear impact, we revert back to a risk-neutral agent with inventory aversion through a running penalty as in all sections, other than Section 6.6, and so the agent's performance criteria is as in (6.20) repeated here for convenience:

$$
H^\nu(t,x,S,q) = \mathbb{E}_{t,x,S,q}\left[ X_T^\nu + Q_T^\nu \left( S_T^\nu - \alpha\, Q_T^\nu \right) - \phi \int_t^T (Q_u^\nu)^2\, du \right],
$$

and the dynamics of $S^\nu$, $X^\nu$ and $Q^\nu$ are also repeated here with the explicit non-linear impact model written in place:

$$
\begin{aligned}
dS_t^\nu &= -b\,\nu_t\, dt + \sigma\, dW_t, \\
dX_t^\nu &= (S_t^\nu - f(\nu_t))\,\nu_t\, dt, \\
dQ_t^\nu &= -\nu_t\, dt.
\end{aligned}
$$

As usual, the DPP suggests that the value function

$$
H(t,x,S,q) = \sup_{\nu\in\mathcal{A}} H^\nu(t,x,S,q),
$$

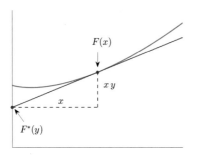

**Figure 6.3** Graphical representation of the Legendre Transform $F^*(y)$ of function $F(x)$. The point at which the tangent hits the vertical axis is the value of the transform evaluated at the slope at the tangent point.

should satisfy the DPE

$$0 = \left(\partial_t + \tfrac{1}{2}\sigma^2\, \partial_{SS}\right) H - \phi\, q^2 + \sup_\nu \left\{\left(\nu\, (S - f(\nu))\, \partial_x - b\, \nu\, \partial_S - \nu\, \partial_q\right) H\right\},$$

with terminal condition $H(T, x, S, q) = x + q\, (S - \alpha\, q)$. Applying the usual ansatz, $H(t, x, S, q) = x + q\, S + h(t, q)$, which separates out the book value of cash in hand and inventory from the value of optimally trading the remaining shares, we have the following non-linear PDE for $h$:

$$0 = \partial_t h - \phi\, q^2 + \sup_\nu \left\{-\nu\, f(\nu) - (b\, q + \partial_q h)\, \nu\right\},$$

with terminal condition $h(T, q) = -\alpha\, q^2$.

To proceed, let us denote $F(\nu) = \nu\, f(\nu)$, and assume that $\nu\, f(\nu)$ is convex. The implication is that the net cost (and not the price impact alone) of trading at a rate of $\nu$ is convex. This certainly holds true for the linear price impact model, for which $f(\nu) = k\, \nu$ and so $F(\nu) = \nu^2$. It also holds for the popular power law price impact models $f(\nu) = k\, \nu^a$ where $a > 0$. Under this convexity assumption, the supremum term becomes

$$\sup_\nu \left\{-\nu\, f(\nu) - (b\, q + \partial_q h)\, \nu\right\} = F^*\left(-(b\, q + \partial_q h)\right),$$

where $F^*$ is the **Legendre transform** of the function $F$ defined as

$$F^*(y) = \sup_x \left(x\, y - F(x)\right).$$

The Legendre transform is a mapping from the graph of a function to the set of its tangents, and can be best understood from Figure 6.3. The figure shows that the Legendre transform $F^*(y)$ of the function $F$ equals the value at which the tangent at a point intersects the vertical axis, and the argument $y$ is the slope of the function at that tangent point. Since the function is convex, the slope is increasing and therefore for each slope $y$, there exists only one point with that slope. Hence, the mapping is one-to-one.

For example, in a power law impact model we write $f(x) = k\, x^a$, and so $F(x) = k\, x^{1+a}$. Then

$$F^*(y) = \sup_x \left(x\, y - k\, x^{1+a}\right).$$

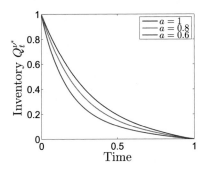

**Figure 6.4** The effect of non-linear impact on the optimal strategy in the case of a power law temporary impact function with power parameter $a$. The model parameters are $b = k = 10^{-4}$, $\phi = 10\,k$, $\alpha = 100\,k$, and $T = 1$.

We can find the optimal point $x^*$ from the first order condition

$$y - k\,(1+a)\,(x^*)^a = 0 \quad \Rightarrow \quad x^* = \left(\frac{y}{(1+a)\,k}\right)^{\frac{1}{a}},$$

and so

$$F^*(y) = \xi\,y^{1+\frac{1}{a}}, \qquad \xi = \frac{a\,k}{((1+a)\,k)^{1+\frac{1}{a}}},$$

and the optimal trading speed in feedback form is

$$\nu^* = \left(-\frac{b\,q + \partial_q h}{(1+a)\,k}\right)^{\frac{1}{a}}. \tag{6.37}$$

We can then write the non-linear PDE for $h$ as

$$\boxed{\partial_t h - \phi\,q^2 + F^*\left(-(b\,q + \partial_q h)\right) = 0, \quad \text{and} \quad h(t,q) = -\alpha\,q^2.} \tag{6.38}$$

In general, this equation cannot be solved analytically, and one must resort to numerical PDE techniques.

In Figure 6.4, we show the effect of the strength of the power in the power law parameter $a$ on the inventory path from following the optimal strategy $Q_t^{\nu^*}$. These curves are obtained by numerically solving (6.38) with a finite difference scheme, substituting the solution into (6.37), and then numerically integrating $dQ_t^{\nu^*} = -\nu_t^*\,dt$, with $Q_0^{\nu^*} = 1$, to obtain $Q_t^{\nu^*}$. The striking result is that as the power law parameter decreases, so that orders of the same size have less and less of an impact, the agent liquidates faster. The intuition here is that since trading does not impact prices as much, the agent prefers to liquidate shares early and reduce her inventory risk, and doing so does not cause her to lose too much from temporary market impact. In some sense, the agent behaves as if she has a larger urgency parameter, but still uses a linear impact model.

## 6.8       **Bibliography and Selected Readings**

Bertsimas & Lo (1998), Large (2007), Obizhaeva & Wang (2013), Bayraktar

& Ludkovski (2011), Schied (2013), Almgren (2003), Almgren, Thum, Hauptmann & Li (2005), Almgren & Chriss (2000), Cont, Kukanov & Stoikov (2013), Lorenz & Almgren (2011), Kharroubi & Pham (2010), Alfonsi, Fruth & Schied (2010), Gatheral (2010), Gatheral, Schied & Slynko (2012), Alfonsi, Schied & Slynko (2012), Schied & Schöneborn (2009), Guéant & Lehalle (2013), Bayraktar & Ludkovski (2012), Guo, De Larrard & Ruan (2013), Gatheral & Schied (2013), Graewe, Horst & Qiu (2013), Graewe, Horst & Séré (2013), Li & Almgren (2014), Almgren (2013), Frei & Westray (2013), Curato, Gatheral & Lillo (2014), Jaimungal & Nourian (2015).

## 6.9    Exercises

E.6.1 The agent wishes to liquidate $\mathfrak{N}$ shares between $t$ and $T$ using MOs. The value function is

$$H(t, S, q) = \sup_{\nu \in \mathcal{A}_{t,T}} \mathbb{E}_{t,S,q} \left[ \int_t^T (S_u - k v_u) \, \nu_u \, du - Q_T^\nu \left( S_T - \alpha Q_T^\nu \right)^2 \right],$$

where $k > 0$ is the temporary market impact, $\nu_t$ is the speed of trading, $\alpha \geq 0$ is the liquidation penalty, and $dS_t = \sigma dW_t$.

(a) Show that the value function $H$ satisfies

$$0 = -\left( \partial_q H - S \right)^2 - 4k \partial_t H - 2k\sigma^2 \partial_{SS} H.$$

(b) Make the ansatz

$$H(t, S, q) = h_2(t) q^2 + h_1(t) q + h_0(t) + q S \tag{6.39}$$

and show that the optimal liquidation rate is

$$\nu_t^* = \frac{Q_t^{\nu^*}}{T - t + \frac{k}{\alpha}}. \tag{6.40}$$

(c) Let $\alpha \to \infty$ and show that (6.40) converges to (6.12). Moreover, discuss the intuition of the strategy when $\alpha \to 0$.

E.6.2 This exercise is similar to that above but with a slightly different setup. The agent wishes to liquidate $\mathfrak{N}$ shares and her objective is to maximise expected terminal wealth which is denoted by $X_T^\nu$ (in the exercise above we wrote terminal wealth as $\int_t^T (S_u - k v_u) \, \nu_u \, du$). The value function is

$$H(t, x, S, q) = \sup_\nu \mathbb{E}_{t,x,S,q} \left[ X_T^\nu - Q_T^\nu \left( S_T - \alpha Q_T^\nu \right)^2 \right], \tag{6.41}$$

where

$$dX_t^\nu = (S_t - k\,\nu_t)\,\nu_t dt\,. \tag{6.42}$$

(a) Show that the HJB satisfied by the value function $H(t, S, q, x)$ is

$$0 = \left(\partial_t + \tfrac{1}{2}\sigma^2\partial_{SS}\right) H + \sup_\nu \left\{ \left(-\nu\partial_q + (S - k\nu)\,\nu\partial_x\right) H \right\}, \tag{6.43}$$

and the optimal liquidation rate in feedback form is

$$\nu_t^* = \frac{\partial_q H - S\partial_x H}{-2k\,\partial_x H}\,. \tag{6.44}$$

(b) To solve (6.43), use the terminal condition $H(T, x, S, q) = x + q\,S - \alpha q^2$ to propose the ansatz

$$H(t, S, x, q) = x + h(t)\,q^2 + q\,S\,, \tag{6.45}$$

where $h(t)$ is a deterministic function of time. Show that

$$h(t) = -\frac{k}{T - t + \frac{k}{\alpha}}\,, \tag{6.46}$$

and

$$\nu_t^* = \frac{Q_t^{\nu^*}}{T - t + \frac{k}{\alpha}}\,.$$

E.6.3 Let the stock price dynamics satisfy

$$dS_t = \mu\,dt + \sigma\,dW_t\,,$$

where $\sigma > 0$, $\mu$ is a constant and $W_t$ is a standard Brownian motion. The agent wishes to liquidate $\mathfrak{N}$ shares and her trades create a temporary adverse move in prices so the price at which she transacts is

$$\hat{S}_t^\nu = S_t - k\nu_t\,,$$

with $k > 0$ and the inventory satisfies

$$dQ_t^\nu = -\nu_t\,dt\,,$$

where $\nu_t$ is the liquidation rate. Any outstanding inventory at time $T$ is liquidated at the midprice and picks up a penalty of $\alpha\,Q_T^2$ where $\alpha \geq 0$ is a constant.

The agent's value function is

$$H(t, S, q) = \sup_\nu \mathbb{E}_{t,S,q}\left[ \int_t^T (S_u - k\nu_u)\,\nu_u\,du - Q_T^\nu\,(S_T - \alpha Q_T^\nu)^2 \right]\,. \tag{6.47}$$

(a) Show that the optimal liquidation rate in feedback form is

$$\nu^* = \frac{\partial_q H - S}{-2k}. \tag{6.48}$$

(b) Use the ansatz $H(t, S, q) = q\,S + h(t, S, q)$ to show that the optimal liqui- dation rate is given by

$$\nu_t^* = \frac{Q_t^{\nu^*}}{(T - t) + \frac{k}{\alpha}} - \frac{1}{4k}\,\mu\,(T - t)\,\frac{(T - t) + 2\frac{k}{\alpha}}{(T - t) + \frac{k}{\alpha}}.$$

Comment on the magnitude of $\mu$ and the sign of the liquidation rate.

(c) Let $\alpha \to \infty$ and show that the inventory along the optimal strategy is given by

$$Q_t^{\nu^*} = (T - t)\left(\frac{\mathfrak{N}}{T} + \frac{\mu}{4k}t\right).$$

# 7 Optimal Execution with Continuous Trading II

## 7.1 Introduction

In the previous chapter we studied the problem of optimal execution for an agent who aims to liquidate/acquire a considerable proportion of the average daily volume (ADV) of shares. There we saw how the agent trades off the impact on prices that her trades would have if she traded quickly, with the uncertainty in prices she would receive/pay if she traded slowly. We find that the agent's optimal strategy is to trade quickly initially (ensuring that she receives a price close to the arrival price, but with a non-trivial impact) and then slow down as time goes by (to reduce her overall impact, but increase price uncertainty). Surprisingly, the optimal strategies we obtain are deterministic and in particular are independent of the midprice process – regardless of the level of urgency required to complete her trade. In this chapter, we incorporate a number of other important aspects of the problem that the agent may wish to include in her optimisation decision, and explore how her trading behaviour adjusts to account for them.

Specifically, we look at three distinct aspects of the optimal execution problem.

i. An upper price limit: In Section 7.2 we study the problem of an agent wishing to acquire a large position, who has an upper price limit on what she is willing to pay. We find that the optimal strategy in this case is no longer independent of the midprice, beyond the obvious change that the agent stops trading when the upper limit price is breached.

ii. Informative order flow: In Section 7.3 we study the problem of an agent wishing to liquidate a large position, taking into account that the order flow from other traders in the market also impacts the midprice. We show that the agent alters her strategy so that when the net effect of other market participants is to trade in her direction, she increases her trading speed; conversely, if the net effect of other agents is to trade in the opposite direction, she decreases her trading speed.

iii. Dark pools: In Section 7.4, the agent has access to a (standard) lit market and also to a dark pool. Trading in the dark pool exposes her to execution risk, but removes some of the price impact. We find that the optimal strategy is still deterministic: initially the agent trades in the lit market at speeds below that dictated by Almgren & Chriss (2000) (AC), and posts the whole of the remaining order in the dark pool, in the hope of it being filled there.

After a while, if the order has not been filled in the dark pool, the agent's speed of trading in the lit pool increases above that of AC.

Throughout this chapter we use the same notation as in Chapter 6.

## 7.2 Optimal Acquisition with a Price Limiter

In this section we solve the problem for an agent whose target is to acquire $\mathfrak{N}$ shares over a trading horizon of $T$, with a cap on the price at which she acquires shares equal to $\overline{S}$. If the midprice reaches this limit price before $T$, all remaining shares are immediately purchased and the acquisition programme stops. We assume that the midprice dynamics follow (6.1b) with $g(\nu_t) = b\nu_t$, $b \geq 0$, and the execution price is as in (6.1c) with linear price impact $f(\nu_t) = k\nu_t$, $k > 0$. The agent will stop trading if any one of the following events occur:

a. the agent's inventory reaches the target level $\mathfrak{N}$,
b. the terminal time $T$ is reached,
c. the midprice $S_t$ reaches the upper *limit price* $\overline{S}$.

These define the following stopping time, $\tau$:

$$\tau = T \wedge \inf\{t : S_t = \overline{S}\} \wedge \{t : Q_t = \mathfrak{N}\}.$$

When either of events (b) or (c) occur, the agent acquires the remaining $\mathfrak{N} - Q_\tau^\nu$ units of the security and pays $S_\tau + \alpha\,(\mathfrak{N} - Q_\tau^\nu)$ per unit, where $\alpha > 0$. To simplify notation we let $Y_t^\nu = \mathfrak{N} - Q_t^\nu$ denote the remaining shares to be acquired, satisfying

$$dY_t^\nu = -\nu_t\,dt\,,$$

where $\nu_t$ is the (positive) rate of trading.

To complete the setup of the problem, we write the agent's performance criteria as

$$H^\nu(t, S, y) = \mathbb{E}_{t,S,y}\left[\int_t^\tau (S_u + k\nu_u)\,\nu_u\,du + Y_\tau\,(S_\tau + \alpha\,Y_\tau) + \phi \int_t^\tau y_u^2\,du\right],$$

$$(7.1)$$

where $\phi\int_t^\tau y_u^2\,du$ with $\phi \geq 0$ is a running inventory penalty of the remaining shares to be acquired (as discussed in the previous chapter this penalty is not a financial penalty), and her value function is

$$H(t, S, y) = \inf_{\nu \in \mathcal{A}} H^\nu(t, S, y)\,,$$

for all $0 \leq t \leq T$, $S \leq \overline{S}$, $0 \leq y \leq Q$, and $\mathcal{A}$ is the admissible set of trading strategies in which $\nu$ is non-negative and uniformly bounded from above.

Applying the dynamic programming principle (DPP), the value function should satisfy the dynamic programming equation (DPE)

$$
\partial_t H + \tfrac{1}{2}\sigma^2 \partial_{SS} H + \phi y^2
$$
$$
+ \min_{\nu} \left\{ -\nu\, \partial_y H + b\,\nu\, \partial_S H + (S + k\nu)\,\nu \right\} = 0 \,,
\tag{7.2a}
$$

subject to the terminal and boundary conditions

$$
H(T, S, y) = (S + \alpha\, y)\, y \,,
\tag{7.2b}
$$
$$
H(t, \overline{S}, y) = (\overline{S} + \alpha\, y)\, y \,,
\tag{7.2c}
$$
$$
H(t, S, 0) = 0 \,.
\tag{7.2d}
$$

The terminal and boundary conditions reflect the fact that the agent acquires the remaining shares at the stopping time. Note that when her inventory equals the target $\mathfrak{N}$ at $t < T$, she stops acquiring and there is no penalty, hence the value function equals zero along $y = 0$.

From the first order conditions, we obtain the optimal acquisition strategy in feedback form as

$$
\boxed{\; \nu^*(t, S, y) = -\tfrac{1}{2k}\left( b\,\partial_S H - \partial_y H + S \right) \,, \;}
$$

and upon substituting back into the DPE above, the value function $H$ then solves the non-linear PDE

$$
\partial_t H + \tfrac{1}{2}\sigma^2 \partial_{SS} H - \tfrac{1}{4k}\left( b\,\partial_S H - \partial_y H + S \right)^2 + \phi y^2 = 0 \,,
\tag{7.3}
$$

subject to the terminal and boundary conditions in (7.2).

### Dimensionality Reduction without Permanent Price Impact

In general, the DPE (7.3) will have to be solved numerically; however, in practice it is normally the case that the effect of permanent impact is much smaller than the temporary impact from walking the LOB, so to reduce the dimension of the problem we set $b = 0$.

In this case, due to the form of the DPE (7.3) and its terminal and boundary conditions in (7.2), it is possible to solve for the dependence in $q$ exactly by using the ansatz

$$
H(t, S, y) = y\, S + y^2\, h(t, S) \,,
$$

in which case, the function $h$ satisfies the Fisher-type PDE

$$
\partial_t h + \tfrac{1}{2}\sigma^2 \partial_{SS} h - \tfrac{1}{k} h^2 + \phi = 0 \,, \quad (t, S) \in [0, T) \times \left( -\infty, \overline{S} \right) \,,
\tag{7.4a}
$$

subject to the terminal and boundary conditions

$$
h(T, S) = \alpha \,, \quad S \le \overline{S} \,,
\tag{7.4b}
$$
$$
h(t, \overline{S}) = \alpha \,, \quad t \le T \,,
\tag{7.4c}
$$

and, furthermore, the optimal acquisition strategy $\nu^*$ reduces to

$$\nu^*(t, S, y) = \frac{1}{k} \, y \, h(t, S) \, .$$

(7.5)

In particular, it follows from (7.5) that as inventory $Q$ increases (so that $Y$ decreases), all else being equal, the optimal rate of acquisition slows down.

## Numerical Solution I: Crank-Nicolson

The (1+1)-dimensional terminal boundary value problem (7.4) can be efficiently solved numerically using a Crank-Nicolson scheme, by placing the problem on a grid in the domain $[0, T] \times [\underline{S}, \overline{S}]$ and treating the quadratic term $h^2$ explicitly.

By introducing this grid, we now have a new boundary at $\underline{S}$, and, in order to have a well-posed problem, we need to specify a boundary condition along $S = \underline{S} \ll \overline{S}$. A usual approach is to specify the boundary condition as $\partial_{SS} h|_{S=\underline{S}} = 0$. In order to evaluate whether this is a reasonable boundary or not we determine what this condition implies about the behaviour of the optimal strategy at the lower boundary.

Denote $\chi(t) = h(t, \underline{S})$. Combining (7.4) with $\partial_{SS} h(t, \underline{S}) = 0$, we see that $\chi(t)$ satisfies the ordinary differential equation (ODE)

$$\begin{cases} \partial_t \chi - \frac{1}{k} \chi^2 + \phi & = 0, \quad t \in [0, T), \\ \chi(T) & = \alpha \, . \end{cases}$$

As this is a Riccati equation, its solution follows from standard methods (as outlined in Section 6.5) and is given explicitly by

$$\chi(t) = \begin{cases} \sqrt{k \, \phi} \, \dfrac{\zeta \, e^{2\gamma(T-t)} + 1}{\zeta \, e^{2\gamma(T-t)} - 1} \, , & \phi > 0 \, , \\[4mm] \left( \dfrac{1}{\alpha} + \dfrac{T-t}{k} \right)^{-1} , & \phi = 0 \, , \end{cases}$$

(7.6)

where the constants are

$$\gamma = \sqrt{\frac{\phi}{k}} \quad \text{and} \quad \zeta = \frac{\alpha + \sqrt{k \, \phi}}{\alpha - \sqrt{k \, \phi}} \, .$$

Note that $0 \le \chi(t) \le \alpha$ and that $\chi$ is increasing in $t$ and takes on its maximum at $t = T$.

Moreover, $q \, S + q^2 \chi(t)$ is precisely the value function in which there is no limit price, as studied in Section 6.5 with permanent impact $b$ set to zero. Therefore, the boundary condition $\partial_{SS} h(t, \underline{S}) = 0$ leads to a solution which has an optimal trading speed equal to the no limit price trading speed.

Thus, imposing the boundary condition $\partial_{SS} h(t, \underline{S}) = 0$ generates a reasonable optimal behaviour on the boundary. But, if we accept this boundary condition because we think the boundary behaviour reasonable, we can equivalently, and more straightforwardly, restrict admissible strategies to those that generate the desired boundary behaviour: $\nu$ must equal the no limit price strategy when $S =$

$\underline{S}$. We therefore impose that the agent ignores the limit price when the midprice is sufficiently far away from it.

Although we cannot obtain an explicit solution to (7.4), we can obtain upper and lower bounds on the function $h(t, S)$ by appealing to the maximum principle. First, from standard results in PDEs, a unique solution to (7.4) exists. Hence, by treating the non-linear term $h^2 = h \cdot \bar{h}$, where the term $\bar{h}$ is the solution to the PDE, we can invoke a standard maximum principle.

THEOREM 7.1 (Maximum Principle)   *Suppose that $\Omega \subset \mathbb{R}$ is a bounded connected set and that $u : \Omega \times [0, \infty) \to \mathbb{R}$ satisfies*

$$\partial_t u + \partial_{xx} u = 0, \qquad\qquad x \in \Omega, \ t > 0,$$
$$u(t, x) = p(t), \qquad\qquad x \in \partial\Omega, \ t > 0,$$
$$u(0, x) = q(x), \qquad\qquad x \in \partial\Omega, \ t > 0.$$

*Then*

$$\max_{\overline{\Omega} \times [0,T]} u \le \max \left( p(t) \, ; \, \max_{\overline{\Omega}} q \right) \quad and \quad \min_{\overline{\Omega} \times [0,T]} u \ge \min \left( p(t) \, ; \, \min_{\overline{\Omega}} q \right).$$

As noted earlier, $\chi(t)$ is increasing in $t$ and $0 \le \chi \le \alpha$. Hence a direct application of this maximum principle to (7.4), supplemented with the boundary condition $h(t, \underline{S}) = \chi(t)$ (or as discussed, equivalently by restricting the trading strategy to equal the no limit price strategy along $S = \underline{S}$) provides us with upper and lower bounds on $h$ as follows:

$$\boxed{\chi(t) \le h(t, S) \le \alpha.} \qquad\qquad (7.7)$$

Since the optimal trading rate $\nu^*(t, S, y) = \frac{1}{k} y \, h(t, S)$, the above inequality implies that at each inventory level, the agent with the limit price constraint trades at least as fast as the agent without the limit price, and attains a maximal speed of trading of $\frac{\alpha}{k} y$. This implies that the agent with the limit price constraint will have acquired more shares than the agent without the constraint at any given fixed point in time.

Finally, it is important to point out that the inventory held by the agent at any given time can be obtained in terms of the path of the midprice process by solving for $Y_t$, where $Y_t$ satisfies the SDE

$$dY_t^* = -\nu_t^* \, dt = -\frac{1}{k} Y_t \, h(t, S_t) \, dt \,,$$

hence

$$Q_t^* = \left( 1 - \exp\left\{ -\frac{1}{k} \int_0^t h(u, S_u) \, du \right\} \right) \mathfrak{N}, \quad t \le \tau.$$

This has the same form as in the case without the limit price, but here the path that the midprice takes plays an important role in determining how much inventory the agent has at any given time. Below, after discussing an alternative numerical solution and the case $T \to \infty$, we show some examples of the optimal strategy for different midprice paths.

### Numerical Solution II: Iterative Scheme

An alternative approach to solving the non-linear terminal boundary value (TBV) problem (7.4) is to use an exact iterative scheme, rather than resorting to finite-difference methods. The essential idea is to take an approximate solution to the problem at the $m^{th}$-iteration, denote it by $h^{(m-1)}(t, S)$, and use it to linearise the TBV to obtain an updated approximation $h^{(m)}(t, S)$ which solves the *linear* TBV

$$\left( \partial_t + \tfrac{1}{2}\sigma^2 \partial_{SS} - \tfrac{1}{k} h^{(m-1)}(t, S) \right) h^{(m)}(t, S) + \phi = 0 \tag{7.8a}$$

subject to

$$h^{(m)}(t, \overline{S}) = \alpha, \quad h^{(m)}(t, \underline{S}) = \chi(t), \quad \forall t \in [0, T), \tag{7.8b}$$

and

$$h^{(m)}(T, S) = \alpha, \quad \forall S \in (\underline{S}, \overline{S}). \tag{7.8c}$$

At each iteration, the boundary and terminal conditions are respected, so all that changes is the behaviour of the solution in the interior. The solution to the PDE (7.8c) can be obtained in closed-form up to a Laplace transform by first introducing an integrating factor and writing

$$h^{(m)}(t, S) = e^{\frac{1}{k} \int_0^t h^{(m-1)}(u,S)\, du} \, g^{(m)}(t, S),$$

so that $g^{(m)}(t, S)$ satisfies the linear PDE

$$\left( \partial_t + \tfrac{1}{2}\sigma^2 \partial_{SS} \right) g^{(m)}(t, S) + \phi\, \ell^{(m-1)}(t, S) = 0, \tag{7.9a}$$

subject to

$$g^{(m)}(t, \overline{S}) = \alpha\, \ell^{(m-1)}(t, \overline{S}), \quad \text{and} \tag{7.9b}$$

$$g^{(m)}(t, \underline{S}) = \chi(t)\, \ell^{(m-1)}(t, \underline{S}), \quad \forall t \in [0, T), \tag{7.9c}$$

and

$$h^{(m)}(T, S) = \alpha\, \ell^{(m-1)}(T, S), \quad \forall S \in (\underline{S}, \overline{S}), \tag{7.9d}$$

where

$$\ell^{(m)}(t, S) = e^{-\frac{1}{k} \int_0^t h^{(m)}(u,S)\, du}.$$

The system of linear PDEs requires an initial guess to begin the iteration. One simple approach is to use a linear interpolation (in $S$) between $h(t, \underline{S}) = \chi(t)$ and $h(t, \overline{S}) = \alpha$ at each point in time. This initial guess can be written as

$$h^{(0)}(t, S) = \frac{S - \underline{S}}{\overline{S} - \underline{S}} \chi(t) + \frac{\overline{S} - S}{\overline{S} - \underline{S}} \alpha.$$

Armed with the initial guess, the above system of linear PDEs can in principle be solved using Laplace transform techniques. The final steps are to show that the mapping $h^{(k-1)} \mapsto h^{(k)}$ is indeed a contraction mapping on a suitable space of functions, and then show that the sequence of solutions converges to the solution of the original non-linear PDE.

## The Perpetual Case

As mentioned earlier, the acquisition problem with the price limit may always be solved numerically, e.g. via a Crank-Nicolson scheme. However, if we assume that the agent's terminal date $T \to \infty$, i.e. the agent trades until she acquires all inventory or the limit price is reached, then the problem reduces to a simpler one that we can solve analytically. In this case, $h$ is independent of time and equation (7.4) reduces to a boundary value problem for the ODE

$$\tfrac{1}{2}\sigma^2 \partial^2_{SS} h - \tfrac{1}{k} h^2 + \phi = 0, \quad S \in (\underline{S}, \overline{S}), \tag{7.10a}$$

with boundary conditions

$$h(\overline{S}) = \alpha, \quad \text{and} \quad h(\underline{S}) = \sqrt{k\,\phi}. \tag{7.10b}$$

The boundary condition $h(\underline{S}) = \sqrt{k\,\phi}$ arises by recalling that we impose the condition $h(t, \underline{S}) = \chi(t)$ and from (7.6) we have $\lim_{T \to \infty} \chi(t) = \sqrt{k\,\phi}$.

The general solution of equations of the above kind are elliptic functions – see e.g., Chapter 18 in Abramowitz & Stegun (1972). In our specific case (with $\phi > 0$), the solution can be written in terms of the Weierstrass $\wp$-function as follows:

$$h(S) = 3\,k\,\sigma^2\,\wp\left(S + C_1;\ \frac{4}{3}\frac{\phi}{k\sigma^4},\ C_2\right), \tag{7.11}$$

where the last two arguments denote the invariants of the elliptic function and the constants $C_1$ and $C_2$ must be determined numerically to match the boundary conditions in (7.10b).

If $\phi = 0$ and $\underline{S} \to -\infty$ the solution can be obtained in terms of elementary functions by solving the ODE (7.10a) using the ansatz $h(S) = \beta_1 (S + \beta_2)^{\beta_3}$, where $\beta_1, \beta_2, \beta_3$ are constants to be determined. Two of the constants can be obtained from the boundary conditions, and the third is obtained by ensuring the ansatz closes the ODE. We leave it as an exercise to show that the solution indeed reduces to the simple equation

$$h(S) = \frac{\sigma^2 k}{2}\left(\frac{\overline{S} - S}{\sqrt{6}} + \sqrt{\frac{\sigma^2 k}{2\alpha}}\right)^{-2}. \tag{7.12}$$

Recall that the optimal trading speed $\nu^* = \tfrac{1}{k} y\, h(t, S)$. Hence, the perpetual solution shows that the agent's speed of trading increases (for the same fixed amount of inventory) as the limit price $\overline{S}$ is approached from below. This is natural, since the agent observes the price getting closer to the limit and therefore wishes to acquire as many shares as possible prior to breaching the price cap without paying too much in immediate impact costs. Note, however, as inventory is acquired, the speed of trading slows down. The net result of these two opposing effects will depend on which of the two is stronger.

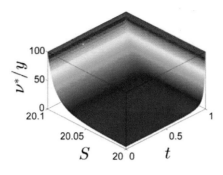
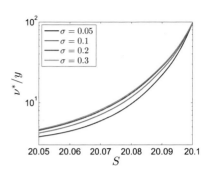

**Figure 7.1** Left panel: The rate of acquisition relative to remaining inventory as a function of time and fundamental price. Right panel: the rate of acquisition relative to inventory as a function of $S$ at $t = 0$ for various volatility levels.

### Simulations of the Strategy with Price Limiter

We now illustrate several aspects of the agent's optimal behaviour when she imposes a price limiter in her strategy.

Throughout, we use the following parameters

$$T = 1\,\text{day}, \quad k = 10^{-4}, \quad b = 0,$$
$$\alpha = 100k, \quad \phi = 10^{-3}, \quad \sigma = 0.1.$$

We also normalise the acquisition target to $\mathfrak{N} = 1$ (this can be viewed as a percentage of the ADV, and we assume the agent trades once per second). Recall that the ratio $\alpha/k$ sets the maximum trading speed, hence our choice for $\alpha$. We set the limiting price to $\overline{S} = S_0 + \sigma$ with initial price $S_0 = 20$, and our choice of volatility $\sigma = 0.1\,(\text{day})^{-\frac{1}{2}}$ corresponds to annualised relative volatility of $0.1/20 \times \sqrt{255} \sim 8\%$.

First, in the left panel of Figure 7.1 we depict the optimal execution strategy $\nu^*$ relative to the current remaining inventory, i.e. $\nu^*/y$. As maturity approaches, the trading rate increases, because for a fixed number of shares remaining the agent must acquire faster to avoid the terminal penalty cost. Moreover, as the fundamental price approaches the limit price, the speed of trading also increases to avoid the terminal penalty. The bounds on the rate of trading implied by the maximum principle (7.7) can also be seen in the figure. In the case with no limit price, the agent's optimal strategy is independent of the asset's volatility. However, with the limit price, as shown in the right panel of Figure 7.1, the agent's behaviour does indeed depend on volatility. The more volatile the market, the faster the agent trades, but as the price moves far away from the limit price, the effect of volatility diminishes.

The agent's strategy is constantly being updated to reflect changes in her inventory (due to her own trading) and the innovations in the midprice, hence the static views in Figure 7.1 do not tell the full story. To gain additional insight

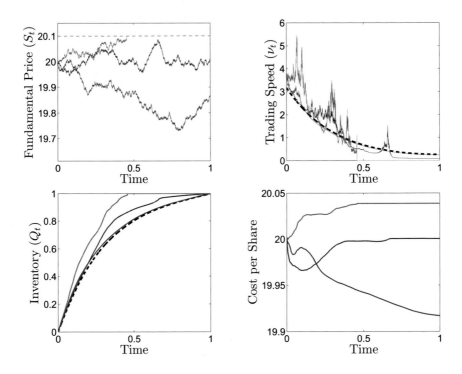

**Figure 7.2** Sample paths of the evolution of the midprice, acquisition rate and inventory. Dashed line in the bottom left panel represents the evolution of inventory in the corresponding AC strategy.

into the dynamic behaviour, in Figure 7.2 we plot three sample paths of the fundamental price together with the rate of acquisition, inventory, and cost per share. In the bottom left panel, using a dashed line, we include the inventories' lower bound, the Almgren-Chriss (AC) strategy, as described by the inequality in (7.7).

As the figures show, the red path is mostly away from the limit price and after some initial noise, the agent's strategy is very close to the deterministic AC strategy. The blue path stays mostly near the arrival price so that the trading speed displays stochastic dynamics. The agent has an increased trading rate between $t = 0.2$ and $t = 0.3$ when the midprice approaches the limit price. Also, around $t = 0.6$ the midprice almost touches the limit price and the agent's trading speed spikes there; however, due to aggressive trading earlier on, by that time she has already acquired a significant proportion of her shares, and the spike is not very large. The green path hits the limit price early on, and as the figure illustrates, the agent trades quickly (relative to the AC strategy) up until the price hits the boundary.

The bottom right panel of Figure 7.2 shows the cost per share along the three paths. Note that the green path, which hits the limit price early on, is more

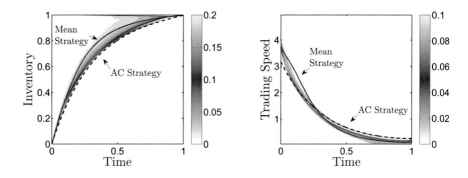

**Figure 7.3** Inventory and trading speed heat-maps from 1,000 simulations of the optimal strategy.

expensive than the other paths for two somewhat interrelated reasons: (i) the midprice was generally higher (and it hit the limit price) during trading; and (ii) since the price generally trended upwards, the agent trades more quickly and hence has a large temporary impact compared with the other paths.

Finally, in Figure 7.3, we show heat-maps of the agent's inventory (left panel) and trading speed (right panel) resulting from 1,000 simulations. We see a thin contribution to the heat-map along an inventory level of 1 as well as along a trading speed of 0. These represent those paths that breached the price limit early. We also show the mean inventory path and the mean trading speed trajectory. As expected, the mean inventory path lies above the AC strategy, since the trading speed for the same level of inventory must lie above the AC strategy as dictated by the maximum principle and encoded in the inequality (7.7).

Inequality (7.7), does not, however, imply that the trading speed will always lie below the AC trading speed. This is because the inventory level varies with the sample path and will not generally equal that of the AC inventory. Indeed, as the right panel of Figure 7.3 shows, the mean trading speed starts above that of the AC, but as time passes, it eventually falls below it. The intuition for this is that since the optimal strategy requires the agent to trade more quickly than the AC, her inventory is generally higher (and closer to the target inventory) than if she were trading according to the AC strategy. Since she eventually has less inventory, her trading speed generally slows down which results in the behaviour observed in the figure.

## 7.3 Incorporating Order Flow

In the previous chapter and in the last section, we assume that in the absence of the agent's trades, the midprice process is a martingale. We also assume that when the agent begins to liquidate (acquire) shares, her actions induce a downward (upward) drift in the midprice process. In other words, the act of her selling (buying) shares induces the market as a whole to adjust prices downwards (upwards). Yet at the same time we are ignoring the trades of other market participants, implicitly assuming that on average their actions even out to yield a net of zero drift. This may be acceptable at an aggregate level, but over short time horizons, there may be order flow imbalance, which very often results in prices trending upwards or downwards over short intervals in time. In this section, we show how to incorporate the order flow from the remainder of the market into the midprice dynamics and how the agent modifies her strategy to adapt to it locally.

## The Model Setup

In addition to the usual state variables and stochastic processes introduced in the previous chapter, we now also model the dynamics of the buy and sell rate of order flow $\mu_t^{\pm}$ and assume that they satisfy the SDE

$$d\mu_t^{\pm} = -\kappa\,\mu_t^{\pm}\,dt + \eta\,dL_t^{\pm}\,, \tag{7.13}$$

where $L_t^{\pm}$ are independent Poisson processes (assumed independent of all other processes as well) with equal intensity $\lambda$. This assumption implies that the buy and sell order flows arrive independently at Poisson times with rate $\lambda$, and induce an increase in the order flow rate by $\eta$ and jumps in order flow rate decay at the speed $\kappa$.

Next, we incorporate the order flow in the midprice process $S_t$, which now satisfies the SDE

$$dS_t^{\nu} = \sigma\,dW_t + (g(\mu_t^+) - g(\mu_t^- + \nu_t))\,dt\,,$$

where $W_t$ is a Brownian motion independent of the Poisson processes and $g$ is an impact function which dictates how the midprice drift is affected by buy/sell order flow. In this manner, the action of the agent's trades and other traders' actions are treated symmetrically. We can define the net order flow $\mu_t = \mu_t^+ - \mu_t^-$ and a short computation shows that

$$
\begin{aligned}
d\mu_t &= -\kappa\left(\mu_t^+ - \mu_t^-\right)dt + \eta\left(dL_t^+ - dL_t^-\right) \\
&= -\kappa\,\mu_t\,dt + \eta\left(dL_t^+ - dL_t^-\right)\,.
\end{aligned}
$$

Hence, if the permanent impact functions $g(x) = b\,x$ are linear (with $b \geq 0$), we can use the net order flow as a state process rather than having to keep track of order flow in both directions separately. Overall, we have

$$dS_t^{\nu} = \sigma\,dW_t + b\left(\mu_t - \nu_t\right)dt\,.$$

The remainder of the agent's optimisation problem is as in Section 6.5. Briefly, the agent's inventory $Q^{\nu}$ is

$$dQ_t^{\nu} = -\nu_t\,dt\,,$$

and her cash process $X^\nu$ satisfies the SDE

$$dX_t^\nu = (S_t^\nu - k\,\nu_t)\,\nu_t\,dt\,,$$

where $k > 0$ is the temporary linear impact parameter. Also, the agent's performance is the usual one so

$$H^\nu(t, x, S, \mu, q) = \mathbb{E}_{t,x,S,\mu,q}\left[X_T^\nu + Q_T^\nu\,(S_T^\nu - \alpha Q_T^\nu) - \phi \int_t^T (Q_u^\nu)^2\,du\right]\,,\quad (7.14)$$

and her value function is

$$H(t, x, S, \mu, q) = \sup_{\nu \in \mathcal{A}} H^\nu(t, x, S, \mu, q)\,.$$

### The Resulting DPE

The DPP for the value function suggests that the value function $H(t, x, S, \mu, q)$ satisfies the DPE (the value function now has an additional state variable, $\mu$)

$$0 = \left(\partial_t + \tfrac{1}{2}\sigma^2\,\partial_{SS}\right) H + \mathcal{L}^\mu H - \phi q^2$$
$$+ \sup_\nu \left\{(\nu\,(S - k\nu)\,\partial_x + b\,(\mu - \nu)\,\partial_S - \nu\,\partial_q)\,H\,,\right\}$$

subject to the terminal condition

$$H(T, x, S, \mu, q) = x + q\,S - \alpha\,q^2\,,$$

where the infinitesimal generator for the net order flow acts on the value function as follows:

$$\mathcal{L}^\mu H(t, x, S, \mu, q) = -\kappa\,\mu\,\partial_\mu H + \lambda\,[H(t, x, S, \mu + \eta, q) - H(t, x, S, \mu, q)]$$
$$+ \lambda\,[H(t, x, S, \mu - \eta, q) - H(t, x, S, \mu, q)]\,.$$
$$(7.15)$$

Inserting the ansatz

$$H(t, x, S, \mu, q) = x + q\,S + h(t, \mu, q)\,,$$

we see that the excess book value function $h(t, \mu, q)$ satisfies the equation

$$\partial_t h + \mathcal{L}^\mu h + b\,\mu\,q - \phi q^2 + \sup_\nu \left\{-k\,\nu^2 - (b\,q + \partial_q h)\,\nu\right\} = 0\,,$$

subject to the terminal condition $h(T, \mu, q) = -\alpha\,q^2$. Recall that $x + q\,S$ represents the cash from the sale of shares so far plus the book value (at midprice) of the shares the agent still holds and aims to liquidate.

The optimal control in feedback form is the same as in (6.22), but the function $h$ satisfies a new equation. More specifically, the first order conditions imply that

$$\nu^* = -\frac{1}{2\,k}\,(b\,q + \partial_q h)\,,$$

and upon substitution back into the previous equation we find that $h$ satisfies the non-linear partial-integral differential equation (PIDE)

$$(\partial_t + \mathcal{L}^\mu)\,h + b\,\mu\,q - \phi q^2 + \frac{1}{4\,k}\,(b\,q + \partial_q h)^2 = 0\,.\quad (7.16)$$

## Solving the DPE

Due to the existence of linear and quadratic terms in $q$ in (7.16), and its terminal conditions, we expect $h(t, \mu, q)$ to be a quadratic form in $q$, and we assume the ansatz

$$h(t, \mu, q) = h_0(t, \mu) + q\, h_1(t, \mu) + q^2\, h_2(t, \mu).$$

Inserting this into (7.16) and collecting like terms in $q$ leads to the following coupled system of PIDEs:

$$(\partial_t + \mathcal{L}^\mu)\, h_0 + \frac{1}{4k}\, h_1^2 = 0, \tag{7.17a}$$

$$(\partial_t + \mathcal{L}^\mu)\, h_1 + b\,\mu + \frac{1}{2k}\, h_1\,(b + 2h_2) = 0, \tag{7.17b}$$

$$(\partial_t + \mathcal{L}^\mu)\, h_2 - \phi + \frac{1}{4k}\, (b + 2h_2)^2 = 0, \tag{7.17c}$$

subject to the terminal conditions

$$h_0(T, \mu) = 0, \quad h_1(T, \mu) = 0, \quad h_2(T, \mu) = -\alpha.$$

Note that since (7.17c) for $h_2$ contains no source terms in $\mu$ and its terminal condition is independent of $\mu$, the solution must be independent of $\mu$, i.e. $h_2$ is a function only of time. In this case, (7.17c) reduces to (6.25) – the equation for $h_2(t)$ in the AC problem. Thus

$$h_2(t, \mu) = \chi(t) - \tfrac{1}{2}\, b, \quad \text{where} \quad \chi(t) = \sqrt{k\phi}\, \frac{1 + \zeta\, e^{2\gamma(T-t)}}{1 - \zeta\, e^{2\gamma(T-t)}},$$

with the constants $\gamma$ and $\zeta$ as defined in (6.26), but repeated here for convenience:

$$\gamma = \sqrt{\frac{\phi}{k}}, \quad \text{and} \quad \zeta = \frac{\alpha - \tfrac{1}{2}b + \sqrt{k\phi}}{\alpha - \tfrac{1}{2}b - \sqrt{k\phi}}.$$

Next, to solve for $h_1$ in (7.17b), we exploit the affine structure of the model for the net order flow and write

$$h_1(t, \mu) = \ell_0(t) + \mu\, \ell_1(t),$$

in which case,

$$\mathcal{L}^\mu h_1 = -\kappa\,\mu\,\ell_1 + \lambda(\eta\,\ell_1) + \lambda(-\eta\,\ell_1) = -\kappa\,\mu\,\ell_1,$$

with terminal conditions $\ell_0(T) = \ell_1(T) = 0$. Therefore, (7.17b) reduces to

$$\left\{ \partial_t \ell_0 + \frac{1}{k}\, \chi(t)\, \ell_0 \right\} + \left\{ \partial_t \ell_1 + \left( \frac{1}{k}\, \chi(t) - \kappa \right) \ell_1 + b \right\} \mu = 0.$$

Since this must hold for every value of $\mu$, each term in the braces must vanish individually and we obtain two simple ODEs for $\ell_0$ and $\ell_1$. Since $\ell_0(T) = 0$ and its ODE is linear in $\ell_0$, the solution is $\ell_0(t) = 0$. For $\ell_1$, due to the source term $b$, the solution is non-trivial and can be written as

$$\ell_1(t) = b \int_t^T e^{-\kappa(s-t)} e^{\frac{1}{k} \int_t^s \chi(u)\, du}\, ds\,. \tag{7.18}$$

As in (6.29), we use the integral

$$\int_0^t \frac{\chi(s)}{k}\, ds = \log \frac{\zeta\, e^{\gamma(T-t)} - e^{-\gamma(T-t)}}{\zeta\, e^{\gamma T} - e^{-\gamma T}}$$

to simplify the expression for $\ell_1$ to

$$\ell_1(t) = b\, \bar{\ell}_1(T - t) \geq 0\,, \tag{7.19}$$

where

$$\bar{\ell}_1(\tau) = \frac{1}{\zeta\, e^{\gamma\tau} - e^{-\gamma\tau}} \left\{ e^{\gamma\tau} \frac{1 - e^{-(\kappa+\gamma)\tau}}{\kappa + \gamma} \zeta - e^{-\gamma\tau} \frac{1 - e^{-(\kappa-\gamma)\tau}}{\kappa - \gamma} \right\}\,,$$

and $\tau = T - t$ represents the time remaining to the end of the trading horizon.

The solution of $h_0$, which satisfies (7.17a), can be obtained in a similar manner, but the optimal speed of trading does not depend on $h_0$ since as we showed earlier, $\nu^* = -(b\,q + \partial_q h)/2\,k$, and $\partial_q h(t, \mu) = h_1(t, \mu) + 2\,q\,h_2(t, \mu)$. Putting these results together we find that the optimal speed of trading is

$$\boxed{\nu_t^* = -\frac{1}{k} \chi(t)\, Q_t^{\nu^*} - \frac{b}{2\,k}\, \bar{\ell}_1(t)\, \mu_t\,.} \tag{7.20}$$

The optimal trading speed above differs from the AC solution by the second term on the right-hand side of (7.20) which represents the perturbations to the trading speed due to excess order flow. Recall that in the limit $\alpha \to \infty$, $\chi \leq 0$, and from the explicit equation above $\bar{\ell}_1 \geq 0$, hence, when the excess order flow is tilted to the buy side ($\mu_t > 0$), the agent slows down trading since she anticipates that excess buy order flow will push the prices upwards – and therefore will receive better prices when she eventually speeds up trading to sell assets later on. Contrastingly, she increases her trading speed when order flow is tilted to the sell side ($\mu_t < 0$), since other traders are pushing the price downwards and she aims to get better prices now, rather than waiting for other traders to push it further down. Another interpretation is that she attempts to hide her orders by trading when order flow moves in her direction. Finally, recall that $\ell_1(t) \xrightarrow{t \to T} 0$, hence, the order flow influences the agent's trading speed less and less as maturity approaches because there is little time left to take advantage of directional trends in the midprice.

Somewhat surprisingly, the volatility of the order flow process $\eta$ does not appear explicitly in the optimal strategy. It does, however, affect the way the agent trades through its influence on the path which order flow takes. When the order flow path is volatile, the optimal trading speed will be volatile as well. It is also interesting to observe that if the jumps $\eta$ in the order flow at the Poisson times were random and not constant, the resulting strategy would be identical, see Exercise E.7.1. Similarly, if we add a Brownian component to the order flow process $\mu_t$, the resulting optimal strategy in terms of $\mu_t$ would be identical, i.e. (7.20) remains true. Naturally, the actual path taken by the order flow, and therefore also that of trading, would be altered by these modifications to the model.

A final point we make about this optimal trading strategy is that $\nu_t$ is not necessarily strictly positive. If the order flow $\mu_t$ is sufficiently positive, then the agent may be willing to purchase the asset to make gains from the increase in asset price (i.e. her liquidation rate becomes negative). This is because the way we have introduced order flow into the model generates predictability in the price process which can be exploited, even if the agent is not executing a trade. In fact, if the agent has liquidated the target $\mathfrak{N}$ at $t < T$ the optimal strategy is not to stop, but to continue trading and exploit the effect of the order flow, and we see this as her inventory can become negative at intermediate times. If there is sufficient selling pressure (i.e. $\mu_t$ is sufficiently negative), then by shorting the asset, she may benefit from the downward price movement.

One approach to avoid such scenarios is to simply restrict the trading strategy in a naive manner, by setting

$$\nu^\dagger = \max\left(-\frac{1}{k}\chi(t)\,Q_t^{\nu^\dagger} - \frac{b}{2\,k}\,\bar{\ell}_t(t)\,\mu_t \; ; \; 0\right) \mathbb{1}_{\left\{Q_t^{\nu^\dagger}>0\right\}}. \tag{7.21}$$

In other words, we can follow the unrestricted optimal solution whenever the trading rate is positive and the agent has positive inventory, otherwise we impose a trading stop. This trading strategy, $\nu^\dagger$, is not the true optimal strategy. To obtain the true optimal strategy we would need to go back to the DPE and impose the constraint $\nu \geq 0$ in the supremum and add an additional boundary condition along $q = 0$. In this case, the DPE will not have an analytical solution, although numerical schemes can be used to solve the problem. Nonetheless, the $\nu^\dagger$ strategy provides a reasonable approximation that is easy to implement.

### Simulations of the Strategy with Order Flow

In this section we perform simulations to show the behaviour of the optimal strategy in this model. Throughout, we use the following parameters:

$$T = 1\,\text{day}, \quad k = 10^{-3}, \quad b = 10^{-4}, \quad \phi = 0.01,$$
$$\lambda = 1000, \quad \kappa = 10, \quad \eta \sim Exp(5), \quad \sigma = 0.1,$$

where $\eta \sim Exp(\eta_0)$ denotes the exponential distribution with mean size $\mathbb{E}[\eta] = \eta_0$.

Figure 7.4 shows three scenarios of the midprice, the order flow, the optimal inventory, and the optimal speed of trading when the agent uses the augmented strategy $\nu^\dagger$ in (7.21). As the figure shows, when the order flow is positive/negative the agent trades more slowly/quickly than the AC trading speed. For example,

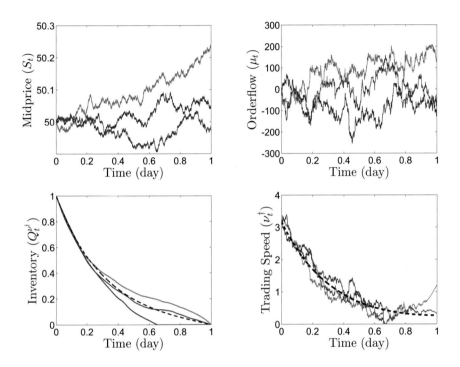

**Figure 7.4** Optimal trading in the presence of order flow. The dashed lines show the classical AC solution.

the large order flow in the buy direction ($\mu_t > 0$), shown by the green path, causes the agent to trade more slowly in the initial stages of the trade. As the end of the trading horizon approaches, the order flow influences her strategy less, but she must speed up her trading since there is little time remaining in which to liquidate the remaining shares. The red path has order flow that fluctuates mostly around zero, and as shown in the diagrams, she follows closely the AC strategy, but locally adjusts her trades relative to the path. Finally, the blue path has a bias towards sell order flow, and the agent adds to this flow by trading more quickly throughout most of the trading horizon and eventually liquidates her shares early.

To gain further insight into the strategy, Figure 7.5 shows heat-maps from 5,000 scenarios of the optimal inventory to hold and the optimal speed of trading. Panel (a) shows the results when $\eta \sim Exp(5)$ as in Figure 7.4, while panel (b) shows the results when $\eta \sim Exp(10)$. As expected, the optimal trading strategy in scenario (b) is more volatile than in scenario (a), despite the optimal strategy (as seen in (7.21)) having no explicit dependence on this volatility.

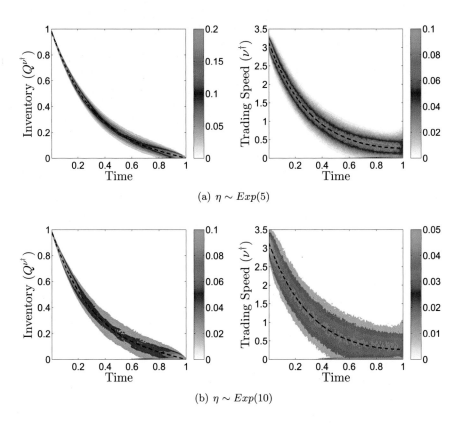

(a) $\eta \sim Exp(5)$

(b) $\eta \sim Exp(10)$

**Figure 7.5** Heat-maps of the optimal trading in the presence of order flow for two volatility levels. The dashed lines show the classical AC solution.

### 7.3.1    Probabilistic Interpretation

Above we studied a particular choice of the dynamics of other market participants' rate of trading $\mu_t$. Here we provide a general solution where we do not assume a particular model for $\mu$, all we specify is the generator $\mathcal{L}^\mu$, which nests (7.15) as a particular case. In this general setup the solution to the problem is very similar to that derived above. We need to solve the system of coupled PIDEs in (7.17) and in particular we must solve for $h_1$ which satisfies equation (7.17b), which we repeat here for convenience:

$$(\partial_t + \mathcal{L}^\mu)\, h_1 + b\,\mu + \frac{1}{2k}\, h_1\,(b + 2h_2) = 0\,, \qquad h_1(T) = 0\,,$$

where

$$h_2(t, \mu) = \sqrt{k\,\phi}\,\frac{1 + \zeta\,e^{2\gamma(T-t)}}{1 - \zeta\,e^{2\gamma(T-t)}} - \tfrac{1}{2}\,b\,.$$

This is a linear PIDE for $h_1$ in which $h_2 + \frac{1}{2}b$ acts as an effective discount rate and $b\,\mu$ is a source term. The general solution of such an equation can be represented using the Feynman-Kac Theorem. Thus we write

$$h_1(t,\mu) = b\, \mathbb{E}_{t,\mu}\left[\int_t^T \exp\left\{\frac{1}{k}\int_t^u \left(h_2(s) + \tfrac{1}{2}b\right)ds\right\}\mu_u\, du\right]$$

and using (6.28) to simplify the exponential term, we obtain

$$h_1(t,\mu) = b\, \mathbb{E}_{t,\mu}\left[\int_t^T \left(\frac{e^{-\gamma(T-u)} - \zeta e^{\gamma(T-u)}}{e^{-\gamma(T-t)} - \zeta e^{\gamma(T-t)}}\right)\mu_u\, du\right]. \qquad (7.22)$$

Recall that the optimal speed of trading is given by

$$\nu^* = -\frac{1}{2k}\left(b\,q + \partial_q h\right),$$

with

$$h(t,\mu,q) = h_0(t,\mu) + q\, h_1(t,\mu) + q^2\, h_2(t,\mu).$$

Hence, we have

$$\nu_t^* = \gamma\, \frac{\zeta\, e^{2\gamma(T-t)} + 1}{\zeta\, e^{2\gamma(T-t)} - 1}\, Q_t^{\nu^*}$$

$$-\frac{b}{2k}\int_t^T \left(\frac{\zeta e^{\gamma(T-u)} - e^{-\gamma(T-u)}}{\zeta e^{\gamma(T-t)} - e^{-\gamma(T-t)}}\right)\mathbb{E}\left[\mu_u \mid \mathcal{F}_t^\mu\right]du. \qquad (7.23)$$

In the limit in which the terminal penalty becomes infinite ($\alpha \to \infty$), so that the agent must completely liquidate her position by the end of the trading horizon, we have $\zeta \to 1$, and the optimal trading speed simplifies to

$$\lim_{\alpha\to\infty}\nu_t^* = \gamma\,\frac{\cosh(\gamma\,(T-t))}{\sinh(\gamma\,(T-t))}\, Q_t^{\nu^*} - \frac{b}{2k}\int_t^T \frac{\sinh(\gamma(T-u))}{\sinh(\gamma(T-t))}\mathbb{E}\left[\mu_u \mid \mathcal{F}_t^\mu\right]du.$$

The first term corresponds to the classical AC solution, while the second corrects the liquidation speed based on the weighted average of the future expected net order flow. If this weighted average future order flow is positive (which would occur, e.g., if the current order flow is positive and hence biased towards buying), then the agent slows down to take advantage of the upward trend in prices that the excess positive order flow will have. The opposite holds if order flow is negative. This dependence on order flow becomes less important as maturity approaches, and the agent instead focuses on completing her execution.

## 7.4 Optimal Liquidation in Lit and Dark Markets

Up until now, the agent has been trading on transparent (lit) markets, where all agents can observe quantities being offered for sale or purchase at different prices, that is, the LOB is visible to all interested parties. We now consider the possibility that the agent can also trade in what are known as **dark pools**. Dark pools are trading venues which, in contrast to traditional (or lit) exchanges, do not display bid and ask quotes to their clients. Trading may occur continuously, as soon as orders are matched, or consolidated and cleared periodically (sometimes referred to as throttling). We focus on a particular kind of dark pool known as a *crossing network* defined by the Securities Exchange Commission (SEC) as (see also Section 3.6)

> "...systems that allow participants to enter unpriced orders to buy and sell securities, these orders are crossed at a specified time at a price derived from another market..."

Typically, the price at which transactions are crossed is the midprice in a corresponding lit trading venue. When a trader places an order in a dark pool, she may have to wait for some time until a matching order arrives so that her order is executed. Thus, on the one hand the trader who sends orders to the dark pool is exposed to execution risk, but on the other hand does not receive the additional temporary price impact of walking the LOB.

Here we analyse the case when the agent trades continuously in the lit market and simultaneously posts orders in the dark pool with the aim to liquidate $\mathfrak{N}$ shares.

## Model Setup

On the lit market, we assume, as before, that the agent is exposed to a temporary market impact from her market orders so when trading $\nu_t dt$ in the lit market, she receives $\hat{S}_t = S_t - k\nu_t$ per share, with $k > 0$, where the midprice $S_t$ is a Brownian motion. In addition to trading in the lit market, the agent posts $y_t \leq q_t$ units of inventory in the dark pool, where $q_t \leq \mathfrak{N}$ are the remaining shares to be liquidated, and she may continuously adjust this posted order. Matching orders in the dark have no price impact because they are pegged to the lit market's midprice, so the agent receives $S_t$ per share for each unit executed in the dark pool which is not necessarily the whole amount $y_t$.

Furthermore, other market participants send matching orders to the dark pool which are assumed to arrive at Poisson times and the volumes associated with the orders are independent. More specifically, let $N_t$ denote a Poisson process with intensity $\lambda$ and let $\{\xi_j : j = 1, 2, \ldots\}$ be a collection of independent and identically distributed random variables corresponding to the volume of the various matching orders which are sent by other market participants into the dark pool. The total volume of buy orders (which may match the agent's posted sell order) placed in the dark pool up to time $t$ is the compound Poisson process

$$V_t = \sum_{n=1}^{N_t} \xi_n .$$

When a matching order arrives, it may be larger or smaller than the agent's posted sell order, hence the agent's inventory (accounting for both the continuous trading in the lit market and her post in the dark pool) satisfies the SDE

$$dQ_t^{\nu,y} = -\nu_t \, dt - \min\left(y_t, \, \xi_{1+N_{t-}}\right) dN_t \,,$$

and recall that the agent's aim is to liquidate $\mathfrak{N}$ shares on or before the terminal date $T$. In the equation above the first term on the right-hand side represents the shares that the agent liquidates using MOs in the lit market and the second represents the orders she sends to the dark pool.

We assume that the agent is at the front of the sell queue in the dark pool, so that she is first to execute against any new orders coming into that market. The model can be modified to account for the agent not being at the front. This can be done by introducing another random variable representing the volume of orders in front of the agent. This, however, complicates but does not alter the approach in a fundamental way, so we leave the interested reader to try this, see Exercise E.7.2.

Hence, the agent's cash process $X_t^{\nu,y}$ satisfies the SDE

$$dX_t^{\nu,y} = (S_t - k\,\nu_t)\,\nu_t \, dt + S_t \, \min\left(y_t, \, \xi_{1+N_{t-}}\right) dN_t \,.$$

Her performance criteria is, as usual, given by

$$H^{\nu,y}(t,x,S,q) = \mathbb{E}_{t,x,S,q}\left[X_\tau + Q_\tau^{\nu,y}\left(S_\tau - \alpha\,Q_\tau^{\nu,y}\right) - \phi\int_t^\tau \left(Q_u^{\nu,y}\right)^2 du\right] \,,$$

where $\mathbb{E}_{t,x,S,q}\left[\cdot\right]$ denotes expectation conditional on $X_{t-} = x$, $S_t = S$, $Q_{t-} = q$, and the stopping time

$$\tau = T \wedge \inf\{t \; : \; Q_t = 0\} \,,$$

represents the time until the agent's inventory is completely liquidated, or the terminal time has arrived. The value function is

$$H(t,x,S,q) = \sup_{\nu,y\in\mathcal{A}} H^{\nu,y}(t,x,S,q) \,,$$

where the set of admissible strategies consists of $\mathcal{F}$-predictable processes bounded from above, and her posted volume in the dark pool is at most her remaining inventory, i.e. $y_t \le Q_t^{\nu,y}$.

## The Resulting DPE

Applying the DPP shows that the value function should satisfy the DPE

$$\partial_t H + \tfrac{1}{2}\sigma^2 \partial_{SS} H - \phi\,q^2$$
$$+ \sup_\nu \left\{(S - k\,\nu)\,\nu\,\partial_x H - \nu\,\partial_q H\right\}$$
$$+ \sup_{y\le q} \left\{\lambda\,\mathbb{E}\left[H\left(t, x + S\,\min(y,\xi), \, S, \, q - \min(y,\xi)\right) - H\right]\right\} = 0 \,,$$

subject to the terminal condition

$$H(T,x,S,q) = x + q\left(S - \alpha\,q\right) \,.$$

In the above, the expectation represents an expectation over the random variable $\xi$ and the various terms in the DPE carry the following interpretations:

- the term $\partial_{SS}$ represents the diffusion of the midprice,
- the $-\phi q^2$ term represents the running penalty which penalises inventories different from zero,
- the $\sup_\nu \{\cdot\}$ term represents optimising over continuous trading in the lit market,
- the $\sup_{y \le q}$ term represents optimising over the volume posted in the dark pool, and the expectation is there to account for the fact that buy volume coming into the dark pool from other traders is random.

The terminal condition once again suggests the ansatz $H(t, x, S, q) = x + q S + h(t, q)$. Recall that $x + q S$ represents the cash from sales so far, in both lit and dark markets, plus the book value (at midprice) of the shares the agent still holds and aims to liquidate. Hence, $h$ represents the value of optimally trading beyond the book value of cash and assets. The DPE then reduces to a simpler equation for $h$:

$$\partial_t h - \phi q^2 + \sup_\nu \left\{ -k \nu^2 - \nu \partial_q h \right\}$$
$$+ \lambda \sup_{y \le q} \mathbb{E} \left[ h \left( t, q - \min(y, \xi) \right) - h(t, q) \right] = 0, \tag{7.24}$$

subject to the terminal condition $h(T, q) = -\alpha q^2$. Next, the first order condition for $\nu$ implies that the optimal speed to trade in feedback control form is

$$\boxed{ \nu^* = -\frac{1}{2k} \partial_q h, } \tag{7.25}$$

so

$$\sup_\nu \left\{ -k \nu^2 - \nu \partial_q h \right\} = \frac{1}{4k} (\partial_q h)^2 .$$

To determine the optimal over $y$ (i.e. the optimal volume to post in the dark pool), we need to either resort to numerics or place more structure on the random variable $\xi$.

### 7.4.1 Explicit Solution when Dark Pool Executes in Full

To obtain an explicit solution to the problem we assume that the agent's desired execution (the liquidation order) is small relative to the volume coming into the dark pool, $\xi_i \ge \mathfrak{N}$ (for all $i = 1, 2, \dots$). This assumption ensures that when a matching buy order arrives in the dark pool, the agent's order is executed in full, as the incoming buy order is larger than the amount posted in the agent's sell order – the agent's inventory at any point in time is at most $\mathfrak{N}$ (the initial amount she must liquidate). As the agent's posts are always filled entirely, in (7.24) $\min(\xi_1, y) = y$.

We hypothesise that the ansatz is a polynomial in $q$. Before proposing the ansatz, note that the DPE contains an explicit $q^2$ penalty, the optimum over $\nu$ is quadratic in $\partial_q h$, and the terminal condition is $-\alpha\, q^2$. Thus this suggests the following ansatz for $h(t, q)$:

$$h(t, q) = h_0(t) + h_1(t)\, q + h_2(t)\, q^2 \, ,$$

with terminal conditions $h_0(T) = h_1(T) = 0$ and $h_2(T) = -\alpha$. The supremum over $y$ becomes

$$\sup_{y \leq q} \mathbb{E}\left[h\left(t, q - \min(y, \xi)\right) - h(t, q)\right]$$

$$= \sup_{y \leq q} \left[h\left(t, q - y\right) - h(t, q)\right]$$

$$= \sup_{y \leq q} \left[-y\, h_1 + (y^2 - 2\, q\, y)\, h_2\right]$$

$$= -\frac{1}{4\, h_2}\, (h_1 - 2\, q\, h_2)^2 \, ,$$

and the optimal dark pool volume in feedback form is

$$y^* = q + \frac{1}{2}\frac{h_1}{h_2} \, .$$

From the terminal condition, $h_2(t) < 0$. It remains to be seen that $h_1(t) \geq 0$ so that indeed $y^* \leq q$ and the admissibility criteria are satisfied.

Furthermore, the optimal speed of trading, in feedback form, simplifies to

$$\nu^* = -\frac{1}{2k}\, (h_1 + 2q\, h_2) \, .$$

Notice that both $y^*$ and $\nu^*$ are independent of $h_0$, so while $h_0$ is important in determining the value function, it is irrelevant for obtaining the optimal strategy.

Inserting the above feedback controls into the DPE (7.24), collecting terms in powers of $q$, and setting each to zero, leads to the coupled system of ODEs

$$\partial_t h_2 - \phi - \lambda\, h_2 + \tfrac{1}{k} h_2^2 = 0 \, , \tag{7.26a}$$

$$\partial_t h_1 + \left(\lambda + \tfrac{1}{k}\, h_2\right) h_1 = 0 \, , \tag{7.26b}$$

$$\partial_t h_0 + \tfrac{1}{k}\, h_1^2 - \frac{\lambda}{4}\frac{h_1^2}{h_2} = 0 \, . \tag{7.26c}$$

Since $h_1$ vanishes at $T$ and its ODE in (7.26b) is linear in $h_1$, the solution is $h_1(t) = 0$ and it is also trivial to see that $h_0(t) = 0$. If there was a drift in the midprice, these terms would not vanish, see Exercise E.7.3. Then, overall, we are left with only the $h_2$ equation, which is modified somewhat from the no dark pool case ((6.25) with $b = 0$) by the term $-\lambda\, h_2$. This term represents a "leakage" of inventory resulting from the possibility that the order posted in the dark pool is fully executed. Clearly, we see that if there is no dark pool, that is $\lambda = 0$, the problem reduces to that of optimal liquidation already discussed above in Chapter 6. For instance, see that for $\lambda = 0$, ODE (7.26a) is the same as (6.25) and both have the same boundary condition.

The equation for $h_2$ is of Riccati type and can be solved explicitly. Let $\zeta^\pm$ denote the roots of the polynomial $\phi + \lambda p - \frac{1}{k} p^2 = 0$, then write (7.26a) as

$$\partial_t h_2 = -\frac{1}{k}\left(h_2 - \zeta^+\right)\left(h_2 - \zeta^-\right),$$

where

$$\zeta^\pm = \tfrac{1}{2}k\lambda \pm \sqrt{\tfrac{1}{4}k^2\lambda^2 + k\phi}.$$

Cross multiplying and writing as partial fractions, we have

$$\partial_t h_2 \left(\frac{1}{h_2 - \zeta^+} - \frac{1}{h_2 - \zeta^-}\right) = -\frac{1}{k}\left(\zeta^+ - \zeta^-\right),$$

and integrating from $t$ to $T$ leads to

$$\log\left(\frac{h_2 - \zeta^-}{h_2 - \zeta^+}\right) - \log\left(\frac{\alpha + \zeta^-}{\alpha + \zeta^+}\right) = -\frac{1}{k}\left(\zeta^+ - \zeta^-\right)(T - t),$$

where we have used the terminal condition $h_2(T) = -\alpha$. Re-arranging the equation, we finally obtain

$$h_2(t) = \frac{\zeta^- - \zeta^+\,\beta\,e^{-\gamma(T-t)}}{1 - \beta\,e^{-\gamma(T-t)}},$$

where the constants are

$$\beta = \frac{\alpha + \zeta^-}{\alpha + \zeta^+} \quad \text{and} \quad \gamma = \tfrac{1}{k}(\zeta^+ - \zeta^-).$$

Therefore, the optimal trading strategy is

$$\boxed{\nu_t^* = -\frac{1}{k}\,h_2(t)\,Q_t^{\nu^*,y^*} \quad \text{and} \quad y_t^* = Q_t^{\nu^*,y^*}.} \qquad (7.27)$$

As before, we can obtain the optimal inventory to hold, up to the arrival of matching order in the dark pool, by solving

$$dQ_t^{\nu^*,y^*} = -\nu_t^*\,dt = \frac{1}{k}\,h_2(t)\,Q_t^{\nu^*,y^*}\,dt,$$

so that

$$Q_t^{\nu^*,y^*} = Q_0 \exp\left\{\tfrac{1}{k}\int_t^T h_2(u)\,du\right\},$$

and therefore by direct integration

$$\boxed{Q_t^{\nu^*,y^*} = e^{(\zeta^-/k)\,t}\left(\frac{1 - \beta\,e^{-\gamma(T-t)}}{1 - \beta\,e^{-\gamma T}}\right)\mathfrak{N}.} \qquad (7.28)$$

In the limit in which the terminal penalty $\alpha$ is very large, i.e. $\alpha \to \infty$ (so that the agent guarantees full execution by the end of the trading horizon), $\beta \to 1$ and hence,

$$Q_t^{\nu^*,y^*} \xrightarrow{\alpha\to\infty} e^{\left(\frac{\zeta^-}{k}+\frac{\gamma}{2}\right)t}\,\frac{\sinh\left(\frac{\gamma}{2}(T-t)\right)}{\sinh\left(\frac{\gamma}{2}T\right)}\mathfrak{N}.$$

Furthermore, in the limit $\lambda \to 0$, $\zeta^- \to -\sqrt{k\phi}$ and $\gamma \to 2\sqrt{\phi/k}$ and thus

$$Q_t^{\nu^*,y^*} \xrightarrow{(\alpha,\lambda)\to(\infty,0)} \frac{\sinh\left(\sqrt{\frac{\phi}{k}}(T-t)\right)}{\sinh\left(\sqrt{\frac{\phi}{k}}T\right)} \mathfrak{N},$$

which recovers the results from the AC case without the dark pool.

### Liquidation Strategy with Dark Pool

It is clear that the optimal amount to send to the dark pool is always what remains to be liquidated. This makes sense because in our model there is no market impact in the dark pool so the agent obtains the midprice for orders that are crossed in the dark pool. The more interesting part of the liquidation strategy is how much the agent should send to the lit markets now that she has access to a dark pool. To answer this question it is useful to compare the lit market liquidation rate $\nu^*$ in (7.27) with the optimal liquidation strategy when there is no dark pool, i.e. $\lambda = 0$. Recall that when $\lambda = 0$ the optimal speed of trading in the lit market is that given by the AC solution, see for instance (6.27) with $b = 0$ or simply use (7.27) with $\lambda = 0$. It is not immediately clear whether the modified rate at which the agent is trading in the lit market is larger or smaller than the liquidation rate when the agent does not have access to a dark pool. Also, it is not clear whether the trading rate is decreasing as in the AC case.

In Figure 7.6 we plot the optimal liquidation rate in the lit market given in (7.27) for different levels of the rate of arrival of matching orders in the dark pool. The figure shows that the trading rate may be larger or smaller than the AC case, and it may be increasing or decreasing. Also, the optimal inventory to hold (up to the time at which a matching order arrives) may be either convex or concave or neither.

In particular, the top two panels of Figure 7.6 show the optimal inventory path and optimal speed of trading where we also assume that the liquidation penalty at time $T$ is $\alpha \to \infty$ and the paths shown are prior to the order posted in the dark pool being executed. Other model parameters are $k = 0.001$ and $\phi = 0.01$.

The bottom two panels of the figure show the case where the order in the dark pool was executed at time $t = 0.6$ and the agent's inventory drops to zero. Since the execution of the orders in the dark pool occur according to a Poisson process with intensity $\lambda$, the time at which this occurs is exponentially distributed with mean $1/\lambda$. Thus, when $\lambda = 0$ there are no executions in the dark pool and the liquidation strategy corresponds to the AC solution. When $\lambda > 0$ the agent starts trading slower than the AC speed in the lit market, to allow for the potential of dark pool execution, but then as time runs out and no execution occurs, her rate of trading increases to compensate for the initially slow trading. Interestingly, the optimal trading curve ceases to be convex, and its convexity changes signs. In the limiting case when $\lambda \to \infty$, the agent does not trade at all in the lit market, since execution in the dark pool is guaranteed. In this case, the optimal inventory path flows along $Q_t^* = \mathfrak{N}\, \mathbb{1}_{t<T}$, but is then infinitely fast at $T$ to rid herself of the assets.

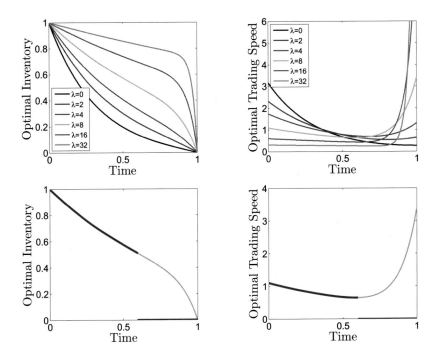

**Figure 7.6** The top panels show optimal inventory path and speed of trading prior to a matching order in the dark pool. The bottom panels show the optimal inventory and trading speed where we assume that the dark pool matching order arrives at $t = 0.6$, right after which inventory drops to zero.

## 7.5     Bibliography and Selected Readings

Laruelle, Lehalle & Pagès (2011), Buti, Rindi & Werner (2011b), Buti, Rindi & Werner (2011a), Crisafi & Macrina (2014), Iyer, Johari & Moallemi (2014), Moallemi, Saglam & Sotiropoulos (2014), Bechler & Ludkovski (2014), Cartea & Jaimungal (2014b).

## 7.6     Exercises

E.7.1 Use the same setup as in Section 7.3 but allow for $\eta$ to be random so that

$$d\mu_t^{\pm} = -\kappa\,\mu_t^{\pm}\,dt + \eta\,dL_t^{\pm}\,, \qquad (7.29)$$

where $\eta$ are i.i.d. with finite first moment. Find the optimal speed of liquidation and compare this result to (7.20).

E.7.2 Assume the setup in Section 7.4. Instead of the agent's orders always being at the front of the queue in the dark pool, assume that ahead of her order there are other market participants' orders which have priority. Model this volume as a random variable and derive the optimal speed of trading in the lit market and the number of shares that the agent sends to the dark pool.

E.7.3 The setup is as in section 7.4.1 and let the midprice satisfy

$$dS_t = \mu dt + \sigma dW_t \,,$$

where $\mu$ is constant. Derive the DPE and propose an ansatz to specify the optimal speed of trading. Compare your results to the case where $\mu = 0$.

# 8 Optimal Execution with Limit and Market Orders

## 8.1 Introduction

In the previous two chapters we focused on execution strategies which relied on market orders (MOs) only. One of the advantages of sending MOs is that execution is guaranteed. The execution price, however, is generally worse than the midprice due to both the existence of non-zero spread and the fact that orders may walk the book. In practice, the agent also employs limit orders (LOs) because instead of picking up liquidity-taking fees and incurring market impact costs, the prices at which LOs are filled are better than the midprice, but there is no guarantee that a matching order will arrive.

To address these issues, this chapter looks at optimal execution problems when the agent employs LOs and possibly also MOs. In Sections 8.2 and 8.3, the agent is only allowed to use LOs. In Section 8.4, the agent is allowed to trade with both LOs and MOs, and in Section 8.5, the agent aims to track a given schedule using LOs and MOs.

In all cases, when the agent posts LOs to liquidate a position, she posts a limit sell order for a fixed volume (e.g., some percentage of the average size of an MO, or a fixed amount of, say, 10 shares) at a price of $S_t + \delta_t$, where $S_t$ is the midprice. Hence, $\delta$ is a premium the agent demands for providing liquidity to the market. The larger $\delta$, the larger the premium, but the probability that an order arrives and walks the limit order book (LOB), up to the posted depth, decreases with $\delta$. The strategy used by the agent relies on speed to post-and-cancel LOs. At every instant in time: the agent reassesses market conditions, cancels any LO resting in the book, posts a new LO at the optimal level, and so on. To do this, requires software, hardware, and connection to the exchange so that the strategy does not have stale quotes in the LOB and can quickly process information.

The probability of being filled when posting at a given depth $\delta$, conditional on the arrival of an MO, is called the **fill probability** which we denote by the function $P(\delta)$. Naturally, $P$ must be decreasing, it changes throughout the day, and it is sensitive to the current status of the LOB. To see this, consider the left panel of Figure 8.1 which shows a block-shaped LOB together with (i) a post at $\delta = 10$ (the dashed line); (ii) the depth to which an MO of volume 700 lifts sell orders (dark green region); and (iii) the depth to which an MO of volume 1,500 lifts sell LOs (dark plus light green region). The deeper the LO is posted (i.e.

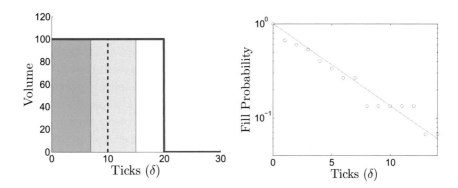

**Figure 8.1** (Left) A flat (or block shaped) LOB. (Right) Empirical fill probabilities for NFLX on June 21, 2011 for the time interval 12:55pm to 1:00pm using 500 millisecond resting times. The straight line shows the fit to an exponential function.

further away from the midprice), the less likely it is that MOs large enough walk the LOB up to that price level. Hence, the probability of being filled decreases as $\delta$ increases.

If we assume that the volume of individual MOs, denoted by $V$, is exponentially distributed with mean volume of $\eta$, and that the LOB is block shaped with height $A$, i.e. the posted volume at a price of $S + \delta$ is equal to a constant $A$ out to a maximum price level of $S + \bar{\delta}$, then the probability of fill is exponential. That is, conditional on the arrival of an MO of volume $V$, the probability that the sell LO is lifted is given by

$$\mathbb{P}(\text{order posted at depth } \delta \text{ is lifted}) = \mathbb{P}(V > A\delta) = \exp\left\{-\frac{A}{\eta}\delta\right\}. \quad (8.1)$$

One could in principle also use power law fill probabilities; however, to keep the analysis consistent and self-contained we use the exponential fill probability throughout.

## 8.2 Liquidation with Only Limit Orders

Chapters 6 and 7 looked at the optimal execution problem for an agent who places only MOs. In this section, the agent posts only LOs and the setup of the problem is similar to that in Chapter 6. Now we must track not only the agent's inventory, but also the arrival of other traders' MOs, which is what will (possibly) lift the agent's posted sell LOs. We summarise the model ingredients and the notation here:

- $\mathfrak{N}$ is the amount of shares that the agent wishes to liquidate,
- $T$ is the terminal time at which the liquidation programme ends,
- $S = (S_t)_{0 \leq t \leq T}$ is the asset's midprice with $S_t = S_0 + \sigma W_t$, $\sigma > 0$, and $W = (W_t)_{0 \leq t \leq T}$ is a standard Brownian motion,

- $\delta = (\delta)_{0 \leq t \leq T}$ denotes the depth at which the agent posts limit sell orders, i.e. the agent posts LOs at a price of $S_t + \delta_t$ at time $t$,
- $M = (M_t)_{0 \leq t \leq T}$ denotes a Poisson process (with intensity $\lambda$) corresponding to the number of market buy orders (from other traders) that have arrived,
- $N^\delta = (N_t^\delta)_{0 \leq t \leq T}$ denotes the (controlled) counting process corresponding to the number of market buy orders which lift the agent's offer, i.e. MOs which walk the sell side of the book to a price greater than or equal to $S_t + \delta_t$,
- $P(\delta) = e^{-\kappa \delta}$ with $\kappa > 0$ is the probability that the agent's LO will be lifted when a buy MO arrives,
- $X^\delta = (X_t^\delta)_{0 \leq t \leq T}$ is the agent's cash process and satisfies the SDE

$$dX_t^\delta = (S_t + \delta_t)\, dN_t^\delta , \qquad (8.2)$$

- $Q_t^\delta = \mathfrak{N} - N_t^\delta$ is the agent's inventory which remains to be liquidated.

Note that whenever the process $N$ jumps, the process $M$ must also jump, but when $M$ jumps, $N$ will jump only if the MO is large enough to walk the book and lift the agent's posted LO. Moreover, conditional on an MO arriving (i.e. $M$ jumps), $N$ jumps with probability $P(\delta_t) = e^{-\kappa\,\delta_t}$; however $N$ is not a Poisson process since its activity reacts to the depth at which the agent posts. Moreover, in contrast to the setup in Chapters 6 and 7, here, when the agent's orders are executed, she receives better than midprices.

Finally, the filtration $\mathcal{F}$ on which the problem is setup is the natural one generated by $S$, $N$ and $M$. Moreover, the agent's depth postings (or strategy) $\delta$ will be $\mathcal{F}$-predictable and in particular will be left-continuous with right limits.

## The Agent's Optimisation Problem

The agent wishes to maximise the profit from liquidating $\mathfrak{N}$ shares, but also requires that most, if not all, of the shares are sold by the terminal time $T$. If the agent has inventory remaining at the end of the trading horizon, she liquidates it using an MO for which she obtains worse prices than the midprice. As argued in Chapter 6, a linear impact function on MOs is a reasonable first order approximation of market impact, hence the agent's optimisation problem is to find

$$H(x, S) = \sup_{\delta \in \mathcal{A}} \mathbb{E}\left[ X_\tau^\delta + Q_\tau^\delta \left( S_\tau - \alpha\, Q_\tau^\delta \right) \mid X_{0-}^\delta = x,\, S_0 = S, Q_{0-}^\delta = \mathfrak{N} \right] , \quad (8.3)$$

where $\alpha \geq 0$ is the liquidation penalty (linear impact function). Moreover, the admissible set $\mathcal{A}$ consists of strategies $\delta$ which are bounded from below, and the stopping time

$$\tau = T \wedge \min\{ t : Q_t^\delta = 0 \}$$

is the minimum of $T$ or the first time that the inventory hits zero, because then no more trading is necessary.

The corresponding value function is

$$H(t, x, S, q) = \sup_{\delta \in \mathcal{A}} \mathbb{E}_{t,x,S,q} \left[ X_\tau^\delta + Q_\tau^\delta \left( S_\tau - \alpha Q_\tau^\delta \right) \right], \tag{8.4}$$

where the notation $\mathbb{E}_{t,x,S,q}[\cdot]$ represents expectation conditional on $X_{t-}^\delta = x$, $S_t = S$, and $Q_{t-}^\delta = q$. In this setup, the agent does not have any urgency, i.e. does not penalise inventories different from zero as discussed in Section 6.5. Indeed, one can add in such a penalty and we leave this as an exercise for the reader, see Exercise E.8.1.

### The Resulting DPE

The dynamic programming principle (DPP) suggests that the value function solves the following dynamic programming equation (DPE):

$$\begin{cases} \partial_t H + \frac{1}{2}\sigma^2 \partial_{SS} H \\ \quad + \sup_\delta \left\{ \lambda e^{-\kappa\delta} \left[ H(t, x + (S+\delta), S, q-1) - H(t, x, S, q) \right] \right\} = 0, \\ \qquad\qquad\qquad\qquad\qquad H(t, x, S, 0) = x, \\ \qquad\qquad\qquad\qquad\qquad H(T, x, S, q) = x + q\,(S - \alpha\,q). \end{cases}$$

We have an optimal trading problem where the state variables jump and the resulting DPE results in a non-linear partial integral differential equation (PIDE) rather than a non-linear PDE. Below we elaborate on the interpretation of the various terms of the PIDE.

(i) The operator $\partial_{SS}$ corresponds to the generator of the Brownian motion which drives the midprice.
(ii) The supremum takes into account the agent's ability to control the depth of her sell LOs.
(iii) The term $\lambda e^{-\kappa\delta}$ represents the rate of arrival of other market participants' buy MOs which lift the agent's posted sell LO at price $S + \delta$.
(iv) The difference (jump) term $H(t, x + (S+\delta), S, q-1) - H(t, x, S, q)$ represents the change in the agent's value function when an MO fills the agent's LO – the agent's cash increases by $S + \delta$ and her inventory decreases by 1.

The terminal condition at $t = T$ represents the cash the agent has acquired up to that point in time, plus the value of liquidating the remaining shares at the worse than midprice of $(S - \alpha q)$ per share – recall that at $T$ she must execute an MO to complete her trade and as a result walks the book. The boundary condition along $q = 0$ represents the cash the agent has at that stopping time and since $q = 0$ there is no liquidation value, and the agent simply walks away with $x$ in cash.

The terminal and boundary conditions suggest that the ansatz for the value function is

$$H(t, x, S, q) = x + q\,S + h(t, q), \tag{8.5}$$

for a yet to be determined function $h(t,q)$. This ansatz has three terms. The first term is the accumulated cash, the second term denotes the book value of the remaining inventory which is marked-to-market using the midprice, and finally the function $h(t,q)$ represents the added value to the agent's cash from optimally liquidating the remaining shares. With this ansatz, upon substitution into the DPE above, we find that $h(t,q)$ satisfies the coupled system of non-linear ODEs

$$\begin{cases} \partial_t h + \sup_{\delta} \left\{ \lambda e^{-\kappa\delta} \left[ \delta + h(t,q-1) - h(t,q) \right] \right\} &= 0, \\ h(t,0) &= 0, \\ h(T,q) &= -\alpha\,q^2. \end{cases} \tag{8.6}$$

The optimal depth can be found in feedback form by focusing on the first order conditions for the supremum. This provides us with the following:

$$\begin{aligned} 0 &= \partial_\delta \left\{ \lambda e^{-\kappa\delta} \left[ \delta + h(t,q-1) - h(t,q) \right] \right\} \\ &= \lambda \left( -\kappa\,e^{-\kappa\delta} \left[ \delta + h(t,q-1) - h(t,q) \right] + e^{-\kappa\delta} \right) \\ &= \lambda e^{-\kappa\delta} \left( -\kappa \left[ \delta + h(t,q-1) - h(t,q) \right] + 1 \right), \end{aligned}$$

and hence the optimal strategy $\delta^*$ in feedback control form is given by

$$\boxed{\delta^*(t,q) = \frac{1}{\kappa} + \left[\, h(t,q) - h(t,q-1) \,\right].} \tag{8.7}$$

This form for the optimal depth has an interesting interpretation. Consider the first term $\frac{1}{\kappa}$. It stems from optimising the instantaneous expected profits from selling one share. The profit is given by the revenue of selling one share at $(S+\delta)$, minus the cost $S$, which results in $\delta$. Hence, the expected profit is $\delta\,P(\delta)$, and when the fill probability is $P(\delta) = e^{-\kappa\delta}$, the maximum is attained at $\delta^\dagger = \frac{1}{\kappa}$, see also the discussion in 2.1.4.

The difference term $h(t,q) - h(t,q-1)$ can be viewed as the agent's correction to this static optimisation taking into account her future optimal behaviour. In particular, it represents a **reservation price**, which is defined as the price $p$ such that $H(t,x+p,S,q-1) = H(t,x,S,q)$, i.e. it is the additional wealth the agent demands for selling the asset such that her value function remains unchanged.

We expect that $\delta^*(t,q)$ is decreasing in $q$, since the more inventory the agent has, the more urgent she should be in getting rid of her holdings and hence the closer to the midprice she should post. It may be that for large enough $q$ the optimal depth becomes negative and the solution to the control problem is no longer financially meaningful. We should instead solve the constrained problem, where $\delta^* \geq 0$ is enforced in the set of admissible strategies. One naive approach, which avoids solving the constrained problem, is to view negative depths as an indicator that the agent should execute a market order instead of positing a limit order. The sound approach to addressing the optimal posting of LOs versus MOs is investigated later in Section 8.4.

Inserting the optimal depth in feedback control form into (8.6) provides a non-linear coupled system of ODEs for $h(t, q)$

$$\partial_t h + \frac{\tilde{\lambda}}{\kappa} \exp\left\{ -\kappa\left[h(t, q) - h(t, q - 1)\right] \right\} = 0, \tag{8.8}$$

where $\tilde{\lambda} = \lambda e^{-1}$ and the same terminal and boundary conditions as in (8.6) apply. This system of ODEs can be solved exactly by making the substitution $h(t, q) = \frac{1}{\kappa} \log w(t, q)$ and writing a new equation for $w(t, q)$, in which case,

$$
\begin{aligned}
0 &= \partial_t h + \frac{\tilde{\lambda}}{\kappa} \exp\left\{ -\kappa\left[h(t, q) - h(t, q - 1)\right] \right\} \\
&= \frac{1}{\kappa} \frac{\partial_t w(t, q)}{w(t, q)} + \frac{\tilde{\lambda}}{\kappa} \frac{w(t, q - 1)}{w(t, q)},
\end{aligned}
$$

which implies that

$$0 = \partial_t w(t, q) + \tilde{\lambda} w(t, q - 1), \tag{8.9}$$

and the terminal and boundary conditions are now

$$w(T, q) = e^{-\kappa \alpha q^2} \quad \text{and} \quad w(t, 0) = 1,$$

respectively.

### Solving the DPE

The coupled system of ODEs (8.9), which the DPE reduces to, can be solved explicitly (see Exercise E.8.2) resulting in the expression

$$w(t, q) = \sum_{n=0}^{q} \frac{\tilde{\lambda}^n}{n!} e^{-\kappa \alpha (q-n)^2} (T - t)^n. \tag{8.10}$$

This solution provides the function $h(t, q)$ which can then be substituted into the equation for the optimal depth (8.7) to find

$$\delta^*(t, q) = \frac{1}{\kappa} \left[ 1 + \log \frac{\displaystyle\sum_{n=0}^{q} \frac{\tilde{\lambda}^n}{n!} e^{-\kappa \alpha (q-n)^2} (T - t)^n}{\displaystyle\sum_{n=0}^{q-1} \frac{\tilde{\lambda}^n}{n!} e^{-\kappa \alpha (q-1-n)^2} (T - t)^n} \right], \tag{8.11}$$

for $q > 0$. The optimal depth at which to post is a decreasing function of time for any model parameter, a decreasing function of the agent's inventory $q$, and increases the rate of arrival of MOs. The increasing behaviour in activity rate is intuitive since as market order arrival rates increase, the agent is willing to post deeper in the book so that her effective rate of filled LOs remains essentially constant, while reaping more profits if a matching arrives.

In Figure 8.2, the optimal depths are shown as a function of time for several inventory levels as well as penalty parameter $\alpha$. MOs arrive at the rate of $50/\min$

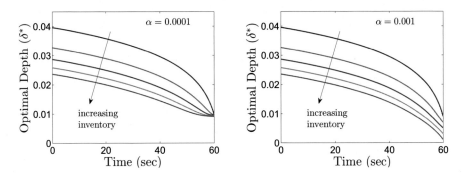

**Figure 8.2** The optimal depths $\delta^*$ at which the agent posts LOs as a function of time and current inventory. The parameters are $\lambda = 50/\min$, $\kappa = 100$, and $\mathfrak{N} = 5$ with the penalty $\alpha$ shown in each panel. The lowest depth corresponds to $q = 5$ and the highest depth to $q = 0$.

and the agent is attempting to liquidate $\mathfrak{N} = 5$ shares – and is hence 10% of the average market volume. These plots show several interesting features of the optimal depths, as described below.

(i) The depths are decreasing in inventory. This is natural, as if the agent's inventory is large, she is willing to accept a lower premium $\delta$, for providing liquidity, to increase the probability that her order is filled. At the same time, this ensures that she may complete the liquidation of the $\mathfrak{N}$ shares by end of the time horizon and avoid crossing the spread (i.e. using MOs) and paying a terminal penalty. However, if inventories are low, the agent is willing to hold on to it in exchange for large $\delta$, because with low inventory the terminal penalty she picks up when crossing the spread will be moderate.

(ii) For fixed inventory level, the depths all decrease in time. Once again, this is due to the agent becoming more averse to holding inventories as the terminal time approaches, due to the penalty they will receive from crossing the spread.

(iii) As the penalty parameter $\alpha$ increases, all depths decrease because increasing the penalty induces the trader to liquidate her position faster, but at lower prices. We point out that if $\alpha$ or $q$ is large then the optimal depths can become negative. In practice one cannot post LOs which improve the best quote on the other side of the LOB, so one may want to interpret this as the agent being very keen to get her LO filled, but here we do not allow the agent to submit MOs, we do this below in Section 8.4.

(iv) The depths keep increasing as one moves further from the end of the trading horizon. The reason is that the agent is only being penalised by her terminal inventory, so far from terminal time, there is no incentive to liquidate her position. If the agent instead penalises inventories through time, the strategies will become asymptotically constant far from maturity. For this case, see Exercise E.8.4.

Far from the terminal time, i.e. when $\tau = T - t \gg 1$, the ratio appearing in the logarithm above (i.e. $w(t,q)/w(t,q-1)$) is to $o\left((T-t)^{-1}\right)$ given by the ratio of the two terms $n = q - 1$ and $n = q$ in the numerator to the term $n = q - 1$ in the denominator. Therefore we can write

$$\frac{w(t,q)}{w(t,q-1)} = \frac{\frac{\tilde{\lambda}^{q-1}}{(q-1)!}e^{-\kappa\,\alpha\,(q-(q-1))^2}\tau^{q-1} + \frac{\tilde{\lambda}^q}{q!}e^{-\kappa\,\alpha\,(q-q)^2}\tau^q}{\frac{\tilde{\lambda}^{q-1}}{(q-1)!}e^{-\kappa\,\alpha\,(q-1-(q-1))^2}\tau^{q-1}} + o\left(\tau^{-1}\right)$$

$$= e^{-\kappa\,\alpha} + \frac{\tilde{\lambda}}{q}\tau + o\left(\tau^{-1}\right).$$

Therefore, far from the terminal time, the agent posts at depths that grow logarithmically as follows:

$$\delta^*(t,q) = \frac{1}{\kappa}\left[1 + \log\left(e^{-\kappa\,\alpha} + \frac{\lambda e^{-1}}{q}\tau\right)\right] + o\left(\tau^{-1}\right).$$

In this expression, the dependence of the optimal depth on the parameters becomes clear: it is increasing in activity rate and decreasing in inventory, time, fill probability and terminal penalty.

### Numerical Experiments

In this section we carry out a simulation study to explore the optimal execution strategy. Throughout we use the following parameters:

$$T = 60\,\mathrm{sec}\,, \quad \lambda = 50/\min\,, \quad \kappa = 100\$^{-1}\,, \quad \alpha = 0.001\,\$/\mathrm{share}\,, \quad \mathfrak{N} = 5\,,$$
$$S_0 = \$30.00\,, \quad \text{and} \quad \sigma = \$\sec^{-1/2} 0.01\,,$$

so that the agent is trading 10% of the market over this time interval.

Figure 8.3 shows three simulated sample paths for the midprice (panel a), the optimal depth (panel b), the resulting inventory (panel c) and the average price per share (panel d) computed as $X_t/(\mathfrak{N} - q_t)$. In the average price per share panel, TWAP, which is given by

$$A_{TWAP} = \frac{1}{T}\int_0^T S_u\,du\,,$$

is often used as a benchmark for comparison purposes so we include it as well.

Panel (c) shows that the algorithm may sometimes acquire all assets early, e.g. along the blue and green paths, or may need to execute MOs at the end of the interval, e.g., as in the red path. When we combine this panel with panel (b), which shows the inventory path, it illustrates how immediately after the agent's LO is filled, the agent increases the posted depth, but if the agent's LO is not filled, she posts closer to the midprice. Panel (d) illustrates how the algorithm (solid lines) outperforms TWAP (dashed lines). The key reason is that the agent mostly uses LOs to achieve her goal of liquidating $\mathfrak{N}$ shares, which provides profits in excess of the midprice. However, some paths that do not completely

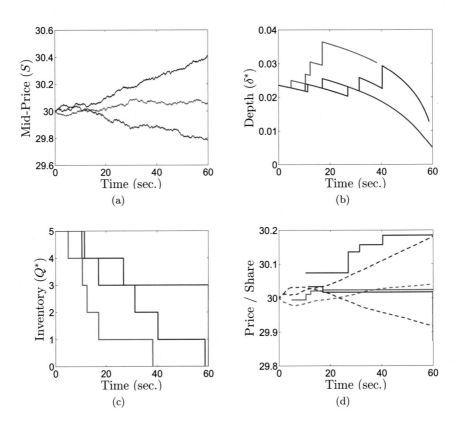

**Figure 8.3** Three sample paths for the agent following the optimal strategy. In panel (d), the dashed line indicates the TWAP curve.

liquidate prior to maturity, such as the red one, force the agent to execute MOs at the end of the trading horizon. Doing so causes her to lose some of the premia, as measured by the depth $\delta$, earned by executing LOs throughout the strategy.

The left panel of Figure 8.4, which shows the histogram of the number of executed MOs over 10,000 scenarios, demonstrates that only a small number of paths require executing MOs at the terminal time. The right panel shows a heat-map of the agent's inventory through time for the same 10,000 scenarios. The dashed line is the mean inventory at each point in time – notice that it is almost linear and reduces to almost, but not equal to, zero at the end of the trade horizon.

To illustrate the savings provided by the algorithm, Figure 8.5 shows the histogram of the difference between the price per share from the algorithm and the TWAP over the 10,000 scenarios.

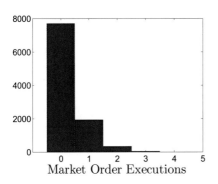
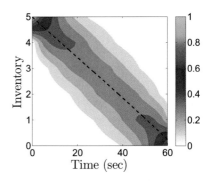

**Figure 8.4** Left panel shows the histograms (from 10,000 scenarios) of executed MOs at the end of trade execution. Right panel shows a heat-map of inventory through time.

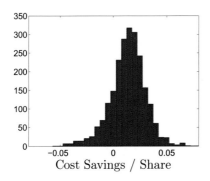

**Figure 8.5** Histogram of the cost savings per share relative to TWAP.

## 8.3    Liquidation with Exponential Utility Maximiser

In the previous section the agent is indifferent to uncertainty in the value of her sales and so her objective is to maximise expected proceeds from selling $\mathfrak{N}$ shares. A more realistic setup is one in which the agent includes a running inventory penalty. Exercise E.8.4 shows that in this case the agent's performance criteria becomes

$$G^\delta(t, x, S, q) = \mathbb{E}_{t,x,S,q}\left[X_\tau^\delta + Q_\tau^\delta(S_\tau - \alpha Q_\tau) - \phi \int_0^\tau \left(Q_s^\delta\right)^2 ds\right], \qquad (8.12)$$

and, compared to the case with no running inventory penalty (i.e. $\phi = 0$), the optimal strategy is modified to become more aggressive (sell faster) earlier on and then less aggressive towards the end of the trading horizon. This allows her to control the distribution of the value of the total sales, as well as how fast inventory is liquidated. This is akin to the approach taken in Chapter 6, where the agents have a running inventory penalty, or urgency, constraint. Some agents, however, may instead wish to penalise uncertainty in their sales directly. Here, we show that if the agent uses exponential utility as a performance measure, her strategy is identical (up to a constant and a re-scaling of parameters) to the one implied by (8.12).

To demonstrate this equivalence, first let us consider the agent who sets preferences based on expected utility of terminal wealth with exponential utility $u(x) = -e^{-\gamma x}$. In this case, her performance criteria is

$$H^\delta(t, x, S, q) = \mathbb{E}_{t,x,S,q}\left[ -\exp\left\{ -\gamma \left( X_T^\delta + Q_T^\delta(S_T - \alpha\, Q_T^\delta) \right) \right\} \right],$$

and proceeding as usual, her value function

$$H(t, x, S, q) = \sup_{\delta \in \mathcal{A}} H^\delta(t, x, S, q)$$

should satisfy the DPE

$$\partial_t H + \tfrac{1}{2}\sigma^2 \partial_{SS} H$$
$$+ \sup_\delta \left\{ \lambda e^{-\kappa\delta}\, [H(t, x + (S + \delta), S, q - 1) - H(t, x, S, q)] \right\} = 0,$$

subject to the terminal and boundary conditions

$$H(T, x, S, q) = -e^{-\gamma(x + q(S - \alpha q))} \quad \text{and} \quad H(t, x, S, 0) = -e^{-\gamma x}.$$

We leave it as an exercise for the reader to show that ansatz

$$H(t, x, S, q) = -e^{-\gamma(x + q\,S + h(t,q))} \tag{8.13}$$

leads to the following equation for $h(t, q)$:

$$\partial_t h - \tfrac{1}{2}\sigma^2\,\gamma\, q^2 + \sup_\delta \lambda e^{-\kappa\delta} \frac{1 - e^{-\gamma\,[\delta + h(t, q-1) - h(t,q)]}}{\gamma} = 0, \tag{8.14}$$

with terminal and boundary conditions

$$h(T, q) = -\alpha\, q^2 \quad \text{and} \quad h(t, 0) = 0.$$

The interpretation of the ansatz (8.13) is similar to that of (8.5), in particular, the right-hand side of equation (8.13) is the utility derived from the sum of: accumulated cash, the book value of the remaining inventory which is marked-to-market using the midprice, and, finally, the function $h(t, q)$ representing the added value to the agent's utility from optimally liquidating the remaining shares.

If one takes the limit in which $\gamma \to 0$, we see that $h$ satisfies the PDE

$$\partial_t h + \sup_\delta \left\{ \lambda e^{-\kappa\delta}[\delta + h(t, q - 1) - h(t, q)] \right\} = 0, \tag{8.15}$$

which is precisely the equation that the excess value function $h$ satisfied in the previous section when the agent has linear utility (see (8.6)).

Going back to the general case $\gamma > 0$, we obtain, from the first order condition, the optimal depth in feedback control form as

$$\boxed{\delta^* = \frac{1}{\gamma} \log\left(1 + \frac{\gamma}{\kappa}\right) + [\,h(t, q) - h(t, q - 1)\,].} \tag{8.16}$$

This form is very similar to, but slightly differs from, the optimal depth in the previous section provided in (8.7). The $h$ functions may differ and the base line level $\kappa^{-1}$ is modified to $\hat{\kappa}^{-1} = \frac{1}{\gamma} \log \left(1 + \frac{\gamma}{\kappa}\right)$. This modification can be seen as a risk aversion bias. Indeed, in the limit of zero risk-aversion

$$\hat{\kappa} \xrightarrow{\gamma \downarrow 0} \kappa \,,$$

and the result from the previous section is recovered. Furthermore, we can view the contribution $\frac{1}{\gamma} \log \left(1 + \frac{\gamma}{\kappa}\right)$ as stemming from the agent maximising her utility, from selling at a price of $(S + \delta)$ and immediately repurchasing at $S$ (i.e. measuring relative to midprice):

$$\max_{\delta} \left\{ u(x + \delta) \, P(\delta) + u(x) \, (1 - P(\delta)) \right\} = \frac{1}{\gamma} \log \left(1 + \frac{\gamma}{\kappa}\right) \,.$$

Substituting the feedback form of the optimal depth (8.16) into the DPE (8.14), we now find the non-linear system of coupled ODEs for $h$ to be

$$\partial_t h - \tfrac{1}{2}\sigma^2 \gamma \, q^2 + \frac{\hat{\lambda}}{\kappa} \, \exp\left\{-\kappa \left[h(t, q) - h(t, q - 1)\right]\right\} = 0 \,, \tag{8.17}$$

where,

$$\hat{\lambda} = (\kappa/(\kappa + \gamma))^{1+\kappa/\gamma} \, \lambda \,.$$

In the limit of zero risk-aversion $\hat{\lambda} \xrightarrow{\gamma \downarrow 0} e^{-1}\lambda = \tilde{\lambda}$ and once again we recover the parameter that appears in (8.8). The above ODE is in fact identical in structure to (8.8), except that it contains the additional term $-\frac{1}{2}\sigma^2 \gamma \, q^2$. As shown in Exercise E.8.4, when the value function $G$ for the running penalty performance criteria (8.12) is written as $G = x + q \, S + g(t, q)$, then $g$ satisfies the system of coupled ODEs (where we write the parameters in the running penalty model with a subscript 0)

$$\partial_t g - \phi \, q^2 + \frac{e^{-1}\lambda_0}{\kappa_0} \, \exp\left\{-\kappa_0 \left[g(t, q) - g(t, q - 1)\right]\right\} = 0 \,,$$

and the optimal strategy is

$$\delta_0^* = \frac{1}{\kappa_0} + g(t, q) - g(t, q - 1) \,.$$

Hence, with

$$\phi = \tfrac{1}{2}\sigma^2 \gamma \,, \quad \lambda_0 = e^{+1} \, \hat{\lambda} \,, \quad \text{and} \quad \kappa_0 = \kappa \,,$$

we see that the $h(t, q)$ and $g(t, q)$ coincide and the optimal strategies satisfy the relation

$$\boxed{\delta^* = \delta_0^* + \left(\hat{\kappa}^{-1} - \kappa_0^{-1}\right) \,.} \tag{8.18}$$

In other words, with a re-scaling of model parameters, the optimal strategy for the utility maximising agent is the same, up to a constant shift, as that of the agent who only penalises running inventory – with an appropriate choice of risk-aversion level and a re-scaling of arrival rates.

In addition to the relationship between the optimal strategies, the value functions can be written in terms of one another. Since $h(t,q) = g(t,q)$, we have

$$G(t,x,S,q) = -\frac{1}{\gamma} \log\left(-H(t,x,S,q)\right),$$

or writing the value function in its original control form

$$\begin{aligned}
&\sup_{\delta \in \mathcal{A}} \mathbb{E}^0_{t,x,S,q}\left[X^\delta_\tau + Q^\delta_\tau(S_\tau - \alpha Q_\tau) - \phi \int_0^\tau (Q^\delta_s)^2\, ds\right]\\
&= -\frac{1}{\gamma} \log\left(-\sup_{\delta \in \mathcal{A}} \mathbb{E}_{t,x,S,q}\left[-\exp\{-\gamma\left(X^\delta_\tau + Q^\delta_\tau(S_\tau - \alpha Q_\tau)\right)\}\right]\right),
\end{aligned} \tag{8.19}$$

where $\mathbb{E}^0[\cdot]$ represents expectation under a probability measure where the arrival rate is $\lambda_0$. This relationship between the value functions is in fact part of a more general result that relates optimisation problems with exponential utility and optimisation problems with penalties (see the further readings section).

## 8.4 Liquidation with Limit and Market Orders

In the previous two sections, the agent considers posting only LOs and, as shown, posts more aggressively (i.e. depth $\delta$ decreases so LOs are posted nearer the midprice) as maturity approaches when her inventory is held fixed. Here, we consider the situation in which the agent is allowed to post MOs in addition to LOs. In this case, when she is far behind schedule, i.e. when maturity is approaching but she still has many shares to liquidate, then she could be willing to execute an MO in order to place her strategy back on target. In this case, the agent searches for both an optimal control and a sequence of optimal stopping times at which to execute MOs.

### The Agent's Optimisation Problem

To formalise the problem, we now need to keep track of the agent's posted MOs, in addition to other traders' MOs, and her executed LOs. Below we list the additional stochastic processes and changes to the cash process to account for executing MOs. All other stochastic processes, including the midprice $S$, other trader's MOs $M$, and the agent's filled LOs $N$, posted at depth $\delta$, remain unaltered in their definition.

- $M^a = (M^a_t)_{0 \leq t \leq T}$ denotes the counting process for the agent's MOs.
- The corresponding increasing sequence of stopping times at which the agent executes MOs is denoted by $\tau = \{\tau_k : k = 1, \ldots, K\}$, with $K \leq \mathfrak{N}$, so that $M^a_t = \sum_{k=1}^K \mathbb{1}_{\tau_k \leq t}$. Note that the agent may place fewer, but never more, than $\mathfrak{N}$ MOs.

- $\xi$ denotes the half-spread, i.e. half-way distance between the best ask and best bid.

- $X = (X_t)_{0 \leq t \leq T}$ denotes the agent's cash process and satisfies the SDE

$$dX_t^{\tau, \delta} = (S_t + \delta_{t-}) \, dN_t^{\delta} + (S_t - \xi) \, dM_t^{a, \tau}.$$

The first term on the right-hand side of the cash process denotes the cash received from having an LO lifted, and the second term is the cash received from selling a share using an MO. Note that when the agent executes a sell MO she crosses the spread, which is why the proceeds from selling one unit of the asset is the midprice minus the half-spread, i.e. the best bid. Furthermore, we assume that the size of the MOs is small enough not to walk the LOB.

We assume that the agent is averse to holding inventory throughout the strategy – unlike in Section 8.2 where she wishes to rid herself of inventory due only to the terminal penalty. To achieve this, we apply an urgency penalty to her performance criteria, much like in Section 6.5, and more specifically equation (6.20). Hence, her performance criteria is

$$
\begin{aligned}
H^{(\tau, \delta)} &(t, x, S, q) \\
&= \mathbb{E}_{t,x,S,q} \left[ X_T^{\tau, \delta} + Q_T^{\tau, \delta} \, S_T - \ell \left( Q_T^{\tau, \delta} \right) - \phi \int_t^T \left( Q_u^{\tau, \delta} \right)^2 \, du \right],
\end{aligned}
\tag{8.20}
$$

where as usual $\mathbb{E}_{t,x,S,q}[\cdot]$ denotes expectation conditional on $X_{t-}^{\tau, \delta} = x$, $S_{t-} = S$, $Q_{t-}^{\tau, \delta} = q$, and the terminal liquidation penalty

$$\ell(q) = q \, (\xi + \alpha \, q).$$

The terminal liquidating cost per share of the shares remaining at the end is written as $(S_T - \xi - \alpha \, Q_T)$ because the agent must cross the spread and then walk the LOB to liquidate the remaining shares – recall that we assumed that the MOs sent during the liquidation strategy before the terminal date did not walk the LOB. And, since the agent may execute MOs, her inventory is reduced each time an LO is filled or an MO is executed, so that

$$Q_t^{\tau, \delta} = \mathfrak{N} - N_t - M_t^a.$$

The set of admissible strategies $\mathcal{A}$ now includes seeking over all $\mathcal{F}$-stopping times in addition to the set of $\mathcal{F}$-predictable, bounded from below, depths $\delta$. In this case, the value function is

$$H(t, x, S, q) = \sup_{(\tau, \delta) \in \mathcal{A}} H^{(\tau, \delta)}(t, x, S, q).$$

In the following we omit the dependence on $(t, x, S, q)$ when there is no confusion.

### The Resulting DPE

Now, the DPP implies that the value function should satisfy the quasi-variational-inequality (QVI), rather than the usual non-linear PDE,

$$
0 = \max\left\{ \partial_t H + \tfrac{1}{2}\sigma^2 \partial_{SS} H - \phi q^2 \right.
$$

$$
+ \sup_\delta \lambda\, e^{-\kappa\delta} \left[ H(t, x + (S + \delta), S, q - 1) - H(t, x, S, q) \right];
$$

$$
\left. \left[ H(t, x + (S - \xi), S, q - 1) - H(t, x, S, q) \right] \right\},
$$

with boundary and terminal conditions

$$
H(t, x, S, 0) = x, \qquad \text{and}
$$
$$
H(T, x, S, q) = x + q\,S - \ell(q).
$$

Note that the first part of the maximisation above is identical to the previous section where we have limit orders only. The various terms in the QVI may be interpreted as described below.

(i) The overall max operator represents the agent's choice to either post an LO (the continuation region) resulting in the first term in the max operator, or to execute an MO (the stopping region) resulting in a value function change of $[H(t, x + (S - \xi), S, q - 1) - H(t, x, S, q)]$ – the agent's cash increases by $S - \xi$ and inventory decreases by 1 upon executing an MO.

(ii) Within the continuation region where the agent posts LOs (the first term in the max):

  (a) the operator $\partial_{SS}$ corresponds to the generator of the Brownian motion which drives midprice,

  (b) the term $-\phi q^2$ corresponds to the contribution of the running inventory penalty,

  (c) the supremum over $\delta$ takes into account the agent's ability to control the posted depth,

  (d) the $\lambda\, e^{-\kappa\delta}$ coefficient represents the arrival rate of MOs which fill the agents posted LO at the price $S + \delta$,

  (e) the difference term $[H(t, x + (S + \delta), S, q - 1) - H(t, x, S, q)]$ represents the change in the value function when an MO fills the agent's LO – the agent's cash increases by $S + \delta$ and her inventory decreases by 1.

As before, the terminal and boundary conditions suggest the ansatz for the value function $H(t, x, S, q) = x + q\,S + h(t, q)$. Making this substitution, we find that $h(t, q)$ satisfies the much simplified QVI

$$\max \left\{ \partial_t h - \phi q^2 + \sup_\delta \lambda e^{\kappa \delta} \left[ \delta + h(t, q-1) - h(t, q) \right] \right. ;$$
$$\left. -\xi + h(t, q-1) - h(t, q) \right\} = 0, \tag{8.22a}$$

with terminal and boundary conditions

$$h(T, q) = -\ell(q), \qquad q = 1, \ldots, \mathfrak{N}, \quad \text{and} \tag{8.22b}$$
$$h(t, 0) = 0. \tag{8.22c}$$

Focusing on the supremum term, through the same computations as in the LO only case leading to (8.7), the optimal posting in feedback control form is

$$\delta^* = \frac{1}{\kappa} + [h(t, q) - h(t, q-1)] . \tag{8.23}$$

In this feedback control form, the optimal posting is identical to the one without MOs (see (8.7)), but the precise function $h(t, q)$ which enters into its computation is different. The first term $\frac{1}{\kappa}$ has the same interpretation as before: it is the optimal depth to post to maximise the expected instantaneous profit from a round-trip liquidated at midprice, i.e. the $\delta$ that maximises $\delta P(\delta)$ (and recall that $P(\delta)$ is the probability of the LO being filled conditional on an MO arriving). The difference term is the correction to this static optimisation to account for the agent's ability to optimally trade.

The timing of MO executions also have a simple feedback form. From (8.22a), we see that an MO will be executed at time $\tau_q$ whenever

$$h(\tau_q, q-1) - h(\tau_q, q) = \xi . \tag{8.24}$$

This can be interpreted as executing an MO whenever doing so increases the value function by the half-spread. Combining this observation with the feedback form for the optimal depth above, we can place a simple lower bound on $\delta^*$ of

$$\delta^* \geq \frac{1}{\kappa} - \xi .$$

Thus, it is clear that if we require $\delta > 0$, so that the strategy never posts sell LOs below the midprice, we must require that $\xi < \frac{1}{\kappa}$.

Upon substituting the optimal control in feedback form into the simplified QVI (8.22), we find that $h(t, q)$ satisfies

$$\max \left\{ \partial_t h - \phi q^2 + \frac{e^{-1}\lambda}{\kappa} e^{-\kappa [h(t,q)-h(t,q-1)]} \right. ;$$
$$\left. -\xi + h(t, q-1) - h(t, q) \right\} = 0. \tag{8.25}$$

At maturity, the agent is forced to execute an MO and pay a cost of $\xi + \alpha q$ per share. However, the agent may execute MOs an instant prior to maturity at a cost of $\xi$ per share. Hence, it is never optimal to wait until $T$ to execute an MO to liquidate remaining inventory. As a result, the left-limit of the value function is not equal to its value at maturity, and instead we have

$$h(T^-, q) = -\xi + h(T^-, q-1)$$

for every $q > 0$, so that $h(T^-, q) = -q\xi$. This feature of having the left-limit of the solution different from the terminal condition is sometimes referred to as *face-lifting*.

A further reduction of the DPE can be made by using the transformation

$$h(t, q) = \tfrac{1}{\kappa} \log w(t, q),$$

which, after some algebra, leads to the following coupled system of QVIs for $w(t, q)$:

$$\max \Big\{ \partial_t w(t, q) - \kappa \phi q^2 w(t, q) + \tilde{\lambda} w(t, q-1) \,;$$
$$e^{-\kappa \xi} w(t, q-1) - w(t, q) \Big\} = 0,$$

where $\tilde{\lambda} = e^{-1}\lambda$ and the terminal and boundary conditions are

$$w(T, q) = e^{-\kappa q(\xi + \alpha q)}, \quad \text{and} \quad w(t, 0) = 1,$$

for $q = 1, \ldots, \mathfrak{N}$.

The intuition behind the system of equations is that one first solves for $w(t, 1)$, knowing the $q = 0$ condition $w(t, 0) = 1$. The $q = 1$ solution then feeds into the $q = 2$ solution, and so on.

## Solving the DPE

We now illustrate how one can in principle first solve the QVI analytically and then provide a simple numerical implementation using an explicit finite-difference scheme for its solution.

### Constructing the Analytic Solution

**The $q = 1$ case:** Let us begin by considering $q = 1$, in which case (since $w(t, 0) = 1$) $w(t, 1)$ satisfies the equation

$$\max \Big\{ \partial_t w(t, 1) - \kappa \phi w(t, 1) + \tilde{\lambda} \,;\, e^{-\kappa \xi} - w(t, 1) \Big\} = 0, \tag{8.26a}$$

$$w(T, 1) = e^{-\kappa(\xi + \alpha)}. \tag{8.26b}$$

As pointed out above, it is optimal to execute MOs for all inventories greater than zero an instant prior to maturity, and we therefore have $w(T^-, q) = e^{-q\kappa \xi}$. Next, the solution to the ODE

$$\partial_t g_1(t) - \kappa \phi g_1(t) + \tilde{\lambda} = 0, \qquad g_1(T^-) = e^{-\kappa \xi},$$

is given by

$$g_1(t) = e^{-\kappa \xi} e^{-\kappa \phi(T-t)} + \tilde{\lambda} \frac{1 - e^{-\kappa \phi(T-t)}}{\kappa \phi}. \tag{8.27}$$

The solution to the QVI therefore has two distinct behaviours depending on the relative sizes of the parameters. First, if

$$\phi \leq \frac{\tilde{\lambda} e^{\kappa \xi}}{\kappa} \,,$$

then $g_1(t) \geq e^{-\kappa \xi}$ for all $t \in (0, T)$ (i.e. the continuation value is always larger than the execution value), hence, the solution to the QVI (8.26) is

$$w(t, 1) = g_1(t) \, \mathbb{1}_{t < T} + e^{-\kappa (\xi + \alpha)} \, \mathbb{1}_{t=T} \,,$$

and it is never optimal to execute an MO except for an instant prior to maturity. Moreover, during the trade horizon, the optimal LO depth (see (8.23)) is

$$\delta^*(t, 1) = \tfrac{1}{\kappa} + \tfrac{1}{\kappa} \log \left( e^{-\kappa \xi} e^{-\kappa \phi (T-t)} + \tfrac{\tilde{\lambda}}{\kappa \phi} \left( 1 - e^{-\kappa \phi (T-t)} \right) \right) \,,$$

for $t < T$. In this parameter regime, the depths narrow as maturity approaches.

If on the other hand

$$\phi > \frac{\tilde{\lambda} e^{\kappa \xi}}{\kappa} \,,$$

then $g(t) < e^{-\kappa \xi}$ for all $t \in (0, T)$ (i.e. the continuation value is always less than the execution value), hence, the solution to the QVI (8.26) is

$$w(t, 1) = e^{-\kappa \xi} \, \mathbb{1}_{t < T} + e^{-\kappa (\xi + \alpha)} \, \mathbb{1}_{t=T} \,,$$

and it is always optimal to execute an MO at all points in time, i.e. $\tau_1 = 0$.

The financial interpretation of this result is that if the running penalty is small enough, the agent is willing to post LOs and wait all the way until maturity before executing an MO. If on the other hand, the running penalty is large enough, it is always optimal to immediately execute an MO because the agent's urgency outweighs any potential gain in waiting for the possibility of filling her posted LO.

**The $q = 2$ case:** This case has more structure in its solution and we must solve the equation

$$\max \left\{ \partial_t w(t, 2) - 4 \kappa \phi w(t, 2) + \tilde{\lambda} w(t, 1) \; ; \; e^{-\kappa \xi} w(t, 1) - w(t, 2) \right\} = 0 \,,$$

$$w(T, 2) = e^{-2\kappa (\xi + 2\alpha)} \,.$$

As mentioned in the previous section, although there is an explicit terminal condition on $w(t, 2)$, the optimal strategy will force the agent to execute one single MO an instant prior to maturity, so that the terminal condition is face-lifted to $w(T^-, 2) = e^{-\kappa \xi} w(T^-, 1) = e^{-2\kappa \xi}$.

Here we only consider the case where $\phi < (\tilde{\lambda} e^{\kappa \xi})/\kappa$ so that it is optimal to post LOs when $q = 1$. In this case, we must determine the time $\tau_2$ at which the solution to the QVI "peels away" from its immediate execution value of $w(t, 2) = e^{-\kappa \xi} w(t, 1)$. This point is determined by ensuring that $w(t, 2)$ and its derivative are continuous at that time, i.e. such that

$$w(\tau_2^-, 2) = w(\tau_2, 2), \quad \text{and} \quad \partial_t w(\tau_2^-, 2) = \partial_t w(\tau_2, 2) = e^{-\kappa \xi} \partial_t w(\tau_2, 1) \,.$$

From the equation that $w(t, 2)$ must satisfy in the continuation region, we have

$$
\begin{aligned}
0 &= \partial_t w(\tau_2, 2) - 4\,\kappa\,\phi\,w(\tau_2, 2) + \tilde{\lambda}\,w(\tau_2, 1) \\
&= e^{-\kappa\xi}\,\partial_t w(\tau_2, 1) - 4\,\kappa\,\phi\,e^{-\kappa\xi} w(\tau_2, 1) + \tilde{\lambda}\,w(\tau_2, 1) \\
&= e^{-\kappa\xi}\left(\kappa\,\phi\,w(\tau_2, 1) - \tilde{\lambda}\right) - 4\,\kappa\,\phi\,e^{-\kappa\xi}\,w(\tau_2, 1) + \tilde{\lambda}\,w(\tau_2, 1) \qquad \text{(from (8.26))} \\
&= \left[-3\,\kappa\,\phi\,e^{-\kappa\xi} + \tilde{\lambda}\right] w(\tau_2, 1) - e^{-\kappa\xi}\,\tilde{\lambda},
\end{aligned}
$$

where the third equality follows from (8.26), i.e. in the continuation region for $q = 1$ (which is $t \in (0, T)$) we have $\partial_t w(t, 1) = \phi\,w(t, 1) - \tilde{\lambda}$. Hence, the optimal time at which to execute an MO when the agent has two units of inventory, solves

$$
w(\tau_2, 1) = e^{-\kappa\xi}\,\tilde{\lambda}\left[\tilde{\lambda} - 3\,\kappa\,\phi\,e^{-\kappa\xi}\right]^{-1}. \tag{8.28}
$$

The above can be solved explicitly for $\tau_2$ since $w(t, 1) = g_1(t)$ where $g_1(t)$ is provided in (8.27). Alternatively, a numerical zero finder can be used.

Once again, there are two parameter regimes which have differing behaviour. First, if

$$
\phi < \frac{\tilde{\lambda}\,e^{\kappa\xi}}{3\,\kappa},
$$

then a solution to (8.28) exists. Therefore, when the agent holds two units of inventory, she posts LOs in the interval $[0, \tau_2)$ and at $\tau_2$ immediately executes an MO. At that point in time, she will post an LO up until her order is executed, or she arrives an instant prior to maturity. Second, if on the other hand,

$$
\phi \geq \frac{\tilde{\lambda}\,e^{\kappa\xi}}{3\,\kappa},
$$

no solution exists, and the agent never posts LOs and instead immediately executes an MO if holding two units of inventory.

The financial intuition for these two cases is as before: if the running penalty is too high, there is no incentive to post LOs and hope for them to be matched, rather the agent aims to liquidate her position quickly. If, on the other hand, the penalty is low enough, she will be patient and post LOs up until the critical time $\tau_2$ is reached when she executes an MO.

In the remainder of the section, we assume that $\phi < \left(\tilde{\lambda}e^{\kappa\xi}\right)/3\kappa$. Given the optimal time to execute an MO, we can write the full solution for $w(t, 2)$ by solving the continuation equation from $\tau_2$ backwards. For this, the solution to the ODE

$$
\partial_t g_2(t) - \kappa\,\phi\,g_2(t) + \tilde{\lambda}\,g_1(t) = 0, \qquad g_2(\tau_2) = \Upsilon,
$$

where $\Upsilon = e^{-\kappa\xi}w(\tau_2, 1) = e^{-2\kappa\xi}\,\tilde{\lambda}\left[\tilde{\lambda} - 3\kappa\,\phi e^{-\kappa\xi}\right]^{-1}$, is

$$g_2(t) = \Upsilon\,e^{-4\phi(\tau_2-t)} + \tilde{\lambda}\int_t^{\tau_2} e^{-4\phi(u-t)}\,g_1(u)\,du$$

$$= \Upsilon\,e^{-4\kappa\,\phi(\tau_2-t)} + \tilde{\lambda}\left\{e^{-\kappa\xi}\frac{1 - e^{-5\kappa\,\phi(\tau_2-t)}}{5\kappa\,\phi}\right.$$

$$\left. + \frac{\tilde{\lambda}}{\phi}\left(\frac{1 - e^{-4\kappa\,\phi(\tau_2-t)}}{4\kappa\,\phi} - \frac{1 - e^{-5\kappa\,\phi(\tau_2-\tau_2)}}{5\kappa\,\phi}\right)\right\}.$$

Hence, the solution to the QVI for $w(t, 2)$ is

$$w(t, 2) = g_2(t)\,\mathbb{1}_{t<\tau_2} + g_1(t)\,\mathbb{1}_{t\geq\tau_2}.$$

Finally, the optimal depths at which the agent posts LOs is given by

$$\delta^*(t, 2) = \frac{1}{\kappa} + \frac{1}{\kappa}\log\frac{w(t, 2)}{w(t, 1)}.$$

The procedure outlined here can be applied recursively to obtain the optimal times at which to execute MOs for all inventory levels, as well as the optimal depths at which to post. The formulae, as one can appreciate, become rather cumbersome fairly quickly. Hence, in the next section we take a numerical approach and solve the QVIs using finite-difference methods to show different aspects of the optimal strategy.

## Numerical Experiments

Here we carry out a simple numerical implementation to compute the optimal execution strategy, including the times at which to post an MO and the depth of posted LOs. Throughout we use the following parameters:

$$T = 1\,\text{min}, \quad \mathfrak{N} = 10, \quad \lambda = 50/\text{min}, \quad \kappa = 100,$$
$$S_0 = \$30.00, \quad \sigma = \$0.01, \quad \xi = 0.005, \quad \text{and} \quad \alpha = 0.001,$$

so that the agent is trading 20% of the market over this time interval. The running penalty parameter $\phi$ will be varied to illustrate its effect on the optimal execution strategy.

Figure 8.6 shows the optimal depths as well as the optimal time at which the agent should execute an MO. The TWAP schedule is also shown for comparison purposes. As before, the optimal depth at which to post an LO is decreasing in inventory and time, i.e. the agent becomes more aggressive, and posts closer to the midprice, when there is more inventory to unwind or maturity approaches.

The right panels show when it is optimal to execute an MO and can be understood as follows. If the agent holds inventory at a point in time that lies to the right of a dot, then she must immediately execute an MO. Prior to this execution time, she will post LOs at the corresponding depth shown in the left panels. For example, when $\phi = 10^{-4}$, at the start $t = 0$, the agent will immediately execute

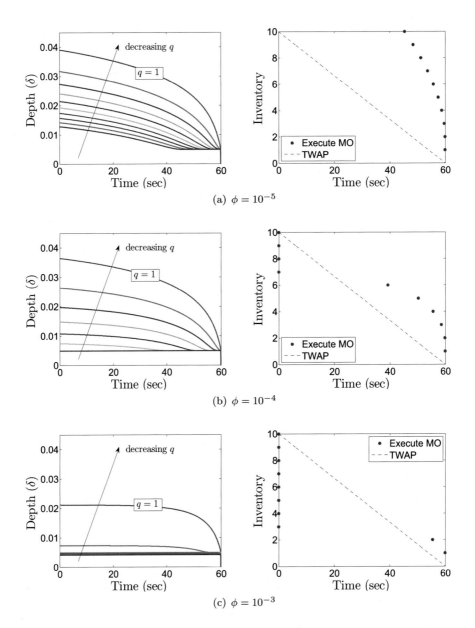

**Figure 8.6** The optimal execution strategy showing: (left panels) the optimal depths $\delta^*$ at which an agent posts LOs as a function of time and current inventory; (right panels) the times at which to execute an MO if LO has not been filled.

a sequence of four MOs reducing her inventory from 10 to 6. She will then post an LO at a depth of about 0.007 and slowly decrease it towards 0.005 until either an MO arrives and lifts her order, or if she is not matched by about $t \sim 40 \, \text{sec}$ she executes an MO, dropping her inventory to 5. Her posts then jump up to

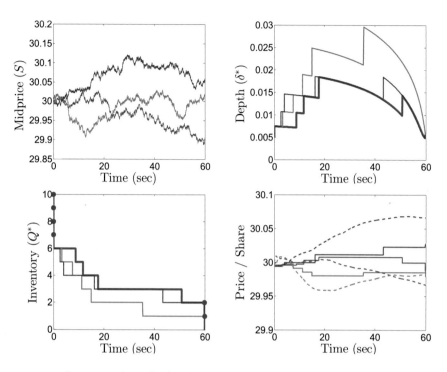

**Figure 8.7** Some sample paths for the agent following the optimal strategy. In the inventory path (bottom left panel) the blue dots indicate that the agent executed an MO. The dashed lines indicate the TWAP for that path. Here, $\phi = 10^{-4}$.

about 0.007 and she will keep decreasing it until her order is lifted, or $t \sim 50\,\mathrm{sec}$ at which point she would execute another MO, and so on.

To provide additional insight into the dynamical behaviour of the optimal strategy, we next perform a simulation of the trading strategy using a running penalty of $\phi = 10^{-4}$. Figure 8.7 shows the midprice, depth, inventory and cost per share for three simulated paths. In the inventory path (bottom left panel) the blue dots indicate when the agent executed an MO. In all scenarios, the agent immediately executes four MOs, and every time an LO is lifted or an MO is executed, the optimal depth instantly jumps upwards and then decays with time until the next LO arrives, or the agent executes an MO.

Finally, in the left panel of Figure 8.8 we show a heat-map of the agent's inventory through time as well as the mean inventory at each point in time. The agent immediately executes four MOs in all scenarios (hence the drop to $Q_0 = 6$), then on average she slowly liquidates the remaining inventory by varying the depth at which she posts her LOs. In most scenarios she ends with one or zero shares just prior to the end of the trading horizon. If she has any inventory an instant prior to maturity, she executes MOs in sequence to unwind all shares before reaching the terminal date and picking up the terminal penalty as a result of walking the LOB. It is also instructive to compare the performance of the

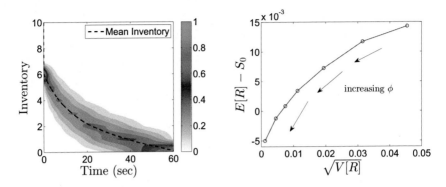

**Figure 8.8** The heat-map of the optimal inventory through time and the risk-reward profile as $\phi$ varies over $\{5 \times 10^{-6}, 5 \times 10^{-5}, 1 \times 10^{-4}, 2.5 \times 10^{-4}, 5 \times 10^{-4}, 10^{-3}, 10^{-2}\}$ from right to left.

algorithm as the value of the running penalty $\phi$ varies. This is shown in the right panel of Figure 8.8 which contains a risk-reward plot of the profit and loss (P&L) relative to the arrival price: $R = X_T - Q_T(S_T - \xi - \alpha Q_T)$. Increasing $\phi$ has two effects: (i) it decreases the standard deviation of the revenue; and (ii) it decreases the P&L. The limiting P&L is $-0.005$ which equals the half-spread $\xi$ used in these experiments, and results from large penalties inducing the agent to liquidate her shares immediately by executing MOs that pick up the half-spread cost.

## 8.5     Liquidation with Limit and Market Orders Targeting Schedules

In the previous sections we investigated the optimal strategies followed by an agent who wishes to liquidate $\mathfrak{N}$ shares and who penalises inventory that differs from zero by including the running inventory penalty term $\phi \int_t^T (Q_u)^2 \, du$ in her performance criteria. As pointed out before, this penalty term can be interpreted as representing the agent's urgency in ridding herself of inventory or her aversion to holding too much inventory at any one point in time. Or put another way, the agent's execution strategy is targeting a schedule where at any point in time the strategy should be tracking an inventory schedule of zero and deviations from this target are penalised. How heavy or light the penalty will be depends on the parameter $\phi \geq 0$.

An agent may, however, have a particular target schedule in mind that her strategy should track as part of the liquidation programme. For example, she may be interested in liquidating shares but also in tracking the inventory schedule followed by TWAP or a schedule such as those that were solved for in a continuous trading model in Chapter 6.

Here, we illustrate how the agent can achieve that goal. To this end, let $\mathfrak{q}_t$ denote the (deterministic) schedule she wishes to target. To account for her desire to target this schedule, we can easily extend the methodology of the previous section by replacing the penalty term

$$\phi \int_t^T \left(Q_u^{\tau,\delta}\right)^2 du \rightsquigarrow \phi \int_t^T \left(Q_u^{\tau,\delta} - \mathfrak{q}_u\right)^2 du.$$

Making this replacement clearly penalises strategies that deviate from the target strategy $\mathfrak{q}_t$. Her optimal behaviour will then be modified to track this schedule, and the parameter $\phi$ determines how closely she matches the target schedule. It is trivial to see that if we choose to target $\mathfrak{q}_t = 0$ for all $t$, we obtain the running inventory penalty discussed above.

We leave it as an exercise for the reader to show that making this replacement in the agent's performance criteria, but keeping the ansatz as $H(t,x,S,q) = x + q\,S + h(t,q)$, leads to the usual optimal strategy

$$\delta^* = \frac{1}{\kappa} + [h(t,q) - h(t,q-1)] ,$$

and the optimal timing $\tau_q$ of MOs solves

$$h(\tau_q, q-1) - h(\tau_q, q) = \xi,$$

where $h$ satisfies the following modification of the QVI in (8.25):

$$\max\left\{ \partial_t h - \phi\,(q - \mathfrak{q}_t)^2 + \frac{e^{-1}\lambda}{\kappa} e^{-\kappa\,[h(t,q)-h(t,q-1)]} ; \right.$$

$$\left. -\xi + h(t,q-1) - h(t,q) \right\} = 0,$$

subject to the terminal and boundary conditions

$$h(T,q) = -\ell(q), \quad \text{and} \quad h(t,0) = -\phi \int_t^T \mathfrak{q}_u^2 \, du.$$

The QVI can be linearised as before by making the transformation

$$h(t,q) = \frac{1}{\kappa} \log w(t,q),$$

to reveal that $w(t,q)$ satisfies

$$\max\left\{ \left(\partial_t - \kappa\,\phi\,(q - \mathfrak{q}_t)^2\right) w(t,q) + e^{-1}\,\lambda w(t,q-1) ; \right.$$

$$\left. e^{-\kappa\xi}\,w(t,q-1) - w(t,q) \right\} = 0, \tag{8.29}$$

subject to the terminal and boundary conditions

$$w(T,q) = e^{-\kappa\,\ell(q)}, \quad \text{and} \quad w(t,0) = e^{-\kappa\,\phi \int_t^T \mathfrak{q}_u^2 \, du}.$$

## Numerical Experiments

The QVI in (8.29) can be solved analytically as outlined in the previous section. Here, however, we solve it numerically and investigate the resulting strategy. For this we use the model parameters

$$T = 60\,\text{sec}, \quad \mathfrak{N} = 10, \quad \lambda = 50/\text{min}, \quad \kappa = 100,$$
$$S_0 = \$30.00, \quad \sigma = \$0.01, \quad \xi = 0.005, \quad \alpha = 0.001, \quad \phi = 10^{-3}.$$

The target schedule $q_t$ is the continuous Almgren-Chriss (AC) trading schedule with temporary and permanent impact studied in Section 6.5, see (6.30), which we repeat here for convenience in the form of a target schedule:

$$q_t = \frac{\zeta\, e^{\gamma(T-t)} - e^{-\gamma(T-t)}}{\zeta\, e^{\gamma T} - e^{-\gamma T}}\,\mathfrak{N},$$

where

$$\gamma = \sqrt{\frac{\phi}{k}} \quad \text{and} \quad \zeta = \frac{\alpha - \frac{1}{2}b + \sqrt{k\,\phi}}{\alpha - \frac{1}{2}b - \sqrt{k\,\phi}},$$

and use the following parameters:

$$T = 60\,\text{sec}, \quad \mathfrak{N} = 10, \quad k = 0.001, \quad \phi = 10^{-5}, \quad b = 0, \quad \alpha = +\infty.$$

Figure 8.9 shows (top left) the optimal depth at which the agent posts at each point in time and inventory level, (top right) the optimal time at which to execute an MO at each inventory level, (bottom left) the heat-map from 10,000 simulations of inventory she holds through time, and (bottom right) a histogram of the number of MOs she executes during the strategy. There are several typical features seen here. First, as time evolves the LOs are posted closer to the midprice – as the agent runs out of time, she becomes more aggressive in her posts to match the given target. As before, the less inventory she holds, the deeper she posts to reap additional revenue in exchange for being relatively ahead of schedule.

Second, the times at which she executes an MO occur when her LO posts are at the lower bound $\delta^* = \frac{1}{\kappa} - \xi$.

Third, the optimal time to post MOs (the blue dots) follows the target schedule fairly closely when the schedule changes rapidly, but allows for some slack when the schedule is not changing rapidly. This slack can be removed by increasing the penalty $\phi$.

In these simulations, the agent posts on average $\sim 4.36$ MOs during the execution, which is considerably smaller than the $\mathfrak{N} = 10$ inventory she wishes to liquidate. Most of these posts occur within the first 10 sec of trading, during the time when the target is changing rapidly. In the heat-map there are jumps downward at every stopping time corresponding to an MO execution. Finally, most paths lead to holding one unit up to an instant prior to maturity.

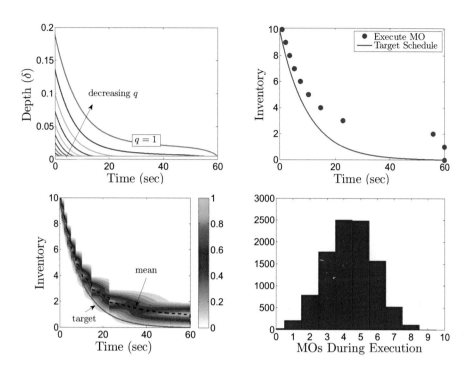

**Figure 8.9** The optimal depth, time at which to execute MOs, heat-map of inventory, and histogram of executed MOs for the agent who targets an AC schedule. The solid blue lines represent the target inventory. The dashed line represents the mean inventory of the strategy.

## 8.6 Bibliography and Selected Readings

Huitema (2013), Guilbaud & Pham (2013), Cartea, Jaimungal & Kinzebulatov (2013), Guéant, Pu & Royer (2013), Guéant, Lehalle & Fernandez Tapia (2012), (Buti & Rindi 2013), Cartea & Jaimungal (2013), Horst & Naujokat (2014).

## 8.7 Exercises

E.8.1 Use the setup provided in Section 8.2. Assume that the agent penalises running inventory so that her value function (8.4) becomes

$$H(t, x, S, q) = \sup_{\delta \in \mathcal{A}} \mathbb{E}_{t,x,S,q} \left[ X_\tau^\delta + Q_\tau^\delta \left( S_\tau - \alpha Q_\tau^\delta \right) - \phi \int_t^\tau \left( Q_u^\delta \right)^2 du \right],$$

where $\phi \geq 0$. Find the optimal depth at which the agent posts the limit sell orders.

E.8.2 Show that (8.10) is indeed the solution to (8.9) by completing the steps below.

(a) Compute $w(t, q)$ for $q = 1, 2, 3$ by explicit integration of (8.9).

(b) Notice that the solutions are all polynomials in $(T - t)$ which increase in order as $q$ increases. Hence, write the ansatz

$$w(t, q) = \sum_{n=0}^{q} a_n^{(q)} (T - t)^n$$

and show that the coefficients $a_n^{(q)}$ satisfy the recursion

$$a_n^{(q)} = \frac{\tilde{\lambda}}{n} a_{n-1}^{(q-1)}, \qquad (8.30)$$

for $n = 1, \ldots, q$, $q = 1, 2, \ldots$ and $a_0^{(q)} = e^{-\kappa \alpha q^2}$.

(c) Prove via induction that the above form of the solution is indeed correct.

(d) Solve the recursion and show that

$$a_n^{(q)} = \frac{\tilde{\lambda}^n}{n!} e^{-\kappa \alpha (q-n)^2}, \qquad (8.31)$$

for $n = 0, \ldots, q$ and $q = 1, 2, \ldots$.

E.8.3 In the optimisation problem (8.3), the terminal penalty is assumed to be $-\alpha q^2$. Suppose instead that terminal penalty is a generic bounded and increasing function of the terminal inventory $\ell(q)$, so that the agent's optimisation problem is

$$H(x, S) = \sup_{\delta \in \mathcal{A}} \mathbb{E}_{0, x, S, \mathfrak{N}} \left[ X_\tau + Q_\tau S_\tau - \ell(Q_\tau) \right]. \qquad (8.32)$$

(a) Derive the corresponding DPE for the associated value function, and solve for the value function and the optimal trading strategy using the same methods as outlined in Section 8.2.

(b) Many markets provide rebates to liquidity providers. This means that each time that an agent posts an LO and it is filled before being cancelled, the agent receives a rebate $\beta$. Account for such rebates in the formulation of the agent's optimisation problem and determine the modified optimal posting strategy.

E.8.4 Suppose that the agent wishes to penalise inventories different from zero not just at the terminal time, but also throughout the entire duration of trading. In this case, the agent adds a running penalty term to the optimisation problem and wishes to optimise

$$G(x, S) = \sup_{\delta \in \mathcal{A}} \mathbb{E}_{0, x, S, \mathfrak{N}} \left[ X_\tau + Q_\tau (S_\tau - \alpha Q_\tau) - \phi \sigma^2 \int_0^\tau Q_s^2 \, ds \right] \qquad (8.33)$$

instead of (8.3), with $\phi \geq 0$. When $\phi = 0$, the agent solves the old optimisation problem, but when $\phi > 0$, the agent modifies her behaviour to reflect her risk preference towards holding inventory.

(a) Show that the corresponding value function can be written as $G(t, x, S, q) = x + q\,S + g(t, q)$, where $g(t, q)$ satisfies the coupled non-linear system of ODEs

$$\begin{cases} \partial_t g + \tilde{\lambda} \exp\left\{-\kappa\left[g(t, q) - g(t, q - 1)\right]\right\} &= \sigma\phi\,q^2\,, \\ g(t, 0) &= 0\,, \\ g(T, q) &= -\alpha\,q^2\,, \end{cases} \tag{8.34}$$

and that the optimal depth $\delta^*$ is still provided in feedback form as

$$\delta^*(t, q) = \frac{1}{\kappa} + [g(t, q) - g(t, q - 1)]\,. \tag{8.35}$$

(b) By writing $g(t, q) = \frac{1}{\kappa} w(t, q)$, solve for $w(t, q)$ and the optimal control $\delta$.

(c) Demonstrate that if $\phi > 0$, then $\lim_{T \to +\infty} \delta^*(t, q)$ is finite for each $q$ and independent of current time $t$.

# 9 Targeting Volume

## 9.1 Introduction

Execution algorithms are designed to minimise the market impact of large orders. As discussed in the previous chapters, slicing and dicing parent orders into child orders is the main principle upon which most algorithms are devised. One source of uncertainty which determines the market impact of each child order is the volume of the child order relative to the volume that the market can bear at that point in time. To see why, consider executing one child order. If it is small, then the order will not walk beyond the best quotes in the limit order book (LOB) and it will have little or no temporary market impact. If the order size is considerable, then it may walk through several layers of the LOB and, therefore, receive poor execution prices relative to the midprice. Furthermore, to complete this description of order size and volume we must also ask whether any other orders are reaching the market at the same time or just prior to the arrival of the child order.

Over short-time scales (seconds), the impact of a market order (MO) depends on many factors where size, relative to what is displayed in the LOB, is key. But what traders see on the LOB might change by the time their orders reach the market. Even traders with access to ultra-fast technology are exposed to the risk of changes in the quantity and prices displayed by limit orders (LOs) because there is a delay between sending an MO and its execution. These changes are due to modifications in the provision of liquidity and the activity of liquidity takers. LOs may be cancelled or more may be added, thus the best quotes and/or depth of the LOB change. Similarly, other MOs may arrive just before the agent's and deplete liquidity that was sitting in the LOB. Thus, the size of the agent's child order is relative to what the LOB can bear when all MOs amalgamate with that of the agent's on the liquidity taking side of the market.

Over long-time scales (minutes/hours), the accumulated orders sent by the agent can exert unusual one-sided pressure which may result in further adverse market impact. Ideally, an agent's strategy may avoid adverse over-tilting of the market order flow by devising algorithms that camouflage her orders. One way to do this is to choose a rate of trading which targets a predetermined fraction of the total volume traded over the time horizon of the strategy.

Here are two strategies that aim at executing a number of shares equivalent to a fraction of:

i. the rate at which other participants are sending MOs; and
ii. the total volume that has been traded over the entire time horizon.

The rate and the total volume quantities are connected because total volume is the sum over the rate of trading, but the optimal execution strategy could be quite different in both cases. One simple approach to targeting (i) is to observe the volume traded over the last several seconds or minutes, and then trade a percentage of this volume over the next several seconds or minutes. Obviously this approach is not optimal because it does not address the problem of market impact when the agent's orders amalgamate with other orders.

Targeting (ii) is difficult because total volume traded, over the planned execution horizon, is not known ahead of time. Naturally, trading a percentage of the volume that has been traded over the last several seconds will also target (ii), although it may not be optimal.

Moreover, neither (i) nor (ii) is entirely compatible with the objective of completing the acquisition or liquidation of an order in full by the end of the trading horizon, because there is no guarantee that the sum of the fractions of volume traded will add up to the number of shares that the agent set out to acquire or liquidate.

Trading algorithms that target benchmarks based on volume are extensively used. One of the most popular benchmarks is the Volume Weighted Average Price, known as **VWAP**. This benchmark consists, as it name clearly suggests, in calculating

$$\text{VWAP}(T_1, T_2) = \frac{\int_{T_1}^{T_2} S_t \, dV_t}{\int_{T_1}^{T_2} dV_t}, \tag{9.1}$$

where $V_t$ is the total volume executed up to time $t$, $S_t$ is the midprice, and $[T_1, T_2]$ is the interval over which VWAP is measured.

Targeting VWAP is challenging for it is difficult to know ahead of time how many shares will be traded over a period of time. Investors target VWAP because of their desire to ensure that when acquiring or liquidating a large position they obtain an average price close to what the market has traded over the same period of time. One way to target VWAP is to follow strategy (i) because targeting a fraction of the rate of trading at every instant in time ensures that the investor is tracking the average price. Ideally, if the investor's strategy smoothes the execution of the number of shares she wishes to execute over the planned time horizon and at the same time adamantly targets a fixed proportion of the rate at which other market participants are trading, then the average cost of the shares she executes will be close to VWAP.

In this chapter we show how to formulate and solve the agent's liquidation problem for (i) and (ii) in a way that is consistent with the overall goal of full

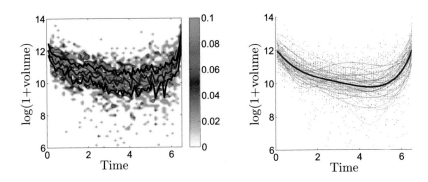

**Figure 9.1** Trading volume, for both buy and sell orders, for INTC for Oct-Nov, 2013 using 5 minute windows.

(or partial) liquidation – the acquisition problem is very similar. Strategies that target (i) are often called percentage of volume (**POV**) and we label strategies that target (ii) as percentage of cumulative volume (**POCV**).

An important source of risk when targeting POV and POCV is that one cannot anticipate the timing and volume of the arrival of other trader's MOs. This uncertainty introduces another dimension of risk into the execution problem. In Figure 9.1 we show the volume of trades (for both buy and sell orders) of INTC (Intel Corporation) using 5 minute windows for every trading day (which consists of 6.5 hours) of the fourth quarter of 2013. The panel on the left shows a heat-map of the data together with the median (second quartile), and first and third quartile estimates – note that we plot $\log(1 + \text{volume})$ because there are 5 minute windows with no trades so volume is zero. The panel on the right shows a functional data analysis (FDA) approach to viewing the data whereby the volumes are regressed against Legendre polynomials (the thin lines). The mean of the regression is then plotted as the solid blue line, which represents the expected (or average) trading volume throughout the day for this ticker. The data are also shown using the dots. From the two pictures one observes that although volume exhibits a 'U'-shape pattern, high volumes at the start and end of the trading day and lower volume in the hours in between, there are days where realised traded volumes deviate from this intraday pattern.

Figure 9.2 uses the same data as Figure 9.1 but instead of the volume it shows the intensity of trades for both buy and sell orders. For each 5 minute window we calculate the intensity $\lambda$ as the number of trades that were made over that time window. The figure shows $\log(1 + \lambda)$ because there are 5 minute intervals where no MOs were sent. The panel on the left shows a heat-map of the data together with the median, and first and third quartile estimates. The panel on the right shows an FDA approach to viewing the data whereby the intensities are regressed against Legendre polynomials (the thin lines). The mean of the regression is then plotted as the solid blue line, which represents the expected (or average) trade intensity through the day. The data are also shown using the

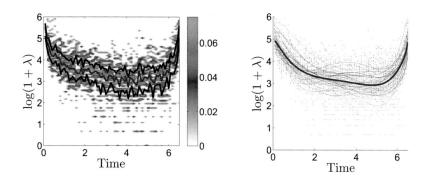

**Figure 9.2** Trading intensity, for both buy and sell orders, for INTC for Oct-Nov, 2013 using 5 minute windows.

dots. As expected, the trading intensity follows a similar pattern to that of the volume shown in Figure 9.1.

We structure this chapter as follows. In Section 9.2 we show how an agent optimally liquidates shares when her strategy targets a fraction of the speed of the rest of the market. We use simulations to show that targeting a fraction of POV can deliver an average execution price which is very close to VWAP. Section 9.3 shows how the agent liquidates shares and her strategy targets POCV. In Section 9.4 the agent modifies her strategy because her own and other market participants' rates of trading have a permanent effect on the midprice. Finally, Section 9.5 shows how to manage price risk through exponential utility when the agent targets POV.

## 9.2 Targeting Percentage of Market's Speed of Trading

In this section we assume that the agent's execution strategy **targets** a percentage of the speed at which other market participants are trading, and we focus on the liquidation strategy with MOs only. The setup for optimal acquisition, as opposed to liquidation, is very similar. In the liquidation problem, the agent searches for an optimal liquidation speed, which we denote by $\nu_t$, to target a fraction $\rho$ of the speed at which the overall market (excluding the agent) is trading. This is different from a strategy which caps the optimal liquidation speed to be at most a fraction of other market participant's speed of trading – this will become clear when we write down the agent's performance criteria. The agent's inventory $Q^\nu$ satisfies the SDE

$$dQ_t^\nu = -\nu_t \, dt\,, \qquad Q_0^\nu = \mathfrak{N}\,.$$

Let $\mu_t$ denote the speed at which all other market participants are selling shares using MOs. This rate of selling can be estimated by summing all shares that are executed over a small time window, and dividing by the time window.

We assume that the agent's speed of liquidation is not taken into account when calculating $\mu_t$. The case when the agent targets a percentage of the total order flow (including her own trades) is left as Exercise E.9.3. Therefore, since the agent's objective is to seek an optimal liquidation speed $\nu_t$ which targets the POV $\rho\mu_t$ at every instant in time, with $0 < \rho < 1$, her performance criteria and value function are

$$H^\nu(t, x, S, \mu, q) = \mathbb{E}_{t,x,S,\mu,q}\left[X_T^\nu + Q_T^\nu(S_T^\nu - \alpha Q_T^\nu) - \varphi\int_t^T (\nu_u - \rho\mu_u)^2 du\right] \quad (9.2)$$

and

$$H(t, x, S, \mu, q) = \sup_{\nu\in\mathcal{A}} H^\nu(t, x, S, \mu, q), \quad (9.3)$$

respectively.

Here $X_T^\nu$ is terminal cash, $\alpha \geq 0$ is a liquidation penalty, and $\varphi \geq 0$ is the target penalty parameter. In this setup, deviations from the target are penalised by $\varphi\int_t^T(\nu_u - \rho\mu_u)^2 du$, but this penalisation does not affect the cash process. High values of $\varphi$ constrain the strategy to closely track the target $\rho\mu_t$ at every instant in time, and low values of $\varphi$ result in liquidation strategies which are more lax about tracking the POV target.

The agent's speed of trading $\nu_t$ has both temporary and permanent impact on the price of the asset. We assume that the impacts are linear in $\nu_t$, so

$$dS_t^\nu = -b\nu_t\,dt + \sigma\,dW_t\,, \qquad S_0^\nu = S\,, \qquad (9.4a)$$

$$\hat{S}_t^\nu = S_t^\nu - k\nu_t\,, \qquad \hat{S}_0^\nu = \hat{S}\,, \qquad (9.4b)$$

$$dX_t^\nu = \hat{S}_t^\nu\,\nu_t\,dt\,, \qquad X_0^v = x\,, \qquad (9.4c)$$

with $b \geq 0$ and $k \geq 0$. In this setup we assume that the order flow $\mu_t$ from other agents does not affect the midprice process. In Section 9.4 we modify this assumption and have the order flow of all agents impacting the midprice.

## 9.2.1    Solving the DPE when Targeting Rate of Trading

We solve the agent's control problem (9.3) assuming that order flow of other agents $\mu_t$ is Markov and independent of all other processes (specifically it is independent of the Brownian motion $W_t$ which drives the midprice), and denote its infinitesimal generator by $\mathcal{L}^\mu$. The dynamic programming principle suggests that the value function should satisfy the DPE

$$0 = \left(\partial_t + \tfrac{1}{2}\sigma^2\partial_{SS} + \mathcal{L}^\mu\right)H$$
$$+ \sup_\nu\left\{(S - k\nu)\nu\,\partial_x H - \nu\,\partial_q H - b\nu\,\partial_S H - \varphi(\nu - \rho\mu)^2\right\}, \quad (9.5)$$

subject to the terminal condition $H(T, x, S, \mu, q) = x + q(S - \alpha q)$, and attains a supremum at

$$\nu^* = \frac{S\partial_x H - \partial_q H - b\partial_S H + 2\varphi\rho\mu}{2(k + \varphi)}. \quad (9.6)$$

To solve (9.5) we make the ansatz

$$H(t, x, S, \mu, q) = x + q\,S + h(t, \mu, q),\qquad(9.7)$$

which can be interpreted as the accumulated cash of the liquidation strategy, the marked-to-market book value of the inventory at the midprice, and the added value obtained from optimally liquidating the remaining shares ($h(t, \mu, q)$).

Upon substituting the ansatz in (9.5) we obtain the following equation satisfied by $h(t, \mu, q)$:

$$0 = (\partial_t + \mathcal{L}^\mu)\,h + \tfrac{1}{4(k+\varphi)}\,(\partial_q h + b\,q - 2\varphi\,\rho\,\mu)^2 - \varphi\,\rho^2\,\mu^2,\qquad(9.8)$$

subject to the terminal condition $h(T, \mu, q) = -\alpha q^2$. By observing that the terminal condition and the DPE (9.8) are at most quadratic in $q$, we use the ansatz

$$h(t,\,\mu,\,q) = h_0(t,\,\mu) + q\,h_1(t,\,\mu) + q^2\,h_2(t,\,\mu).\qquad(9.9)$$

With this ansatz, the optimal trading speed in feedback form reduces considerably to

$$\nu^* = \tfrac{1}{k+\varphi}\left\{\left[\varphi\,\rho\,\mu - \tfrac{1}{2}\,h_1(t,\mu)\right] - \left[\tfrac{1}{2}b + h_2(t,\mu)\right]q\right\}.$$

Moreover, substituting back into the DPE, after straightforward (but tedious) manipulations, collecting terms in $q$, then setting each to zero, we find the problem reduces to solving the coupled system of equations

$$0 = (\partial_t + \mathcal{L}^\mu)\,h_2 + \frac{\left(h_2 + \tfrac{1}{2}b\right)^2}{k + \varphi},\qquad(9.10a)$$

$$0 = (\partial_t + \mathcal{L}^\mu)\,h_1 + \frac{h_1 - 2\,\varphi\,\rho\,\mu}{k + \varphi}\left(h_2 + \tfrac{1}{2}b\right),\qquad(9.10b)$$

$$0 = (\partial_t + \mathcal{L}^\mu)\,h_0 + \frac{1}{4(k + \varphi)}\left(h_1 - 2\varphi\,\rho\,\mu\right)^2 - \varphi\,\rho^2\,\mu^2,\qquad(9.10c)$$

with terminal conditions $h_2(T, \mu) = -\alpha$ and $h_1(T, \mu) = h_0(T, \mu) = 0$. Each equation is in fact a linear PDE with non-linear sources terms given by the solution to the other PDEs. These equations are also dependent in a constructive manner, i.e. $h_2$ is independent of all others, $h_1$ only depends explicitly on $h_2$, while $h_0$ only depends on $h_1$. Therefore, they can be solved sequentially.

Now observe that equation (9.10a) for $h_2$ contains no source terms dependent on $\mu$ and its terminal condition is independent of $\mu$, hence the solution must also be independent of $\mu$, and it is given by

$$h_2(t, \mu) = -\left(\tfrac{T-t}{k+\varphi} + \tfrac{1}{\alpha - \tfrac{1}{2}b}\right)^{-1} - \tfrac{1}{2}b,\qquad(9.11)$$

and since the optimal speed of trading does not depend on $h_0$, we do not need to solve (9.10c). What remains is to solve the PDE for $h_1(t, \mu)$. At this point, we instead simply assume we have solved for it and derive expressions for the optimal trading speed and the resulting optimal inventory trajectory. Once we have these expressions, in the subsections ahead, we make some specific modelling assumptions and compute $h_1$ explicitly as well as provide a probabilistic representation for the general case.

We can then express the optimal speed of trading as

$$v_t^* = \frac{1}{k+\varphi}\left[\varphi\,\rho\,\mu_t - \tfrac{1}{2}\,h_1(t,\mu_t)\right] + \frac{Q_t^{\nu^*}}{(T-t)+\zeta}, \qquad (9.12)$$

where the constant

$$\zeta = \frac{k+\varphi}{\alpha - \tfrac{1}{2}b}.$$

The optimal speed of trading consists of two terms. The second term on the right-hand side of (9.12) is very similar to the strategy discussed in the optimal liquidation problems in Chapter 6. For instance, one can see that if in (6.27) we let the running inventory penalty parameter $\phi = 0$, we obtain the second term on the right-hand side of (9.12) with $\varphi = 0$. Therefore, one can view this term as the TWAP-like liquidation strategy (we refer to it as TWAP-like because only when $\zeta = 0$ do we obtain TWAP). The first term on the right-hand side of (9.12) provides the volume corrections to the TWAP-like strategy. This correction depends on the POV target, $\rho\,\mu_t$, and on the function $h_1(t,\mu)$ which encodes how the optimal strategy behaves given the dynamics of the volume process.

Before specifying the stochastic process for the rate of trading $\mu_t$, we comment on some general features of the POV strategy. The boundary condition $h_1(T,\mu) = 0$ implies that near the end of the trading horizon, $(T-t) \ll 1$, the optimal strategy behaves like

$$v_t^* = \frac{\varphi\,\rho}{k+\varphi}\,\mu_t + \frac{Q_t^{\nu^*}}{(T-t)+\zeta},$$

and we can further examine two interesting limiting cases.

First, when $\alpha \to \infty$, so that the agent must execute all shares by the trading end, $\zeta \to 0$. For a fixed inventory level, the second term on the right-hand side in the equation becomes dominant as $t$ approaches $T$. Hence, the agent ignores the POV constraint and instead focuses on using TWAP when approaching the end of the trading horizon. Note, however, that the inventory also flows as time evolves, so it is not clear if that term truly dominates near maturity. Below, once we solve for $Q_t^{\nu^*}$, we will see that indeed the POV constraint is ignored near maturity.

Second, if $\alpha$ remains finite, but $\varphi \to \infty$, so that the agent heavily penalises trades that deviate from POV, then $\zeta \to \infty$ and the strategy ignores the TWAP-like term, and trades at a rate of $\rho\,\mu_t$ as expected. Naturally, these limits do not commute as they are contradictory objectives – unless the volume to be traded is exactly equal to $\rho$ percentage of the total volume traded over the execution duration.

Armed with the optimal trading speed in (9.12), we can obtain an expression for the optimal inventory to hold by solving for $Q_t^{\nu^*}$. Recall that $dQ_t^{\nu^*} = -\nu_t^* \, dt$, and hence,

$$dQ_t^{\nu^*} = -\tilde{h}_1(t, \mu_t) \, dt - \frac{Q_t^{\nu^*}}{(T-t)+\zeta} \, dt \, ,$$

where

$$\tilde{h}_1(t, \mu) := \frac{1}{k+\varphi} \left[ \varphi \rho \mu - \tfrac{1}{2} h_1(t, \mu) \right] .$$

The ODE above can be solved by introducing an integrating factor, by writing

$$Q_t^{\nu^*} = e^{-\int_0^t ((T-s)+\zeta)^{-1} \, ds} \, q_t^{\nu^*}$$

$$= e^{\log \frac{(T-t)+\zeta}{T+\zeta}} \, q_t^{\nu^*} = \frac{(T-t)+\zeta}{T+\zeta} \, q_t^{\nu^*} \, ,$$

and solving for the unknown process $q_t^{\nu^*}$. By direct computation we have that

$$dq_t^{\nu^*} = e^{\int_0^t ((T-s)+\zeta)^{-1} \, ds} \left\{ ((T-t)+\zeta)^{-1} Q_t^{\nu^*} \, dt + dQ_t^{\nu^*} \right\}$$

$$= -e^{\int_0^t ((T-s)+\zeta)^{-1} \, ds} \, \tilde{h}_1(t, \mu_t) \, dt$$

$$= -\frac{T+\zeta}{(T-t)+\zeta} \, \tilde{h}_1(t, \mu_t) \, dt \, ,$$

and so

$$q_t^{\nu^*} - q_0 = -\int_0^t \frac{T+\zeta}{(T-s)+\zeta} \, \tilde{h}_1(s, \mu_s) \, ds \, .$$

Therefore, we can write

$$Q_t^{\nu^*} \frac{T+\zeta}{(T-t)+\zeta} - \mathfrak{N} = -\int_0^t \frac{T+\zeta}{(T-s)+\zeta} \, \tilde{h}_1(s, \mu_s) \, ds \, ,$$

so that finally

$$\boxed{\begin{aligned} Q_t^{\nu^*} = {} & \left( 1 - \frac{t}{T+\zeta} \right) \mathfrak{N} \\ & - \frac{1}{k+\varphi} \int_0^t \frac{(T-t)+\zeta}{(T-s)+\zeta} \left\{ \varphi \rho \mu_s - \tfrac{1}{2} h_1(s, \mu_s) \right\} ds \, . \end{aligned}} \tag{9.13}$$

The optimal inventory to hold at any point in time has two components. The first term is a TWAP-like strategy. The second term controls for fluctuations in the market's trading rate. Now that we are equipped with this general framework for the optimal trading speed (9.12) and the optimal inventory to hold (9.13), we proceed by specifying the volume dynamics and then solving for explicit formulae for the optimal speed of trading and the optimal inventory path.

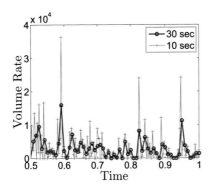

**Figure 9.3** AAPL volume rates per minute from 10.00 to 10.30 on Jan 5, 2011. Estimation using 30 sec and 10 sec time windows.

### 9.2.2 Stochastic Mean-Reverting Trading Rate

Figure 9.3 shows the estimate of the volume rate of trading (per minute) for AAPL on Jan 5, 2011 from 10:00 to 10:30. The estimate is computed by counting the volume traded over 30 sec and 10 sec windows and scaling the counts to one minute. As the figure shows, a reasonable first order model is that traded volume comes in bursts of activity which persists for a while (seconds for instance) and then decays to zero. Thus, for an agent whose objective is to target the rate of trading over short-time horizons, we assume that the sell volume rate $\mu_t$ is a mean reverting process which satisfies the SDE

$$d\mu_t = -\kappa \, \mu_t \, dt + \eta_{1+N_{t-}} \, dN_t, \tag{9.14}$$

where $\kappa \geq 0$ is the mean reversion rate, $N_t$ is a homogeneous Poisson process with intensity $\lambda$, and $\{\eta_1, \eta_2, \dots\}$ are non-negative i.i.d. random variables with distribution function $F$, with finite first moment, independent of $N_t$. The solution to (9.14) is

$$\mu_t = e^{-\kappa t} \mu_0 + \int_0^t e^{-\kappa(t-u)} \eta_{1+N_{u-}} \, dN_u$$

$$= e^{-\kappa t} \mu_0 + \sum_{m=1}^{N_t} e^{-\kappa(t-\tau_m)} \eta_m, \tag{9.15}$$

where $\tau_m$ denotes the time of the $m^{th}$ arrival of the Poisson process. As shown earlier in Figure 9.2, the arrival of trades follows a U-shaped pattern that can be incorporated in (9.14) by introducing a deterministic component in the drift of the process, but here, for simplicity, we assume that the trading rate always mean-reverts to zero. Although the order flow process mean-reverts to zero, its long-run expected value is not zero; indeed from (9.15) we have

$$\mathbb{E}[\mu_t] = e^{-\kappa t} \mu_0 + \int_0^t e^{-\kappa(t-u)} \mathbb{E}[\eta_1] \lambda \, dt$$

$$= e^{-\kappa t} \mu_0 + \frac{\lambda \mathbb{E}[\eta_1]}{\kappa} \left(1 - e^{-\kappa t}\right) \xrightarrow{t \to \infty} \frac{\lambda \mathbb{E}[\eta_1]}{\kappa}.$$

With order flow satisfying (9.14), its infinitesimal generator acts on the value function as follows:

$$\mathcal{L}^\mu H(t, S, \mu, q) = -\kappa \mu \, \partial_\mu H + \lambda \mathbb{E}\left[H(t, S, \mu + \eta, q) - H(t, S, \mu, q)\right], \qquad (9.16)$$

where the expectation is with respect to the random variable $\eta$ with distribution function $F$.

We now find the solution $h_1$ to (9.10b) with this model assumption. Since this model is of the affine type, and (9.10b) is linear in $\mu$, we expect the function $h_1(t, \mu)$ to be linear in $\mu$. Specifically, we write

$$h_1(t, \mu) = \ell_0(t) + \ell_1(t)\,\mu \,,$$

for some deterministic functions of time $\ell_0(t)$ and $\ell_1(t)$. The terminal condition $h(T, \mu) = 0$ implies that we must also impose $\ell_0(T) = \ell_1(T) = 0$. When the generator (9.16) acts on $h_1(t, \mu)$ we obtain

$$\mathcal{L}^\mu h_1(t, \mu) = -\kappa \mu \, \ell_1 + \psi \, \ell_1 \,,$$

where $\psi = \lambda \mathbb{E}\left[\eta\right]$ so that equation (9.10b) becomes

$$0 = \left\{\partial_t \ell_0 + \frac{h_2 + \frac{1}{2}b}{k + \varphi} \ell_0 + \psi \, \ell_1 \right\} + \left\{\partial_t \ell_1 - \kappa \ell_1 + \frac{\ell_1 - 2\varphi\rho}{k + \varphi}\left(h_2 + \tfrac{1}{2}b\right)\right\} \mu \,.$$

To solve for $\ell_0(t)$ and $\ell_1(t)$ in the above equation, we observe that since the equation must hold for all $\mu$, the terms in braces must vanish individually. We first solve the ODE in the second set of braces and then use the expression for $\ell_1$ to solve the ODE contained in the first set of braces.

For $\ell_1$, we solve

$$\partial_t \ell_1 - \kappa \ell_1 + \frac{\ell_1 - 2\varphi\rho}{k + \varphi}\left(h_2 + \tfrac{1}{2}b\right) = 0 \,,$$

using the integrating factor technique by multiplying it by the integrating factor

$$\exp \int \left(\frac{h_2(t) + \frac{1}{2}b}{k + \varphi} - k\right) dt = \left(\frac{T - t}{k + \varphi} + \frac{1}{\alpha - \frac{1}{2}b}\right) e^{-kt} \,,$$

and after simple algebra and integrating between $t$ and $T$, the ODE above reduces to solving

$$\int_t^T d\left(\left(\frac{T - u}{k + \varphi} + \frac{1}{\alpha - \frac{1}{2}b}\right) e^{-ku} \, \ell_1(u)\right) + \frac{2\varphi\rho}{k + \varphi} \int_t^T e^{-ku} \, du = 0 \,.$$

Finally, using the boundary condition $\ell_1(T) = 0$ we obtain

$$\ell_1(t) = 2\,\varphi\,\rho\,\left((T - t) + \zeta\right)^{-1} \frac{1 - e^{-\kappa(T-t)}}{\kappa} \,,$$

where recall that the constant $\zeta = (k + \varphi)/(\alpha - \frac{1}{2}b)$.

We proceed in a similar way to solve the ODE for $\ell_0$ resulting from setting the first set of braces to zero. First substitute the solution for $\ell_1$ into the equation for $\ell_0$, then multiply the ODE through by the integrating factor

$$\exp\int\left(\frac{h_2 + \frac{1}{2}b}{k + \varphi}\right)dt = \left(\frac{T - t}{k + \varphi} + \frac{1}{\alpha - \frac{1}{2}b}\right),$$

and integrate between $t$ and $T$, to obtain

$$\frac{1}{k + \varphi}\int_t^T d\left((T - u + \zeta)\,\ell_0(u)\right) + \frac{2\psi\varphi\rho}{k + \varphi}\int_t^T \frac{1 - e^{-k(T-u)}}{k}\,du = 0,$$

with terminal condition $\ell_0(T) = 0$. Evaluating the explicit integral, we can then write

$$\ell_0(t) = 2\,\varphi\,\rho\,\psi\left((T - t) + \zeta\right)^{-1}\frac{e^{-\kappa(T-t)} - 1 + \kappa(T - t)}{\kappa^2}.$$

Finally, substituting these results back into the optimal speed of trading in (9.12), we have

$$\boxed{\nu_t^* = \frac{1}{k+\varphi}\left[\varphi\,\rho\,\mu_t - \frac{1}{2}\left(\ell_0(t) + \ell_1(t)\,\mu_t\right)\right] + \frac{Q_t^{\nu^*}}{(T - t) + \zeta}\,.} \qquad (9.17)$$

## 9.2.3     Probabilistic Representation

In this section we show how to solve for the optimal liquidation strategy for the most general case, where we do not specify the particular process followed by the trading rate $\mu_t$ and the only assumption is that it is Markov and independent of the Brownian motion driving the midprice. We establish this result by applying the Feynman-Kac Theorem to (9.10b), which is the evolution equation for $h_1(t, \mu)$, and represent $h_1$ as

$h_1(t, \mu)$

$$= -\frac{2\,\varphi\,\rho}{k+\varphi}\,\mathbb{E}_{t,\mu}\left[\int_t^T \exp\left\{\frac{1}{k+\varphi}\int_t^u\left(h_2(s, \mu_s) + \frac{1}{2}b\right)ds\right\}\left(h_2(t, \mu_u) + \frac{1}{2}b\right)\mu_u\,du\right]$$

$$= \frac{2\,\varphi\,\rho}{k+\varphi}\,\mathbb{E}_{t,\mu}\left[\int_t^T \exp\left\{-\frac{1}{k+\varphi}\int_t^u\left(\frac{T-s}{k+\varphi} + \frac{1}{\alpha - \frac{1}{2}b}\right)^{-1}ds\right\}\left[\frac{T-u}{k+\varphi} + \frac{1}{\alpha - \frac{1}{2}b}\right]^{-1}\mu_u\,du\right]$$

$$= \frac{2\,\varphi\,\rho}{k+\varphi}\,\mathbb{E}_{t,\mu}\left[\int_t^T \exp\left\{\log\left(\frac{T-s}{k+\varphi} + \frac{1}{\alpha - \frac{1}{2}b}\right)\Big|_{s=t}^u\right\}\left[\frac{T-u}{k+\varphi} + \frac{1}{\alpha - \frac{1}{2}b}\right]^{-1}\mu_u\,du\right],$$

where $\mathbb{E}_{t,\mu}[\cdot]$ is shorthand notation for the conditional expectation $\mathbb{E}[\cdot \mid \mu_t = \mu]$.

After a series of cancellations in the integrand above, we have the following compact representation for the solution to $h_1(t, \mu)$:

$$\boxed{h_1(t, \mu) = 2\,\varphi\,\rho\,\frac{\int_t^T \mathbb{E}\left[\mu_u \mid \mu_t = \mu\right]\,du}{(T - t) + \zeta}\,.} \qquad (9.18)$$

Recall that the constant $\zeta = (k + \varphi)/(\alpha - \frac{1}{2}b)$. The integral appearing above is precisely the **expected total volume** over the remainder of the trading horizon. This is because it is the integral of the expected trading rate $\mu_t$ between the current time and the end of the strategy's trading horizon. Moreover, the integral term combined with the factor $((T - t) + \zeta)^{-1}$ is approximately the average expected trading rate for the remaining time horizon. This combination would be exactly the average expected trading rate if $k = \varphi = 0$ and/or $\alpha \to \infty$.

Using this general form for $h_1$, the agent's optimal trading speed can be represented as

$$
\nu_t^* = \frac{\varphi\rho}{k + \varphi}\left[\mu_t - \frac{\int_t^T \mathbb{E}\left[\mu_u \mid \mathcal{F}_t^\mu\right] du}{(T - t) + \zeta}\right] + \frac{Q_t^{\nu^*}}{(T - t) + \zeta}, \tag{9.19}
$$

where the conditional expectation is now with respect to the filtration $\mathcal{F}_t^\mu$ generated by $\mu$. By inserting the general result for $h_1$ into the optimal inventory to hold, we also have the compact representation

$$
\begin{aligned}
Q_t^{\nu^*} = &\left(1 - \frac{t}{T + \zeta}\right)\mathfrak{N}\\
&- \frac{\varphi\rho}{k + \varphi}\int_0^t \frac{(T - t) + \zeta}{(T - s) + \zeta}\left[\mu_s - \frac{\int_s^T \mathbb{E}\left[\mu_u \mid \mathcal{F}_s^\mu\right] du}{(T - s) + \zeta}\right] ds .
\end{aligned} \tag{9.20}
$$

To understand the intuition of the strategy (9.19) we start by pointing out that the agent's performance criteria (9.2) includes competing objectives. On the one hand the strategy aims at liquidating $\mathfrak{N}$ shares by $T$, and on the other hand the strategy must track a fraction of other market participants' trading rate. Only when $\mathfrak{N}$ is exactly equal to the desired fraction $\rho$ of the total volume over the execution duration $T$ are these two objectives compatible.

Next, rewrite the optimal trading speed as

$$
\nu_t^* = \frac{\varphi}{k + \varphi}\rho\,\mu_t \tag{9.21a}
$$

$$
+ \frac{1}{(T - t) + \zeta}\left(Q_t^{\nu^*} - \frac{\varphi\rho}{k + \varphi}\int_t^T \mathbb{E}[\mu_u \mid \mathcal{F}_t^\mu] du\right). \tag{9.21b}
$$

The first component of the strategy (9.21a) accounts for the trading rate that must be achieved to meet the POV target, taking into account the trade-off between the POV target penalty $\varphi$ and the costs stemming from temporary price impact $k$. Although the POV target is $\rho\,\mu_t$, the strategy targets a lower amount since $\frac{\varphi\rho}{k+\varphi} \leq \rho$, where equality is achieved if the costs of missing the target are $\varphi \to \infty$ and $k$ remains finite, or there is no temporary impact $k \downarrow 0$.

The second component is a TWAP-like strategy (the first term in the braces in (9.21a)) with a downward adjustment (the second term in the braces) because throughout the trading horizon there is the component targeting the POV. This is why we see that the TWAP-like strategy is applied to the remaining inventory $Q_t^{\nu^*}$ minus the number of shares that are expected to be liquidated as part of the POV target, which will be done by the first term of the strategy (9.21a) over the remaining time of the strategy.

We continue our discussion of the optimal strategy by looking at some limiting cases.

The limiting case in which the agent wishes to always track a fraction $\rho$ of the rate $\mu_t$ is obtained by letting the target penalty parameter $\varphi \to \infty$. In this case, the liquidation speed and inventory path become

$$\nu_t^* \xrightarrow{\varphi \to \infty} \rho \mu_t \qquad \text{and} \qquad Q_t^{\nu^*} \xrightarrow{\varphi \to \infty} \mathfrak{N} - \rho \int_0^t \mu_s \, ds \,,$$

because $\zeta \xrightarrow{\varphi \to \infty} \infty$.

Regardless of what the inventory target is, the strategy liquidates at a rate of $\rho \mu_t$. Clearly, as shown by the inventory path, the strategy could liquidate an amount of shares which exceeds or falls short of the initial target $\mathfrak{N}$. When the strategy reaches $T$ any outstanding shares, short or long, are liquidated with an MO at the midprice and receive a finite penalty of $\alpha q_T^2$ which, in this limit, the agent prefers to picking up the more onerous running penalty which would be infinite if she did not liquidate at the rate $\rho \mu_t$.

The limiting case in which the agent wishes to fully liquidate her inventory leads to a finite trading strategy, with finite inventory paths. In particular, $\zeta \xrightarrow{\alpha \to \infty} 0$, and so we have

$$\nu_t^* \xrightarrow{\alpha \to \infty} \frac{\varphi \rho}{k + \varphi} \left[ \mu_t - \frac{\int_t^T \mathbb{E}[\mu_u \mid \mathcal{F}_t^\mu] \, du}{T - t} \right] + \frac{Q_t^{\nu^*}}{T - t} \,,$$

$$Q_t^{\nu^*} \xrightarrow{\alpha \to \infty} \left( 1 - \frac{t}{T} \right) \mathfrak{N} - \frac{\varphi \rho}{k + \varphi} \int_0^t \frac{T - t}{T - s} \left[ \mu_s - \frac{\int_s^T \mathbb{E}[\mu_u \mid \mathcal{F}_s^\mu] \, du}{T - s} \right] ds \,.$$

Suppose that, in addition to requiring full liquidation, the agent is also very averse to trading at a rate different from $\rho \mu_t$. As $\varphi$ increases, she will target more and more closely the required trading rate; however, due to the constraint that she must fully liquidate, she will not be able to match the required trading rate at all times. Therefore, in the limit in which $\varphi \to \infty$, after we have already taken $\alpha \to \infty$, the value function will become arbitrarily large and negative, and will not be finite. The limiting optimal strategy, however, does remain finite, as does her optimal inventory path, and the net value of liquidating her shares remains finite and well behaved. The optimal speed of trading and inventory position in this second double limiting case are

$$\lim_{\varphi \to \infty} \lim_{\alpha \to \infty} \nu_t^* = \rho \left[ \mu_t - \frac{\int_t^T \mathbb{E}[\mu_u \mid \mathcal{F}_t^\mu] \, du}{T - t} \right] + \frac{Q_t^{\nu^*}}{T - t} \,,$$

$$\lim_{\varphi \to \infty} \lim_{\alpha \to \infty} Q_t^{\nu^*} = \left( 1 - \frac{t}{T} \right) \mathfrak{N} - \rho \int_0^t \frac{T - t}{T - s} \left[ \mu_s - \frac{\int_s^T \mathbb{E}[\mu_u \mid \mathcal{F}_s^\mu] \, du}{T - s} \right] ds \,.$$

Interestingly, when the other agents trade at a constant rate, i.e. $\mu_t$ is a constant,

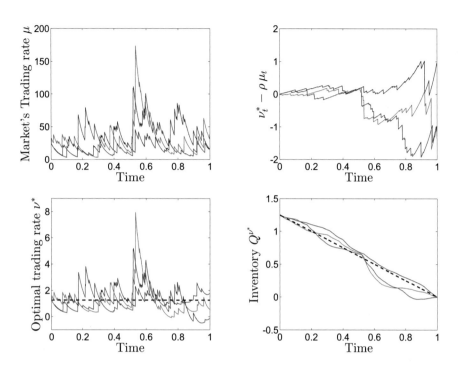

**Figure 9.4** Three sample paths of the market's trading rate, the optimal trading rate, the difference between the optimal trading rate and the targeted rate, and the agent's inventory.

the POV correction terms in the above cancel and the agent's strategy becomes TWAP.

### 9.2.4 Simulations

In this section we provide some simulations of the optimal strategy for the mean-reverting volume model in subsection 9.2.2. We focus on the double limiting case of first ensuring that all inventory is liquidated ($\alpha \to \infty$), and second that the agent wishes to trade very close to POV ($\varphi \to \infty$). In this case, we have

$$
\lim_{\varphi \to \infty} \lim_{\alpha \to \infty} \nu_t^* = \rho \left[ \mu_t - \frac{\frac{1-e^{-\kappa\,(T-t)}}{\kappa} \left( \mu_t - \frac{\psi}{\kappa} \right) + (T-t)\frac{\psi}{\kappa}}{(T-t)} \right] + \frac{Q_t^{\nu^*}}{T-t},
$$

where $\psi = \lambda\,\mathbb{E}[\eta]$ and therefore the long-run expected trading rate is $\mathbb{E}[\mu_t] = \psi/\kappa$.

For the simulations, we use the following modelling parameters:

$$
S_0 = 20\,, \quad \sigma = 0.5\,, \quad T = 1\,,
$$

$$
\mu_0 = \psi/\kappa\,, \quad \eta \sim \mathrm{Exp}(10)\,, \quad \lambda = 50\,, \quad \kappa = 20\,,
$$

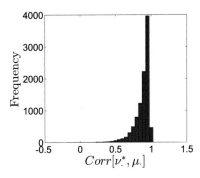
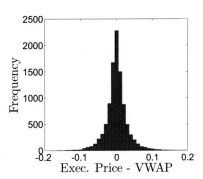

**Figure 9.5** Left: histogram of the correlation between the agent's trading rate $\nu_t^*$ and $\mu_t$. Right: histogram of the difference between the execution price per share and the VWAP.

$$k = 0.1, \quad b = 0.1, \quad \text{and} \quad \rho = 0.05 \,.$$

We assume that the agent is attempting to liquidate $\rho$ percentage of the volume she expects to arrive in the market over her trading horizon, thus $\mathfrak{N} = \rho \, T \, \psi / \kappa = 1.25$. Our parameterisation of the exponential distribution is such that $\mathbb{E}[\eta] = 10$.

In Figure 9.4 we show three sample paths of the trading rate of other market participants $\mu_t$, the optimal trading rate $\nu_t^*$, the difference between the optimal rate and the target rate $\nu_t^* - \rho \, \mu_t$, and the agent's inventory $Q_t^{\nu^*}$. In the bottom left panel, the dotted line is the expected trading rate equal to $\psi / \kappa$ and in the bottom right panel, the dotted line is TWAP. Note that $\nu_t^*$ and $\mu_t$ are strongly correlated. Indeed, in the left panel of Figure 9.5 we show the histogram of the correlation between $\nu_t^*$ and $\mu_t$ viewed as a time series, along 10,000 sample paths. The mean correlation is quite high at 0.88, illustrating the fact that the trading rate tracks the rate of order flow. There are, however, deviations from the targeted rate of $\rho \, \mu_t$. These differences appear most notably towards the end of the trading horizon (as seen in the top right panel in Figure 9.4) where the agent's main concern is to drive her inventory to zero and she is less concerned about targeting other participants' trading rate.

Moreover, in the right panel of Figure 9.5 we show the difference between the executed price per share and the VWAP. As discussed in the introduction, if the agent is closely targeting a fraction $\rho$ of the trading volume then this results in strategies which do indeed target VWAP on average. The deviation around VWAP is symmetric (skewness $= 0.06$) with mean $-1.4 \times 10^{-4}$ and standard error $\pm 3 \times 10^{-4}$.

In Figure 9.6 we illustrate a heat-map of the agent's trading rate and her inventory. The dotted lines here indicate the 5%, 50% and 95% quantiles. Interestingly, the median inventory path is TWAP, while the agent's inventory may deviate both above and below this trajectory in her attempt to match the POV target. The median path of her optimal trading rate is essentially constant through time, although the mode of this trajectory tends to increase towards

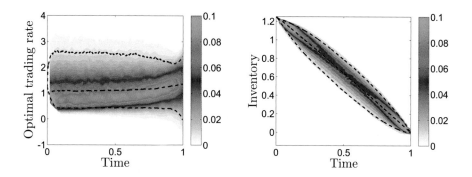

**Figure 9.6** Heat-maps of the optimal trading rate and inventory. The dotted lines show the 5%, 50% and 95% quantiles.

maturity. This suggests there is a slight bias towards first trading a little more slowly compared with TWAP and then trading faster to catch up.

## 9.3 Percentage of Cumulative Volume

In this section we assume that the agent's execution strategy targets a percentage of **cumulative** volume (POCV) and the liquidation strategy relies on MOs only. Here the accumulated volume $V$ of sell orders, excluding the agent's own trades, is given by

$$V_t = \int_0^t \mu_u \, du \,,$$

where as above $\mu_t$ denotes other market participants' rate of trading.

The agent's performance criteria is now modified to

$$H^\nu(t, x, S, \mu, V, q) = \mathbb{E}_{t,x,S,\mu,V,q}\left[X_T^\nu + Q_T^\nu \left(S_T^\nu - \alpha Q_T^\nu\right) \right.$$

$$\left. - \varphi \int_t^T \left((\mathfrak{N} - Q_u^\nu) - \rho V_u\right)^2 du\right], \tag{9.22}$$

where $\mathfrak{N}$ is the number of shares that the agent wishes to liquidate by the terminal date $T$.

The running target penalty $\varphi \int_t^T \left((\mathfrak{N} - Q_u^\nu) - \rho V_u\right)^2 du$ is not a financial cost that the agent incurs. Rather, its purpose is to allow the agent to seek for optimal liquidation rates where the total amount that has been liquidated up to time $t$ is not too far away from a percentage of what the entire market has sold. For example, when the penalty parameter $\varphi \to \infty$, the optimal strategy is forced to liquidate shares so that at any point in time the number of shares that have already been liquidated, $\mathfrak{N} - q_t$, equals $\rho V_t$. In this manner, the agent devises a strategy where the cumulative sum of her own sell MOs is a fraction $0 < \rho < 1$

of the market. As in the last section, in this setup the agent's own trades are not taken into account in the accumulated volume.

As usual, the agent's value function is

$$H(t, x, S, \mu, V, q) = \sup_{\nu \in \mathcal{A}} H^\nu(t, x, S, \mu, V, q), \tag{9.23}$$

and her cash process $X^\nu$, the controlled midprice $S^\nu$, and execution price $\hat{S}^\nu$ satisfy the equations in (9.4). Applying the dynamic programming principle suggests that the value function should satisfy the DPE

$$0 = \left(\partial_t + \mathcal{L}^{S,\mu,V}\right) H - \varphi((\mathfrak{N} - q) - \rho V)^2$$
$$+ \sup_\nu \left\{-b\nu \partial_S H + (S - k\nu) \nu \partial_x H - \nu \partial_q H\right\}, \tag{9.24}$$

subject to the terminal condition $H(T, x, S, y, q) = x + q(S - \alpha q)$. Here, $\mathcal{L}^{S,\mu,V}$ denotes the infinitesimal generator of the joint process $(S_t, \mu_t, V_t)_{0 \le t \le T}$. Moreover, we assume that the volume trading rate $\mu_t$, and therefore the total volume $V_t$, is independent of the asset's midprice.

The first order conditions provide us with the optimal trading rate in feedback form as

$$\nu^* = \tfrac{1}{2k} \left(S \partial_x H - \partial_q H - b \partial_S H\right).$$

The terminal condition suggests that to solve (9.24) we use the ansatz

$$H(t, x, S, \mu, V, q) = x + q S + h(t, \mu, V, q) \tag{9.25}$$

to obtain the following equation satisfied by $h(t, \mu, V, q)$:

$$0 = \left(\partial_t + \mathcal{L}^{\mu,V}\right) h + \frac{1}{4k} \left(\partial_q h + bq\right)^2 - \varphi \left(\mathfrak{N} - q - \rho V\right)^2, \tag{9.26}$$

subject to the terminal condition $h(T) = -\alpha q^2$, where $\mathcal{L}^{\mu,V}$ denotes the generator of the process $(\mu_t, V_t)_{0 \le t \le T}$ which excludes the midprice because the trading rate is independent of the midprice.

The above non-linear equation can be reduced even further by noticing that the equation and its terminal condition are at most quadratic in $q$, hence we use the ansatz

$$h(t, \mu, V, q) = h_0(t, \mu, V) + h_1(t, \mu, V) q + h_2(t, \mu, V) q^2,$$

subject to the boundary conditions

$$h_0(T, \mu, V) = h_1(T, \mu, V) = 0, \quad \text{and} \quad h_2(T, \mu, V) = -\alpha.$$

By substituting this ansatz into equation (9.26), collecting terms with equal powers in $q$, and setting each to zero, we find that $h_0$, $h_1$, and $h_2$ must satisfy the coupled system of PIDEs

$$0 = \left(\partial_t + \mathcal{L}^{\mu,V}\right) h_2 + \tfrac{1}{k} \left(h_2 + \tfrac{1}{2}b\right)^2 - \varphi, \tag{9.27a}$$
$$0 = \left(\partial_t + \mathcal{L}^{\mu,V}\right) h_1 + \tfrac{1}{k} \left(h_2 + \tfrac{1}{2}b\right) h_1 + 2\varphi \left(\mathfrak{N} - \rho V\right), \tag{9.27b}$$
$$0 = \left(\partial_t + \mathcal{L}^{\mu,V}\right) h_0 + \tfrac{1}{4k} \left(h_1\right)^2 - \varphi \left(\mathfrak{N} - \rho V\right)^2. \tag{9.27c}$$

As in the POV section, each equation above is a linear PIDE with non-linear source terms given by the solution to the other PIDEs. These equations are also dependent in a sequential manner, i.e. $h_2$ is independent of all others, $h_1$ only depends explicitly on $h_2$, while $h_0$ only depends on $h_1$. Therefore, they can be solved sequentially.

Now observe that equation (9.27a) for $h_2$ contains no source terms dependent on $V$ and its terminal condition is independent of $V$, hence the solution must also be independent of $V$. Therefore, we must solve the Riccati equation

$$0 = \partial_t h_2 + \frac{1}{k} \left( h_2 + \tfrac{1}{2}b \right)^2 - \varphi,$$

which can be solved explicitly, see (6.25) where we provide detailed steps to solve a similar ODE, resulting in

$$h_2(t) = \sqrt{k\,\varphi}\, \frac{1 + \zeta\, e^{2\xi\,(T-t)}}{1 - \zeta\, e^{2\xi(T-t)}} - \tfrac{1}{2}b, \qquad (9.28)$$

where

$$\xi = \sqrt{\frac{\varphi}{k}} \quad \text{and} \quad \zeta = \frac{\alpha - \tfrac{1}{2}b + \sqrt{k\,\varphi}}{\alpha - \tfrac{1}{2}b - \sqrt{k\,\varphi}}. \qquad (9.29)$$

What remains is to solve for $h_1$, which we defer for now, and instead focus on the form of the optimal trading speed, given its solution. Once we have these expressions, in the subsections ahead, we make some specific modelling assumptions and compute $h_1$ explicitly as well as provide a probabilistic representation for the general case. As before, $h_0$ does not appear in the optimal control and, hence, we do not attempt to solve for it, although a closed-form expression can be obtained.

Substituting this result into the expression for the optimal liquidation rate above, we can now express the optimal speed of trading in feedback form as

$$\boxed{\nu_t^* = -\tfrac{1}{2k} \left\{ h_1(t, \mu_t, V_t) + 2 \left( h_2(t) + \tfrac{1}{2}b \right) Q_t^{\nu^*} \right\},} \qquad (9.30)$$

where $h_1(t, \mu, V)$ is the solution to (9.27b) and $h_2(t)$ is given in (9.28).

Thus, the optimal speed of trading can be decomposed into two terms. The second term is similar to the AC-like solution already seen in (6.27) where the penalty on deviations from POCV plays the same role as the urgency parameter in the AC-like setting. The first term adjusts the strategy to account for the rate and volume of trades up to that point in time. Its specific form depends on the precise modelling assumptions on $\mu$ and $V$. In all cases, however, the importance of this term vanishes as $t \to T$ due to the terminal condition $h_1(T, \mu, V) = 0$, and the agent trades more and more like the AC solution.

Recall that the inventory $Q_t^{\nu^*}$ at time $t$ solves the equation $dQ_t^{\nu^*} = -\nu_t^* \, dt$; hence,

$$dQ_t^{\nu^*} = \tfrac{1}{2k} \left\{ h_1(t, \mu_t, V_t) + 2 \left( h_2(t) + \tfrac{1}{2}b \right) Q_t^{\nu^*} \right\} dt. \qquad (9.31)$$

To explicitly solve this ODE, we first multiply (9.31) through by the integrating factor

$$\exp\left\{-\tfrac{1}{\kappa}\int \left(h_2(t) + \tfrac{1}{2}b\right) dt\right\},$$

and write the ODE as

$$d\left(e^{-\frac{1}{\kappa}\int\left(h_2(t)+\frac{1}{2}b\right)dt}Q_t^{\nu^*}\right) = e^{-\frac{1}{\kappa}\int\left((h_2(t)+\frac{1}{2}b\right)dt}\frac{1}{2k}h_1(t,\mu_t,V_t)\,dt\,. \qquad (9.32)$$

We use (9.28) and recall that $\xi = \sqrt{\varphi/k}$ to calculate explicitly the integral appearing in the integrating factor:

$$\frac{1}{\kappa}\int \left(h_2(t) + \tfrac{1}{2}b\right) dt = \xi\int \frac{1 + \zeta e^{2\xi(T-t)}}{1 - \zeta e^{2\xi(T-t)}}\, dt \qquad (9.33)$$

$$= \xi\int \frac{e^{-2\xi(T-t)}}{e^{-2\xi(T-t)} - \zeta}\, dt + \xi\int \frac{\zeta e^{2\xi(T-t)}}{1 - \zeta e^{2\xi(T-t)}}\, dt$$

$$= \frac{1}{2}\log\left[\left(e^{-2\xi(T-t)} - \zeta\right)\left(1 - \zeta e^{2\xi(T-t)}\right)\right]$$

$$= \frac{1}{2}\log\left[e^{2\xi(T-t)}\left(e^{-2\xi(T-t)} - \zeta\right)^2\right]$$

$$= \log\left[e^{-\xi(T-t)} - \zeta e^{\xi(T-t)}\right], \qquad (9.34)$$

and integrate (9.32) between 0 and $t$ to obtain

$$Q_t^{\nu^*} = \frac{\zeta e^{\xi(T-t)} - e^{-\xi(T-t)}}{\zeta e^{\xi T} - e^{-\xi T}}\,\mathfrak{N}$$

$$+ \frac{1}{2k}\int_0^t \frac{\zeta e^{\xi(T-t)} - e^{-\xi(T-t)}}{\zeta e^{\xi(T-u)} - e^{-\xi(T-u)}}\,h_1\left(u,\mu_u,V_u\right) du\,. \qquad (9.35)$$

In this representation, we can immediately see how the first term represents AC-like optimal holdings, while the second term accounts for fluctuations in trading volume. If we take the limit in which the agent penalises all strategies which do not completely liquidate her inventory, i.e. $\alpha \to +\infty$, then $\zeta \to 1$, and so

$$Q_t^{\nu^*} \xrightarrow{\alpha\to\infty} \frac{\sinh(\xi\,(T-t))}{\sinh(\xi\,T)}\,\mathfrak{N} + \int_0^t \frac{\sinh(\xi\,(T-t))}{\sinh(\xi\,(T-u))}\lim_{\alpha\to\infty}\tfrac{1}{2k}h_1\left(u,\mu_u,V_u\right) du\,.$$

Next we show the optimal speed of liquidation and inventory path for the particular case where volume follows a compound Poisson process. Later, in subsection 9.3.2, we assume that volume increments are due to a trading rate that follows an OU-type process as in (9.14). Finally in subsection 9.3.3 we provide a general solution where we do not require a specific functional form for the volume or trading rate. Thus, the next two subsections are particular cases of the more general solution derived in subsection 9.3.3.

## 9.3.1    Compound Poisson Model of Volume

In subsection 9.2.2, we motivated the volume trading rate by Figure 9.3 which shows the estimate of the volume rate of trading (per minute) for AAPL on Jan 5, 2011 from 10:00 to 10:30. The estimate is computed by counting the volume traded over 30 and 10 second windows and scaling the counts to one minute. As the figure shows, a reasonable first order model is one where traded volume comes in bursts of activity which persists for a while (seconds for instance) and then decays to zero. In this section we model volume of trades as a marked point process:

$$V_t = \sum_{n=1}^{N_t} \eta_n , \tag{9.36}$$

where $N_t$ is a homogenous Poisson process with intensity $\lambda$, and $\{\eta_1, \eta_2, \dots\}$ are non-negative i.i.d. random variables, with distribution function $F$ and finite first moment, independent of $N_t$ and of the Brownian motion driving the midprice. In this case, there is no volume rate process $\mu_t$ so the value function is a function of $t, x, S, V$ and $q$.

With this model for volume, the infinitesimal generator of $V$ acts on the value function as follows:

$$\mathcal{L}^V H (t, x, S, V, q) = \lambda \, \mathbb{E} \left[ H (t, x, S, V + \eta, q) - H (t, x, S, V, q) \right] , \tag{9.37}$$

where $\mathbb{E}$ is the expectation operator with respect to the random variable $\eta$ with distribution function $F$.

In the previous section we derived the general form (9.30) of the optimal strategy, with the only model dependent component stemming from the function $h_1(t, \mu, V)$ which is, under the current model assumptions, now a function only of $V$. Thus, under the compound volume Poisson model, we need to solve

$$0 = \partial_t h_1 + \lambda \, \mathbb{E} \left[ h_1(t, V + \eta) - h_1(t, V) \right] + \tfrac{1}{k} \left( h_2 + \tfrac{1}{2} b \right) h_1 - 2 \varphi \left( \rho \mu - \mathfrak{N} \right) , \tag{9.38}$$

and since the source term is linear in $V$, we make the ansatz

$$h_1(t, V) = \ell_0(t) + \ell_1(t) \, V .$$

The terminal condition $h_1(T, V) = 0$ implies that $\ell_0(T) = \ell_1(T) = 0$, and substituting this ansatz for $h_1$ reduces (9.38) to

$$0 = \left\{ \partial_t \ell_0 + \tfrac{1}{k} \left( h_2 + \tfrac{1}{2} b \right) \ell_0 + \psi \ell_1 + 2 \varphi \mathfrak{N} \right\} + \left\{ \partial_t \ell_1 + \tfrac{1}{k} \left( h_2 + \tfrac{1}{2} b \right) \ell_1 - 2 \varphi \rho \right\} V .$$

Since the above must hold for every $V$, each term in the braces must vanish individually, resulting in two simple ODEs for $\ell_0$ and $\ell_1$.

We first solve the ODE for $\ell_1(t)$. As before, we use the integrating factor technique and find that

$$d \left( \exp \left\{ \tfrac{1}{k} \int \left( h_2 + \tfrac{1}{2} b \right) dt \right\} \ell_1 \right) = 2 \varphi \rho \, \exp \left\{ \tfrac{1}{k} \int \left( h_2 + \tfrac{1}{2} b \right) dt \right\} dt .$$

Next, we use the explicit expression for the integrating factor in (9.34), integrate between $t$ and $T$, and use the boundary condition $\ell_1(T) = 0$ to find

$$\ell_1(t) = \frac{-2\varphi\rho}{\xi\left(\zeta e^{\xi(T-t)} - e^{-\xi(T-t)}\right)} \left(\zeta e^{\xi(T-t)} + e^{-\xi(T-t)} - \zeta - 1\right),$$

where the constants $\zeta$ and $\xi$ are provided in (9.29).

To solve for $\ell_0(t)$ we proceed as above and write

$$d\left(e^{\frac{1}{\kappa}\int\left(h_2(t)+\frac{1}{2}b\right)dt}\,\ell_0(t)\right) = -\left(2\varphi\mathfrak{N} + \psi\ell_1(t)\right)e^{\frac{1}{\kappa}\int\left(h_2(t)+\frac{1}{2}b\right)dt}\,dt. \quad (9.39)$$

Above we solved a similar ODE, so here the only new term we need to integrate is

$$\int_t^T e^{\frac{1}{\kappa}\int\left(h_2(u)+\frac{1}{2}b\right)du}\,\ell_1(u)\,du$$

$$= \frac{2\varphi\rho}{\xi}\int_t^T \left(\zeta e^{\xi(T-u)} + e^{-\xi(T-u)} - \zeta - 1\right)du$$

$$= \frac{2\varphi\rho}{\xi^2}\left(\zeta e^{\xi(T-t)} - e^{-\xi(T-t)} - \xi(\zeta+1)(T-t) + 1 - z\right).$$

Now, putting these results together we obtain

$$\ell_0(t) = \frac{2\varphi\mathfrak{N}}{\xi\left(\zeta e^{\xi(T-t)} - e^{-\xi(T-t)}\right)}\left(\zeta e^{\xi(T-t)} + e^{-\xi(T-t)} - \zeta - 1\right)$$

$$-\frac{2\psi\varphi\rho}{\xi^2}\frac{\left(\zeta e^{\xi(T-t)} - e^{-\xi(T-t)} - \xi(\zeta+1)(T-t) + 1 - \zeta\right)}{\zeta e^{\xi(T-t)} - e^{-\xi(T-t)}}$$

$$= -\frac{\mathfrak{N}}{\rho}\ell_1(t) + \frac{2\psi\varphi\rho}{\xi^2}\frac{\left(\zeta e^{\xi(T-t)} - e^{-\xi(T-t)} - \xi(\zeta+1)(T-t) + 1 - \zeta\right)}{e^{-\xi(T-t)} - \zeta e^{\xi(T-t)}}.$$

To obtain the optimal liquidation rate and inventory path we substitute $h_1$ and $h_2$ into (9.31) and (9.35).

## 9.3.2    Stochastic Mean-Reverting Volume Rate

In this section we adopt the model for trading volume rate developed in subsection 9.2.2. Recall that the cumulative volume of other market participant's trades on the sell side of the market is given by

$$V_t = \int_0^t \mu_u\,du\,, \quad (9.40)$$

and we assume that rate of trading $\mu_t$ is as in (9.14), which we repeat for convenience:

$$d\mu_t = -\kappa\,\mu_t\,dt + \eta_{1+N_t^-}\,dN_t\,.$$

Earlier we derived the general form of the optimal strategy, with the only model dependent component stemming from the function $h_1(t, \mu, V)$ which must satisfy

(9.27b). With our current modelling assumptions, the infinitesimal generator of the joint rate and trading volume acts on $h_1(t, \mu, V)$ as follows:

$$\mathcal{L}^{\mu, V} h_1(t, \mu, V) = (\mu \, \partial_V - \kappa \, \mu \, \partial_\mu) \, h_1(t, \mu, V) + \lambda \, \mathbb{E}[h_1(t, \mu + \eta, V) - h_1(t, \mu, V)],$$

where the expectation is over the random variable $\eta$ with distribution function $F$. Due to the affine nature of this generator and the fact that the terminal condition $h_1(t, \mu, V) = 0$ is a constant, we use the affine ansatz

$$h_1(t) = \ell_0(t) + \ell_1(t) \, V + \ell_2(t) \, \mu, \tag{9.41}$$

with terminal conditions $\ell_i(T) = 0$ for $i = 1, 2, 3$. Therefore, (9.27b) reduces to solving a system of coupled ODEs for $\ell_i(t)$

$$0 = \partial_t \ell_1 + \tfrac{1}{k} \left( h_2 + \tfrac{1}{2} b \right) \ell_1 - 2 \varphi \rho, \tag{9.42a}$$

$$0 = \partial_t \ell_2 + \left( \tfrac{1}{k} \left( h_2 + \tfrac{1}{2} b \right) - \kappa \right) \ell_2 + \ell_1, \tag{9.42b}$$

$$0 = \partial_t \ell_0 + \psi \, \ell_2 + \tfrac{1}{k} \left( h_2 + \tfrac{1}{2} b \right) \ell_0 + 2 \varphi \, \mathfrak{N}, \tag{9.42c}$$

and recall that $h_2$ is given by (9.28).

To solve for $\ell_1$ we once again make use of the integrating factor technique by writing (9.42a) as

$$d \left( e^{\frac{1}{k} \int \left( h_2 + \frac{1}{2} b \right) dt} \ell_1 \right) = 2 \varphi \rho \, e^{\frac{1}{k} \int \left( h_2 + \frac{1}{2} b \right) dt}, \tag{9.43}$$

using (9.33), integrating between $t$ and $T$, and using the terminal condition $\ell_1(T) = 0$ to obtain

$$\ell_1(t) = \frac{2 \varphi \rho}{e^{-\xi(T-t)} - \zeta e^{\xi(T-t)}} \left( e^{-\xi(T-t)} - \zeta e^{\xi(T-t)} + \zeta - 1 \right), \tag{9.44}$$

where $\zeta$ and $\xi$ are given in (9.29).

Similarly, and after straightforward but tedious calculations, we solve (9.42b) to obtain

$$\ell_2(t) = \frac{2 \varphi \rho}{\zeta e^{(\kappa + \xi)(T-t)} - e^{(\kappa - \xi)(T-t)}}$$

$$\times \left( \frac{e^{(\kappa - \xi)(T-t)} - 1}{\kappa - \xi} + \frac{\zeta}{\kappa + \xi} \left( 1 - e^{(\kappa + \xi)(T-t)} \right) + \frac{1 - \zeta}{\kappa} \left( 1 - e^{\kappa(T-t)} \right) \right).$$

Finally, since we have $\ell_2(t)$ we leave it to the reader to solve (9.42c) to obtain $\ell_0(t)$.

### 9.3.3     Probabilistic Representation

In the previous two subsections, we analysed two modelling specifications for the rate of volume arrivals and derived explicit closed-form expressions. It is in fact possible to derive a general form for the function $h_1$ in a quite general setting. First, let us restate the equation satisfied by $h_1$:

$$0 = \left( \partial_t + \mathcal{L}^{\mu, V} \right) h_1 + \tfrac{1}{k} \left( h_2 + \tfrac{1}{2} b \right) h_1 + 2 \varphi \left( \mathfrak{N} - \rho V \right),$$

subject to the terminal condition $h_1(T, \mu, V) = 0$, and recall that $h_2$ is a deterministic function of time given by (9.28). This is a linear PIDE for $h_1$ in which $h_2 + \frac{1}{2}b$ acts as an effective discount rate and $2\varphi\,(\mathfrak{N} - \rho V)$ is a potential (or source) term. The general solution of such an equation can be represented using the Feynman-Kac Theorem. Thus we write

$$h_1(t, \mu, V) = 2\,\varphi\,\mathbb{E}_{t,\mu,V}\left[\int_t^T \exp\left\{\frac{1}{k}\int_t^u \left(h_2(s, \mu_s, V_s) + \frac{1}{2}b\right)\,ds\right\}\,(\mathfrak{N} - \rho\,V_u)\,du\right]$$

$$= 2\,\varphi\,\mathbb{E}_{t,\mu,V}\left[\int_t^T \exp\left\{\sqrt{\frac{\varphi}{k}}\int_t^u \frac{1 + \zeta\,e^{2\xi\,(T-s)}}{1 - \zeta\,e^{2\xi(T-s)}}\,du\right\}\,(\mathfrak{N} - \rho\,V_u)\,du\right].$$

Using (9.34) to compute the integral in the exponent, we can write

$$h_1(t, \mu, V) = 2\,\varphi\,\mathbb{E}_{t,\mu,V}\left[\int_t^T \left(\frac{\zeta\,e^{\xi\,(T-u)} - e^{-\xi\,(T-u)}}{\zeta\,e^{\xi\,(T-t)} - e^{-\xi\,(T-t)}}\right)(\mathfrak{N} - \rho\,V_u)\,du\right]$$

$$= 2\,\varphi\int_t^T g(t, u)\,(\mathfrak{N} - \rho\,\mathbb{E}_{t,\mu,V}[V_u])\,du\,,$$

where

$$g(t, u) = \left(\frac{\zeta\,e^{\xi\,(T-u)} - e^{-\xi\,(T-u)}}{\zeta\,e^{\xi\,(T-t)} - e^{-\xi\,(T-t)}}\right).$$

Upon inserting this representation into the optimal strategy we arrive at the general representation for an agent who targets POCV:

$$\boxed{\begin{aligned}\nu_t^* = {}&\xi\left(\frac{\zeta\,e^{\xi\,(T-t)} + e^{-\xi(T-t)}}{\zeta\,e^{\xi\,(T-t)} - e^{-\xi\,(T-t)}}\right)Q_t^{\nu^*}\\&- \xi^2\int_t^T g(t, u)\,\left(\mathfrak{N} - \rho\,\mathbb{E}\left[V_u\,\big|\,\mathcal{F}_t^{\mu,V}\right]\right)du\,.\end{aligned}} \tag{9.45}$$

Moreover, inserting the expression for $h_1$ into the optimal inventory to hold (9.35) we obtain

$$\boxed{\begin{aligned}Q_t^{\nu^*} = {}&\left(\frac{\zeta\,e^{\xi\,(T-t)} - e^{-\xi(T-t)}}{\zeta\,e^{\xi\,T} - e^{-\xi\,T}}\right)\mathfrak{N}\\&+ \xi^2\int_0^t\int_s^T g(s, u)\,\left(\mathfrak{N} - \rho\,\mathbb{E}\left[V_u\,\big|\,\mathcal{F}_s^{\mu,V}\right]\right)du\,ds\,.\end{aligned}} \tag{9.46}$$

To understand the intuition behind this general result we look at the limiting case in which the agent wishes full execution of all orders, i.e. in the limit $\alpha \to \infty$, $\zeta \xrightarrow{\alpha\to\infty} 1$. In this scenario, the optimal inventory to hold through time simplifies to

$$Q_t^{\nu^*} \xrightarrow{\alpha\to\infty} \frac{\sinh(\xi\,(T-t))}{\sinh(\xi\,T)}\,\mathfrak{N}$$

$$+ \xi^2\int_0^t\int_s^T \frac{\sinh(\xi\,(T-u))}{\sinh(\xi\,(T-s))}\,\left(\mathfrak{N} - \rho\,\mathbb{E}\left[V_u\,\big|\,\mathcal{F}_s^{\mu,V}\right]\right)du\,ds\,.$$

As the expression shows, the first term is the classical AC strategy, and the second term corrects for deviations of the strategy from the POCV.

## 9.4    Including Impact of Other Traders

So far we have assumed that other traders' volume does not move the midprice, but that the liquidating agent's trades do. This is somewhat inconsistent. In this section, we assume that MOs from all market participants, including the agent's, have a permanent effect on the midprice as shown by the SDE

$$dS_t^\nu = b\left(\mu_t^+ - (\nu_t + \mu_t^-)\right)dt + \sigma\, dW_t\,,$$

where $\mu_t^\pm$ denote the rate of trading for buy and sell MOs sent by other traders and $\nu_t > 0$ is the agent's liquidation rate. If the agent was acquiring shares then her trading rate would be added, instead of subtracted, to the drift of the SDE. Moreover, $b > 0$ represents the permanent impact that trading has on the midprice. As before, the agent's execution price $\hat{S}_t^\nu$ is assumed to be linear in her trading rate so that

$$\hat{S}_t^\nu = S_t^\nu - k\nu_t\,.$$

Here, we assume that the agent aims to target the rate of sell trading volume, and we leave the case of targeting cumulative volume to the reader, see Exercise E.9.3. In this case, her performance criteria is given by

$$H^\nu(t,x,S,\boldsymbol{\mu},q) = \mathbb{E}_{t,x,S,\boldsymbol{\mu},q}\left[X_T^\nu + Q_T^\nu(S_T^\nu - \alpha Q_T^\nu) - \varphi\int_t^T (\nu_u - \rho\mu_u^-)^2\, du\right]\,,$$

where $\boldsymbol{\mu} = \{\mu^+, \mu^-\}$, and her value function is

$$H(t,x,S,\boldsymbol{\mu},q) = \sup_{\nu\in\mathcal{A}} H^\nu(t,x,S,\boldsymbol{\mu},q)\,.$$

Applying the dynamic programming principle suggests that the value function should satisfy the DPE

$$0 = \left(\partial_t + \tfrac{1}{2}\sigma^2\partial_{SS} + \mathcal{L}^\mu\right)H$$
$$+ \sup_\nu\left\{(S - k\nu)\,\nu\,\partial_x H - \nu\,\partial_q H\right.\tag{9.47}$$
$$\left. + b((\mu^+ - \mu^-) - \nu)\,\partial_S H - \varphi\,(\nu - \rho\mu^-)^2\right\}\,,$$

where $\mathcal{L}^\mu$ denotes the infinitesimal generator of $\boldsymbol{\mu}$. The only difference between this DPE and the one for the case without impact (see (9.5)), is the term $b\,(\mu^+ - \mu^-)\,\partial_S H$ which does not directly affect the optimisation over the agent's trading $\nu$ in feedback form – it will however alter the value function, and hence the explicit form of $\nu^*$.

We once again use the ansatz

$$H(t,x,S,\boldsymbol{\mu},q) = x + q\,S + h(t,\boldsymbol{\mu},q)\,,$$

where $h$ now depends on both trading rates $\mu^+$ and $\mu^-$, and the terminal condition is $h(T, \boldsymbol{\mu}, q) = -\alpha\, q^2$. Inserting into the DPE above, we find that $h$ must satisfy

$$0 = (\partial_t + \mathcal{L}^\mu)\, h + b\,(\mu^+ - \mu^-)\, q$$
$$+ \sup_\nu \left\{ -k\, \nu^2 - (\partial_q h + b\, q)\, \nu - \varphi\,(\nu - \rho\,\mu^-)^2 \right\}. \tag{9.48}$$

Solving for the first order condition provides the optimal speed of trading in feedback form as

$$\nu^* = \frac{-(\partial_q h + b\, q) + 2\,\varphi\,\rho\,\mu^-}{2(k + \varphi)}, \tag{9.49}$$

and upon inserting into the DPE we find the non-linear equation which $h$ should satisfy:

$$0 = (\partial_t + \mathcal{L}^\mu)\, h + b\,(\mu^+ - \mu^-)\, q + \tfrac{1}{4(k+\varphi)}\, \left(\partial_q h + b\, q - 2\,\varphi\,\rho\,\mu^-\right)^2 - \varphi\,\rho^2\,(\mu^-)^2.$$

Comparing this to the case where there is no impact from other traders shown in (9.8), repeated here for convenience,

$$0 = (\partial_t + \mathcal{L}^\mu)\, h + \tfrac{1}{4(k+\varphi)}\, \left(\partial_q h + b\, q - 2\,\varphi\,\rho\,\mu\right)^2 - \varphi\,\rho^2\,(\mu)^2,$$

we see that the main difference is that the agent must now account for both buy and sell side trading rates, and the additional term $b\,(\mu^+ - \mu^-)\, q$ makes the agent aware of the asset's drift due to net trading in either direction.

To solve the non-linear equation for $h$, we observe once again that the affine nature of the PDE and the terminal condition suggest the ansatz

$$h(t, \boldsymbol{\mu}, q) = h_0(t, \boldsymbol{\mu}) + h_1(t, \boldsymbol{\mu})\, q + h_2(t, \boldsymbol{\mu})\, q^2,$$

subject to the terminal conditions $h_0(T, \boldsymbol{\mu}) = h_1(T, \boldsymbol{\mu}) = 0$ and $h_0(T, \boldsymbol{\mu}) = -\alpha$. On inserting this ansatz, expanding the expression and collecting terms with like powers in $q$, we find that the $h_i$ satisfy the coupled system of equations

$$0 = (\partial_t + \mathcal{L}^\mu)\, h_2 + \frac{\left(h_2 + \tfrac{1}{2} b\right)^2}{k + \varphi}, \tag{9.50a}$$

$$0 = (\partial_t + \mathcal{L}^\mu)\, h_1 + \frac{h_1 - 2\,\varphi\,\rho\,\mu^-}{k + \varphi}\, \left(h_2 + \tfrac{1}{2} b\right) + b\,(\mu^+ - \mu^-), \tag{9.50b}$$

$$0 = (\partial_t + \mathcal{L}^\mu)\, h_0 + \frac{1}{4(k + \varphi)}\, \left(h_1 - 2\,\varphi\,\rho\,\mu^-\right)^2 - \varphi\,\rho^2\,(\mu^-)^2. \tag{9.50c}$$

We see that $h_2$ which satisfies (9.50a), is the same equation as in the case where we ignore the impact of other agents, and its solution is given by (9.11), repeated here for convenience:

$$h_2(t, \boldsymbol{\mu}) = -\left(\tfrac{T-t}{k+\varphi} + \tfrac{1}{\alpha - \frac{1}{2} b}\right)^{-1} - \tfrac{1}{2} b.$$

As before, this is independent of the rate of trading of other agents and corresponds to a TWAP-like trading strategy. In the next section, we focus on computing $h_1$ under general assumptions.

## 9.4.1     Probabilistic Representation

In the general case, we can once again make use of a Feynman-Kac formula and write the solution to (9.50b) as

$$
h_t(t, \boldsymbol{\mu}) = \mathbb{E}_{t,\boldsymbol{\mu}} \left[ \int_t^T e^{\int_t^s \tilde{h}_2(u, \boldsymbol{\mu}_u)\, du} \left\{ -2\, \varphi\, \rho\, \tilde{h}_2(s, \boldsymbol{\mu}_s)\, \mu_s^- + b\left( \mu_s^+ - \mu_s^- \right) \right\} ds \right],
$$

where $\tilde{h}_2(t, \boldsymbol{\mu}_t) := \frac{1}{k+\varphi} \left( h_2(t, \boldsymbol{\mu}_t) + \frac{1}{2}b \right)$. The above expression can be simplified somewhat by noticing that

$$
e^{\int_t^s \tilde{h}_2(u, \boldsymbol{\mu}_u)\, du} = \exp\left\{ -\frac{1}{k+\varphi} \int_t^s \left( \frac{T-s}{k+\varphi} + \frac{1}{\alpha - \frac{1}{2}b} \right)^{-1} ds \right\} = \frac{(T-s)+\zeta}{(T-t)+\zeta},
$$

where

$$
\zeta = \frac{k+\varphi}{\alpha - \frac{1}{2}b}.
$$

Inserting this integral into the expression for $h_1$ above, and interchanging the expectation and outer integral, we arrive at

$$
h_t(t, \boldsymbol{\mu}) = \varphi\, \rho \int_t^T \frac{\mathbb{E}_{t,\boldsymbol{\mu}}\left[ \mu_s^- \right]}{(T-t)+\zeta}\, ds + b \int_t^T \left( \frac{(T-s)+\zeta}{(T-t)+\zeta} \right) \mathbb{E}_{t,\boldsymbol{\mu}}\left[ \mu_s^+ - \mu_s^- \right] ds.
$$

As the above formula shows, the first term accounts for the one-sided trades that move in the same direction as that of the agent (i.e. sell trades), while the second term accounts for the imbalance in buys and sells. Both terms are integrated over the remaining life of the trading horizon and weight the expected selling / imbalance trading rate through time. The first term computes the mean expected future selling rate, while the second weighs the earlier trades more heavily – since those trades have more time in which to impact the midprice.

Recall that the agent's optimal trading speed is given by (9.49) which, in terms of $h_i$, reduces to

$$
\nu_t^* = -\frac{1}{k+\varphi} \left( h_2(t, \boldsymbol{\mu}_t) + \frac{1}{2}b \right) Q_t^{\nu^*} + \frac{1}{2(k+\varphi)} \left( 2\, \varphi\, \rho\, \mu_t^- - h_1(t, \boldsymbol{\mu}_t) \right).
$$

Substituting the above results, we finally obtain

$$
\boxed{
\begin{aligned}
\nu_t^* = {} & \frac{Q_t^{\nu^*}}{(T-t)+\zeta} + \frac{\varphi}{k+\varphi}\, \rho \left\{ \mu_t^- - \frac{\int_t^T \mathbb{E}\left[ \mu_s^- \mid \mathcal{F}_t^{\boldsymbol{\mu}} \right] ds}{(T-t)+\zeta} \right\} \\
& - \frac{b}{k+\varphi} \frac{\int_t^T \left( (T-s)+\zeta \right) \mathbb{E}\left[ (\mu_s^+ - \mu_s^-) \mid \mathcal{F}_t^{\boldsymbol{\mu}} \right] ds}{(T-t)+\zeta}.
\end{aligned}
}
\tag{9.51}
$$

To understand the functioning of the liquidation strategy we see that the first two terms are identical to the case which does not account for the impact of other traders, see (9.19) and discussion that follows. The new term in the second line of (9.51) acts to correct her trading based on her expectations of the net order flow from that point in time until the end of the trading horizon. When there is currently no imbalance ($\mu_t^+ = \mu_t^-$) her liquidation rate is as in (9.19). When there is a surplus of buy trades, however, she slows down her trading rate to allow the midprice to appreciate before liquidating the rest of her order. When there is a surplus of sell orders, she speeds up her trades for two reasons: (i) because she must match the POV on the sell side; and (ii) the action of other sellers in the market will push prices downwards, and therefore degrade her profits if she waits. She therefore attempts to liquidate a larger portion of her orders now rather than later.

### 9.4.2    Example: Stochastic Mean-Reverting Volume

It is helpful to provide a specific modelling example in which we can derive a simple closed-form formula for the optimal liquidation speed. We assume that buy and sell trading volumes arrive independently, where each satisfies the SDE

$$ d\mu_t^\pm = -\kappa\,\mu_t^\pm\,dt + \eta_{1+N_{t^-}^\pm}\,dN_t^\pm\,, $$

where $N_t^\pm$ are independent Poisson processes with intensity $\lambda$, and $\{\eta_1^\pm, \eta_2^\pm, \dots\}$ are i.i.d. random variables, with distribution function $F$, representing jumps in trading volume, and independent of $N_t^\pm$ and of the Brownian motion $W_t$ driving the midprice. In this manner, the model assumes that buy/sell trading rates are two independent jump Ornstein-Uhlenbeck (OU) process.

We can solve the above SDE explicitly, by introducing an integrating factor and writing $\mu_t^\pm = e^{-\kappa t}\,\tilde{\mu}_t^\pm$, so that

$$ d\tilde{\mu}_t^\pm = e^{\kappa t}\,\eta_{1+N_{t^-}^\pm}\,dN_t^\pm\,. $$

Therefore, we can write the trading rates as

$$ \tilde{\mu}_s^\pm - \tilde{\mu}_t^\pm = \int_t^s e^{\kappa u}\,\eta_{1+N_{u^-}^\pm}\,dN_u^\pm\,, $$

and so, for $s > t$, we have

$$ \mu_s^\pm = e^{-\kappa(s-t)}\,\mu_t^\pm + \int_t^s e^{-\kappa(s-u)}\,\eta_{1+N_{u^-}^\pm}\,dN_u^\pm\,. $$

The expression for the optimal trading speed requires only the expected value of the trading rate given its current value. For this purpose, we can compute, for $s > t$,

$$ \mathbb{E}_{t,\mu}\left[\mu_s^\pm\right] = e^{-\kappa(s-t)}\,\mu_t^\pm + \int_t^s e^{-\kappa(s-u)}\,\mathbb{E}[\eta]\,\lambda\,du $$

$$ = e^{-\kappa(s-t)}\,\mu_t^\pm + \frac{\lambda\,\mathbb{E}[\eta]}{\kappa}\left(1 - e^{-\kappa(s-t)}\right) $$

$$ = e^{-\kappa(s-t)}\left(\mu_t^\pm - \frac{\lambda\,\mathbb{E}[\eta]}{\kappa}\right) + \frac{\lambda\,\mathbb{E}[\eta]}{\kappa}\,. $$

Next, substituting this result into the expression for the optimal trading rate

(9.51) and computing the integrals there leads to (with $\tau := T - t$)

$$
v_t^* = \frac{Q_t^{\nu^*}}{\tau + \zeta} + \frac{\varphi}{k + \varphi} \rho \left\{ \mu_t^- - \frac{\int_t^T \left( e^{-\kappa(s-t)} \left( \mu_t^- - \frac{\lambda \mathbb{E}[\eta]}{\kappa} \right) + \frac{\lambda \mathbb{E}[\eta]}{\kappa} \right) ds}{\tau + \zeta} \right\}
$$

$$
+ \frac{b}{k + \varphi} \frac{\int_t^T ((T-s) + \zeta) e^{-\kappa(s-t)} ds}{\tau + \zeta} (\mu_t^+ - \mu_t^-)
$$

$$
= \frac{Q_t^{\nu^*}}{\tau + \zeta} + \frac{\varphi}{k + \varphi} \rho \left\{ \mu_t^- - \frac{\frac{1 - e^{-\kappa\tau}}{\kappa} \left( \mu_t^- - \frac{\lambda \mathbb{E}[\eta]}{\kappa} \right) + \frac{\lambda \mathbb{E}[\eta]}{\kappa} \tau}{\tau + \zeta} \right\}
$$

$$
+ \frac{b}{k + \varphi} \frac{(1 - \kappa\zeta) e^{-\kappa\tau} + \kappa(\tau + \zeta) - 1}{\kappa^2(\tau + \zeta)} (\mu_t^+ - \mu_t^-) \ .
$$

## 9.5 Utility Maximiser

In this section the agent's objective is to maximise expected utility of terminal wealth but to also target POV. The setup is the same as the one described in Section 9.2 and the agent's preferences are described by exponential utility $U(x) = -e^{-\gamma x}$. If the agent ignores the POV objective, her performance criteria is

$$
H^\nu(t, x, S, \mu, q) = \mathbb{E}_{t,x,S,\mu,q} \left[ -e^{-\gamma(X_T^\nu + Q_T^\nu (S_T^\nu - \alpha Q_T^\nu))} \right] .
$$

The question is how to incorporate a POV penalty while maintaining tractability of the problem. One naive answer is to simply add in a penalisation as we did before, e.g. by considering the performance criteria

$$
H^\nu(t, x, S, \mu, q) = \mathbb{E}_{t,x,S,\mu,q} \left[ -e^{-\gamma(X_T^\nu + Q_T^\nu (S_T^\nu - \alpha Q_T^\nu))} - \varphi \int_t^T (\nu_u - \rho \mu_u)^2 \, du \right] ,
$$

where $\mu_t$ is the other agents' (selling) trading rate, $0 < \rho < 1$ is the fraction of the trading rate that the agent targets, and $\varphi \geq 0$ the target penalty parameter.

This approach, however, does not lead to analytically tractable results. The main reason is that the exponential utility and the linear penalty are in a sense incompatible, and, here, even the cash process does not factor out of the problem. Instead, we consider what is sometimes called a recursive intertemporally additive penalty. Stylistically we aim to have a value function defined as the continuous limit of the recursion

$$
H_t = \sup_\nu \mathbb{E} \left[ H_{t+\Delta t} + \varphi \gamma \, H_{t+\Delta t} (\nu_t - \rho \mu_t)^2 \Delta t \right] ,
$$

with $H_T = \mathbb{E} \left[ -e^{-\gamma(X_T^\nu + Q_T^\nu (S_T^\nu - \alpha Q_T^\nu))} \right]$. Alternatively, one can view this as a stochastic differential utility, as in Duffie & Epstein (1992).

By adopting this approach, when the agent wishes to optimally liquidate shares whilst targeting POV, her performance criteria is

$$H^\nu(t, x, S, \mu, q) = \mathbb{E}_{t,S,\mu,q}\left[ - e^{-\gamma(X_T^\nu + Q_T^\nu(S_T^\nu - \alpha Q_T^\nu))} \right.$$

$$\left. + \varphi\gamma \int_t^T (\nu_u - \rho\,\mu_u)^2 \; H^\nu(u, X_u^\nu, S_u^\nu, \mu_u, Q_u^\nu)\, du \right]$$

and her value function is

$$H(t, x, S, \mu, q) = \sup_{\nu \in \mathcal{A}} H^\nu(t, x, S, \mu, q)\,.$$

Note that the penalty term has a positive sign in front of the integral. The reason is that $H$ itself is negative, so that this term is indeed a penalty contribution. Deviations from the target POV are scaled by the value function at that time, so if there is a lot of value at that point in state space, the agent is averse to moving away from POV, while if there is little value at that point in state space, the agent is willing to deviate from POV if it gains her value.

Applying the dynamic programming principle suggests that the value function should satisfy the DPE

$$\left(\partial_t + \tfrac{1}{2}\sigma^2 \partial_{SS} + \mathcal{L}^\mu\right) H$$
$$+ \sup_\nu \left\{ (S - k\,\nu)\,\nu\,\partial_x H - \nu\partial_q H - b\,\nu\,\partial_S H + \varphi\,\gamma\,(\nu - \rho\,\mu)^2\, H \right\} = 0,$$

subject to the terminal condition $H(T, x, S, \mu, q) = -e^{-\gamma(x + q(S - \alpha q))}$, and attains a supremum at

$$\nu^* = \frac{1}{2}\frac{-S\,\partial_x H + \partial_q H + b\,\partial_S H + 2\,\gamma\varphi\,\rho\mu\, H}{-k\,\partial_x H + \gamma\varphi\, H}\,. \tag{9.52}$$

The form of the terminal condition suggests that we use the ansatz

$$H = -e^{-\gamma(x + q\,S + h(t,\mu,q))}\,,$$

which leads to the following equation for $h(t, \mu, q)$:

$$0 = -\,\mathcal{L}^\mu\left(e^{-\gamma h}\right)$$
$$+ \left(\gamma\,\partial_t h - \tfrac{1}{2}\sigma^2\,\gamma^2\,q^2 + \gamma\frac{(b\,q + \partial_q h - 2\,\varphi\,\rho\,\mu)^2}{4(k + \varphi)} - \gamma\varphi\,\rho^2\,\mu^2\right) e^{-\gamma h}\,, \tag{9.53}$$

with terminal condition $h(T, \mu, q) = -\alpha\,q^2$.

## 9.5.1 Solving the DPE with Deterministic Volume

We assume that the rate at which other market participants are selling shares is a deterministic function of time with derivative $d\mu(t) = g(t)\,dt$ so that $\mathcal{L}^\mu H = \partial_\mu H\, g(t)$. In this case, we can view the agent as targeting a predictable trading pattern, which for example, may be taken as the solid blue line in the right panel of Figure 9.2.

Thus we write (9.53) in the form

$$0 = (\partial_t + \mathcal{L}^\mu) \, h + \frac{(\partial_q h + bq - 2\varphi\rho\mu)^2}{4(k+\varphi)} - \varphi\rho^2\mu^2 - \tfrac{1}{2}\sigma^2\gamma q^2 \,. \tag{9.54}$$

Note that (9.54) is the same as (9.8) with the extra term $-\tfrac{1}{2}\sigma^2\gamma q^2$, so we proceed as above and make the ansatz

$$h(t, \mu, q) = h_0(t, \mu) + q \, h_1(t, \mu) + q^2 \, h_2(t, \mu) \,, \tag{9.55}$$

and after straightforward manipulations and collecting terms in $q$ we obtain the coupled system of PIDEs

$$0 = (\partial_t + \mathcal{L}^\mu) \, h_2 + \frac{(h_2 + \tfrac{1}{2}b)^2}{k+\varphi} - \tfrac{1}{2}\sigma^2\gamma \,, \tag{9.56a}$$

$$0 = (\partial_t + \mathcal{L}^\mu) \, h_1 + \frac{h_1 - 2\varphi\rho\mu}{k+\varphi} \, (h_2 + \tfrac{1}{2}b) \,, \tag{9.56b}$$

$$0 = (\partial_t + \mathcal{L}^\mu) \, h_0 + \frac{1}{4(k+\varphi)} \, (h_1 - 2\varphi\rho\mu)^2 - \varphi\rho^2\mu^2 \,, \tag{9.56c}$$

with terminal conditions $h_2(T) = -\alpha$, $h_1(T) = h_0(T) = 0$.

Now observe that equation (9.56a) for $h_2$ contains no source terms dependent on $\mu$ and its terminal condition is independent of $\mu$, hence the solution must also be independent of $\mu$. The equation satisfied by $h_2$ is of Riccati type and can be solved explicitly. We solve for $h_2$, see the solution of the Riccati ODE (6.25), to obtain

$$h_2(t) = \sqrt{(k+\varphi)\,\xi} \, \frac{1 + \zeta \, e^{2\omega \, (T-t)}}{1 - \zeta \, e^{2\omega(T-t)}} - \frac{1}{2}b \,,$$

where

$$\xi = \tfrac{1}{2}\sigma^2\gamma, \quad \omega = \sqrt{\frac{\xi}{k+\varphi}}, \quad \text{and} \quad \zeta = \frac{\alpha - \tfrac{1}{2}b + \sqrt{(k+\varphi)\,\xi}}{\alpha - \tfrac{1}{2}b - \sqrt{(k+\varphi)\,\xi}}. \tag{9.57}$$

Now we turn to solving the PIDE

$$0 = \partial_t h_1 - g \, \partial_\mu h_1 + \frac{h_1 - 2\varphi\rho\mu}{k+\varphi} \, (h_2 + \tfrac{1}{2}b) \,, \tag{9.58}$$

with terminal condition $h_1(T) = 0$. To solve this equation we look at the different components in (9.58) and observe that if we assume an ansatz linear in $\mu$:

$$h_1(t, \mu) = \ell_0(t) + \ell_1(t) \, \mu \,, \tag{9.59}$$

then (9.58) becomes

$$0 = \left\{ \partial_t \ell_0 - g \ell_1 + \frac{h_2 + \tfrac{1}{2}b}{k+\varphi} \ell_0 \right\} + \left\{ \partial_t \ell_1 + \frac{\ell_1 - 2\varphi\rho}{k+\varphi} \, (h_2 + \tfrac{1}{2}b) \right\} \mu \,. \tag{9.60}$$

Now, since this equation must hold for all $\mu$, the terms in braces must vanish individually. Therefore we must solve two uncoupled ODEs:

$$0 = \partial_t \ell_0 - g\ell_1 + \frac{h_2 + \frac{1}{2}b}{k + \varphi}\ell_0 \,, \tag{9.61}$$

$$0 = \partial_t \ell_1 + \frac{\ell_1 - 2\varphi\rho}{k + \varphi}\left(h_2 + \tfrac{1}{2}b\right)\,. \tag{9.62}$$

We first solve (9.62) using the integrating factor technique. First we calculate, in a similar way to (9.33),

$$\frac{1}{k + \varphi}\int\left(h_2(t) + \frac{1}{2}b\right)dt = \sqrt{\frac{\xi}{k + \varphi}}\int\frac{1 + \zeta\,e^{2\omega\,(T-t)}}{1 - \zeta\,e^{2\omega(T-t)}}\,dt$$

$$= \log\left[e^{-\omega(T-t)} - \zeta e^{\omega(T-t)}\right]\,.$$

Now we write

$$d\left(e^{\frac{1}{k+\varphi}\int\left(h_2(t)+\frac{1}{2}b\right)dt}\,\ell_1\right) = \frac{2\varphi\rho}{k + \varphi}e^{\frac{1}{k+\varphi}\int\left(h_2(t)+\frac{1}{2}b\right)dt}\left(h_2(t) + \tfrac{1}{2}b\right)\,,$$

and integrate both sides between $t$ and $T$:

$$-\left(e^{-\omega(T-t)} - \zeta e^{\omega(T-t)}\right)\ell_1(t)$$

$$= 2\varphi\rho\omega\int_t^T\left(e^{-\omega(T-u)} - \zeta e^{\omega(T-u)}\right)\frac{1 + \zeta\,e^{2\omega\,(T-u)}}{1 - \zeta\,e^{2\omega(T-u)}}\,du$$

$$= 2\varphi\rho\left(\zeta e^{\omega(T-t)} - e^{-\omega(T-t)} + 1 - \zeta\right)$$

$$\ell_1(t) = 2\varphi\rho\frac{\zeta e^{\omega(T-t)} - e^{-\omega(T-t)} + 1 - \zeta}{\zeta e^{\omega(T-t)} - e^{-\omega(T-t)}}\,.$$

Similarly, to solve for $\ell_0$ we start as usual by writing

$$d\left(e^{\frac{1}{k+\varphi}\int\left(h_2(t)+\frac{1}{2}b\right)dt}\,\ell_0(t)\right) = e^{\frac{1}{k+\varphi}\int\left(h_2(t)+\frac{1}{2}b\right)dt}\,g(t)\,\ell_1(t)$$

and integrate both sides between $t$ and $T$ to obtain

$$\ell_0(t) = \frac{2\varphi\rho}{e^{-\omega(T-t)} - \zeta e^{\omega(T-t)}}\int_t^T g(u)\left(\zeta e^{\omega(T-u)} - e^{-\omega(T-u)} + 1 - \zeta\right)du\,. \tag{9.63}$$

Finally, after straightforward manipulations, the optimal speed of trading is

$$\boxed{\nu_t^* = -\omega\frac{1 + \zeta\,e^{2\omega\,(T-t)}}{1 - \zeta\,e^{2\omega(T-t)}}Q_t^{\nu^*} + \frac{\varphi\rho - \frac{1}{2}\ell_1(t)}{k + \varphi}\mu_t - \frac{\frac{1}{2}\ell_0(t)}{k + \varphi}\,,} \tag{9.64}$$

where $\omega$, $\zeta$ are constant parameters given in (9.57).

In the optimal speed of liquidation we see that the first term is an AC-type term similar to the one derived in (6.27). The other three terms adjust the speed so that the liquidation rate targets the fraction $\rho$ of POV.

Note that when the risk aversion parameter $\gamma \to 0$, the agent's preferences are as those of a risk-neutral agent and the optimal liquidation strategy would be the same as the one derived in subsection 9.2.1 for this particular choice of $\mathcal{L}^\mu$. It is easy to see that if we set $\gamma = 0$, the system of PIDEs (9.56) is the same as the system (9.10) and $h_1(t)$ and $h_2(t)$ have the same boundary conditions.

## 9.6 Bibliography and Selected Readings

Konishi (2002), Bialkowski, Darolles & Fol (2008), Humphery-Jenner (2011), McCulloch & Kazakov (2012), Frei & Westray (2013), Guéant & Royer (2014), Guéant (2014), Mitchell, Bialkowski & Tompaidis (2013), Cartea & Jaimungal (2014$a$).

## 9.7 Exercises

E.9.1 Assume that the agent's objective is to liquidate $\mathfrak{N}$ shares as in Section 9.2, but the strategy also penalises running inventory so that her performance criteria is

$$H^\nu(t, x, S, q) = \mathbb{E}_{t,x,S,q}\left[ X_T^\nu + Q_T^\nu(S_T^\nu - \alpha Q_T^\nu) \right.$$
$$\left. -\varphi \int_t^T (\nu_u - \rho V_u)^2 \, du - \phi \int_t^T (Q_u)^2 \, du \right].$$

(a) Show that the agent's value function $H$ satisfies

$$0 = \left(\partial_t + \tfrac{1}{2}\sigma^2 \partial_{SS} + \mathcal{L}^V\right) H - \phi q^2$$
$$+ \sup_\nu \left\{ (S - k\nu)\nu - \nu \partial_q H - b\nu \, \partial_S H - \varphi(\nu - \rho V)^2 \right\},$$

subject to the terminal condition $H(T, S, v, q) = q(S - \alpha q)$.

(b) Make the ansatz

$$H(t, S, V, q) = q\, S + h_0(t, V) + q\, h_1(t, V) + q^2\, h_2(t, V)$$

and show that the problem reduces to solving the coupled system of PIDEs

$$0 = \left(\partial_t + \mathcal{L}^V\right) h_2 + \frac{\left(h_2 + \tfrac{1}{2}b\right)^2}{k + \varphi} - \phi, \tag{9.65a}$$

$$0 = \left(\partial_t + \mathcal{L}^V\right) h_1 + \frac{h_1 - 2\varphi\rho V}{k + \varphi}\left(h_2 + \tfrac{1}{2}b\right), \tag{9.65b}$$

$$0 = \left(\partial_t + \mathcal{L}^V\right) h_0 + \frac{1}{4(k + \varphi)}\left(h_1 - 2\varphi\rho V\right)^2 - \varphi\rho^2 V^2. \tag{9.65c}$$

(c) Assuming that $V_t$ satisfies (9.14), find the optimal speed of trading and compare it to (9.17).

**E.9.2** An agent wishes to acquire $\mathfrak{N}$ shares by time $T$. Her performance criteria is

$$H^\nu(t,x,S,y,q) = \mathbb{E}_{t,x,S,y,q}\Big[X_T^\nu + (\mathfrak{N} - Q_T^\nu)(S_T^\nu + \alpha(\mathfrak{N} - Q_T^\nu))$$

$$+ \varphi \int_t^T (Q_u^\nu - \rho Y_u)^2\, du\Big]$$

and her value function is

$$H(t,x,S,y,q) = \inf_{\nu \in \mathcal{A}} H^\nu(t,x,S,y,q)\,, \tag{9.66}$$

where $Y_t$ is the total volume purchased by other market participants, and the acquired inventory $Q_t^\nu$, $S_t^\nu$, and acquisition cost $X_t^\nu$, satisfy

$$\begin{aligned}
dQ_t^\nu &= \nu_t\, dt\,, & Q_0^\nu &= q\,, \\
dS_t^\nu &= b\,\nu_t\, dt + \sigma\, dW_t\,, & S_0^\nu &= S\,, \\
dX_t^\nu &= (S_t^\nu + k\,\nu_t)\,\nu_t\, dt\,, & X_0^\nu &= x\,.
\end{aligned}$$

(a) Show that the value function satisfies the DPE

$$\big(\partial_t + \tfrac{1}{2}\sigma^2\partial_{SS} + \mathcal{L}^{y,V}\big) H + \varphi(q - \rho y)^2$$
$$+ \inf_\nu \big\{ (S^\nu + \nu k)\nu\, \partial_x H + b\nu\partial_S H + \nu\partial_q H \big\} \tag{9.68}$$

subject to terminal and boundary conditions and where $\mathcal{L}^{y,V}$ is the generator of $y$ and $V$.

(b) Show that the optimal acquisition speed is

$$\nu_t^* = \frac{\mathfrak{N} - Q_t^{\nu^*}}{k}\big(h_2(t) - \tfrac{1}{2}b\big) + \frac{h_1(t)}{2k}\,,$$

where

$$h_2(t) = \sqrt{k\varphi}\,\frac{1 + \zeta\, e^{2\xi\,(T-t)}}{1 - \zeta\, e^{2\xi(T-t)}} - \frac{1}{2}b\,,$$

with

$$\xi = \sqrt{\frac{\varphi}{k}} \quad \text{and} \quad \zeta = \frac{\alpha + \tfrac{1}{2}b + \sqrt{k\varphi}}{\alpha + \tfrac{1}{2}b - \sqrt{k\varphi}}\,,$$

and $h_1(t)$ solves the PIDE

$$\big(\partial_t + \mathcal{L}^{y,V}\big) h_1 - \frac{1}{k}\big(h_2(t) - \tfrac{1}{2}b\big) - 2\varphi(\mathfrak{N} - \rho y) = 0\,.$$

Note that if $\varphi = b = 0$ then the optimal acquisition speed is as in (6.17).

**E.9.3** Modify the problem of optimal liquidation described in Section 9.2 so that the liquidation rate $\mu_t$ takes into account the agent's trade. Derive the optimal liquidation rate and the inventory path.

E.9.4 Assume the setup of Section 9.4 where the trading rates of the agent and other traders affect the midprice. Solve the optimal liquidation problem where the agent targets POCV, that is her performance criteria is

$$H^\nu(t, x, S, \boldsymbol{\mu}, q)$$

$$= \mathbb{E}_{t,x,S,\boldsymbol{\mu},q}\left[ X_T^\nu + Q_T^\nu(S_T^\nu - \alpha Q_T^\nu) - \varphi \int_t^T ((\mathfrak{N} - Q_u^\nu) - \rho\, V_u)^2\, du \right],$$

where

$$V_t = \int_0^t \mu_u^- \, du$$

is the total volume traded on the sell side of the market.

# 10 Market Making

## 10.1 Introduction

In this chapter we model how a **market maker** (MM) maximises terminal wealth by trading in and out of positions using limit orders (LOs). The MM provides liquidity to the limit order book (LOB) by posting buy and sell LOs and the control variable is the depth, which is measured from the midprice, at which these LOs are posted. To formalise the problem, we list the relevant variables that we use throughout this section:

- $S = (S_t)_{0 \leq t \leq T}$, denotes the midprice, with $S_t = S_0 + \sigma W_t$, $\sigma > 0$ and $W = (W_t)_{0 \leq t \leq T}$ is a standard Brownian motion,
- $\delta^{\pm} = (\delta_t^{\pm})_{0 \leq t \leq T}$ denote the depth at which the agent posts LOs; sell LOs are posted at a price of $S_t + \delta_t^+$ and buy LOs at a price of $S_t - \delta_t^-$,
- $M^{\pm} = (M_t^{\pm})_{0 \leq t \leq T}$ denote the counting processes corresponding to the arrival of other participants' buy (+) and sell (−) market orders (MOs) which arrive at Poisson times with intensities $\lambda^{\pm}$,
- $N^{\delta, \pm} = (N_t^{\delta, \pm})_{0 \leq t \leq T}$ denote the controlled counting processes for the agent's filled sell (+) and buy (−) LOs,
- conditional on a market order (MO) arrival, the posted LO is filled with probability $e^{-\kappa^{\pm} \delta_t^{\pm}}$, with $\kappa^{\pm} \geq 0$,
- $X^{\delta} = (X_t^{\delta})_{0 \leq t \leq T}$ denotes the MM's cash process and satisfies the SDE

$$dX_t^{\delta} = (S_{t-} + \delta_t^+) \, dN_t^{\delta, +} - (S_{t-} - \delta_t^-) \, dN_t^{\delta, -}, \tag{10.1}$$

which accounts for the cash increase when a sell LO is lifted by a buy MO and the cash outflow when a buy LO is hit by an incoming sell MO,

- $Q^{\delta} = (Q_t^{\delta})_{0 \leq t \leq T}$ denotes the agent's inventory process and

$$Q_t^{\delta} = N_t^{\delta, -} - N_t^{\delta, +}. \tag{10.2}$$

As discussed in Section 8.2, whenever the process $N^{\delta, \pm}$ jumps, the process $M^{\delta, \pm}$ must also jump; but when $M^{\delta, \pm}$ jumps, $N^{\delta, \pm}$ will jump only if the MO is large enough to fill the agent's LO, and $N^{\delta, \pm}$ is not a Poisson process. Moreover, note that the fill rate of LOs can be written as $\Lambda_t^{\delta, \pm} = \lambda^{\pm} e^{-\kappa^{\pm} \delta_t^{\pm}}$, which is the rate of execution of an LO.

To simplify notation, in the rest of this chapter we suppress the superscript $\delta$ in the counting process for filled LOs, cash, and inventory. In Section 10.2 we discuss market making strategies when the MM does not face adverse selection costs, and her strategy takes into account restrictions on the amount of inventory she is willing to hold during the life of the strategy, as well as how costly it is to liquidate outstanding inventory at the terminal date of the strategy. In Section 10.3 the agent maximises terminal utility of cash holdings. Finally, in Section 10.4 we introduce various ways in which the MM faces adverse selection costs and how this affects the market making strategies.

## 10.2 Market Making

In this section we assume that the MM seeks the strategy $(\delta_s^{\pm})_{0 \le s \le T}$ that maximises cash at the terminal date $T$. We also assume that at time $T$ the MM liquidates her terminal inventory $Q_T$ using an MO at a price which is worse than the midprice to account for liquidity taking fees as well as the MO walking the LOB. Finally, the MM caps her inventory so that it is bounded above by $\bar{q} > 0$ and below by $\underline{q} < 0$, both finite, and also includes a running inventory penalty so that the performance criterion is

$$H^\delta(t, x, S, q) = \mathbb{E}_{t,x,q,S}\left[X_T + Q_T^\delta(S_T^\delta - \alpha\, Q_T^\delta) - \phi \int_t^T (Q_u)^2\, du\right],$$

where $\alpha \ge 0$ represents the fees for taking liquidity (i.e. using an MO) as well as the impact of the MO walking the LOB, and $\phi \ge 0$ is the running inventory penalty parameter. The MM's value function is

$$H(t, x, S, q) = \sup_{\delta^{\pm} \in \mathcal{A}} H^\delta(t, x, S, q), \tag{10.3}$$

where $\mathcal{A}$ denotes the set of admissible strategies, i.e. $\mathcal{F}$−predictable, bounded from below.

To solve the optimal control problem, a dynamic programming principle holds and the value function satisfies the DPE

$$0 = \partial_t H + \tfrac{1}{2}\sigma^2 \partial_{SS} H - \phi q^2$$
$$+ \lambda^+ \sup_{\delta^+}\left\{e^{-\kappa^+\delta^+}\left(H(t, x + (S + \delta^+), q - 1, S) - H\right)\right\} \mathbb{1}_{q > \underline{q}}$$
$$+ \lambda^- \sup_{\delta^-}\left\{e^{-\kappa^-\delta^-}\left(H\left(t, x - (S - \delta^-), q + 1, S\right) - H\right)\right\} \mathbb{1}_{q < \bar{q}}, \tag{10.4}$$

where $\mathbb{1}$ is the indicator function, with terminal condition

$$H(T, x, S, q) = x + q(S - \alpha q). \tag{10.5}$$

Recall that the set of admissible strategies imposes bounds on $q_t$, so that when $q_t = \bar{q}$ $(\underline{q})$ the optimal strategy is to post one-sided LOs which are obtained by solving (10.4) with the term proportional to $\lambda^-$ $(\lambda^+)$ absent as enforced by the indicator function $\mathbb{1}$ in the DPE. Alternatively, one can view these boundary cases as imposing $\delta^- = +\infty$ and $\delta^+ = +\infty$ when $q = \bar{q}$ and $\underline{q}$ respectively.

Intuitively, the various terms in the DPE equation represent the arrival of MOs that may be filled by LOs together with the diffusion of the asset price through the term $\frac{1}{2}\sigma^2 \partial_{SS} H$, and the effect of penalising deviations of inventories from zero along the entire path of the strategy which is captured by the term $\phi q^2$. In the second line of the DPE the sup over $\delta^+$ contain the terms due to the arrival of a market buy order (which is filled by a limit sell order), and here we see the change in the value function $H$ due to the arrival of the MO which fills the LO, so that cash increases by $(S+\delta^+)$ and inventory decreases by one unit. Similarly, in the last line in the DPE the sup over $\delta^-$ contain the analogous terms for the market sell orders which are filled by limit buy orders.

To solve the DPE we use the terminal condition (10.5) to make an ansatz for $H$. In particular, we write

$$H(t, x, q, S) = x + q S + h(t, q), \tag{10.6}$$

which has a simple interpretation. The first term is the accumulated cash, the second term is the book value of the inventory marked-to-market (i.e. the value of the shares at the current midprice), and the last term is the added value from following an optimal market making strategy up to the terminal date.

We proceed by substituting the ansatz into (10.4) to obtain

$$\phi q^2 = \partial_t h(t, q) + \lambda^+ \sup_{\delta^+} \left\{ e^{-\kappa^+ \delta^+} \left( \delta^+ + h(t, q-1) - h(t, q) \right) \right\} \mathbb{1}_{q > \underline{q}}$$
$$+ \lambda^- \sup_{\delta^-} \left\{ e^{-\kappa^- \delta^-} \left( \delta^- + h(t, q+1) - h(t, q) \right) \right\} \mathbb{1}_{q < \bar{q}}, \tag{10.7}$$

with terminal condition $h(T, q) = -\alpha q^2$.

Then the optimal depths in feedback form are given by

$$\delta^{+,*}(t, q) = \frac{1}{\kappa^+} - h(t, q-1) + h(t, q), \quad q \neq \underline{q}, \tag{10.8a}$$

$$\delta^{-,*}(t, q) = \frac{1}{\kappa^-} - h(t, q+1) + h(t, q), \quad q \neq \bar{q}, \tag{10.8b}$$

and the boundary cases are $\delta^{+,*}(t, q) = +\infty$ and $\delta^{-,*}(t, q) = +\infty$ when $q = \bar{q}$ and $\underline{q}$ respectively.

To understand the intuition behind the feedback controls we first note that the optimal $\delta^\pm$ can be decomposed into two components. The first component, $1/\kappa^\pm$, is the optimal strategy for a MM who does not impose any restrictions on inventory ($\alpha = \phi = 0$ and $|\underline{q}| = \bar{q} = \infty$) – see below in subsection 10.2.1.

The second component, the term $-h(t, q-1) + h(t, q)$, controls for inventories through time. As expected, if inventories are long, then the strategy consists in posting LOs that increase the probability of limit sell orders being hit. Moreover, the function $h(t, q)$ also induces mean reversion to an optimal inventory level, as a result of penalising accumulated inventories throughout the entire trading horizon and the strategy approaching $T$ as well as the other parameters of the model, including $\phi$.

Substituting the optimal controls into the DPE we obtain

$$\phi q^2 = \partial_t h(t,q) + \frac{e^{-1}\lambda^+}{\kappa^+} e^{-\kappa^+(-h(t,q-1)+h(t,q))} \mathbb{1}_{q>\underline{q}}$$
$$+ \frac{e^{-1}\lambda^-}{\kappa^-} e^{-\kappa^-(-h(t,q+1)+h(t,q))} \mathbb{1}_{q<\overline{q}} . \tag{10.9}$$

## Solving the DPE

It is possible to find an analytical solution to the DPE if the fill probabilities of LOs are the same on both sides of the LOB. In this case, if $\kappa = \kappa^+ = \kappa^-$ then write

$$h(t,q) = \frac{1}{\kappa} \log w(t,q) ,$$

and stack $w(t,q)$ into a vector

$$\boldsymbol{w}(t) = \left[ w(t,\overline{q}), \, w(t,\overline{q}-1), \ldots, \, w(t,\underline{q}) \right]' .$$

Now, let $\mathbf{A}$ denote the $(\overline{q}-\underline{q}+1)$-square matrix whose rows are labelled from $\overline{q}$ to $\underline{q}$ and whose entries are given by

$$\mathbf{A}_{i,q} = \begin{cases} -\phi\kappa q^2, & i=q, \\ \lambda^+ e^{-1}, & i=q-1, \\ \lambda^- e^{-1}, & i=q+1, \\ 0, & \text{otherwise}, \end{cases} \tag{10.10}$$

with terminal and boundary conditions $w(T,q) = e^{-\alpha\kappa q^2}$.

Then (10.9) becomes

$$\partial_t \boldsymbol{w}(t) + \mathbf{A}\,\boldsymbol{w}(t) = \mathbf{0} .$$

The solution of this matrix ODE is straightforward and we finally have,

$$\boxed{\boldsymbol{w}(t) = e^{\mathbf{A}(T-t)}\,\mathbf{z} ,} \tag{10.11}$$

where $\mathbf{z}$ is a $(\overline{q}-\underline{q}+1)$-dim vector where each component is $z_j = e^{-\alpha\kappa j^2}$, $j = \overline{q}, \ldots, \underline{q}$.

## Behaviour of the Strategy

Figure 10.1 shows the behaviour of the optimal depths as a function of time for different inventory levels. In the examples the arrival rate of MOs is $\lambda^\pm = 1$ (there are on average 1 buy and 1 sell MO per second), $\overline{q} = -\underline{q} = 3$, and $\phi = 10^{-5}$ in panel (a), and $\phi = 2 \times 10^{-4}$ in panel (b). In the left of panel (a) we show the optimal sell postings $\delta^+$, i.e. upon the arrival of a market buy order the MM is willing to sell one unit of the asset at the price $S_t + \delta^+$, and in the right of panel (a) we show the optimal buy postings $\delta^+$. For example, when the strategy is far away from expiry and inventories are close to the allowed minimum, the optimal

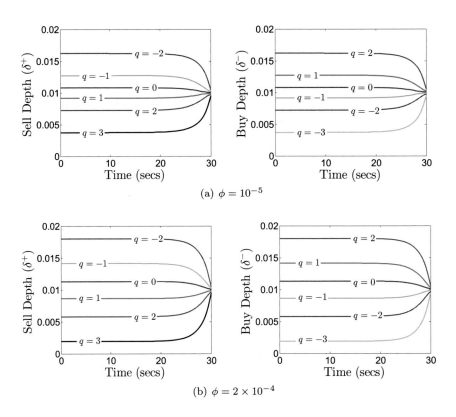

**Figure 10.1** The optimal depths as a function of time for various inventory levels and $T = 30$. The remaining model parameters are: $\lambda^{\pm} = 1$, $\kappa^{\pm} = 100$, $\bar{q} = -\underline{q} = 3$, $\alpha = 0.0001$, $\sigma = 0.01$, $S_0 = 100$.

sell posting is furthest away from the midprice because only at a very 'high' price is the MM willing to decrease her inventories further, and at the same time the optimal buy posting is very close to the midprice because the strategy would like to complete round-trip trades (i.e. a buy followed by a sell or a sell followed by a buy) and push inventories to zero.

We also observe that as the strategy approaches $T$ and $q_t < 0$ ($q_t > 0$), the optimal sell (buy) depth $\delta^+$ ($\delta^-$) decreases (increases). To understand the intuition behind the optimal strategy note that if the terminal inventory $q_T < 0$ is liquidated at the price $S_T - \alpha q_T$, then when $\alpha$ is sufficiently low, as well as being fractions of a second away from expiry, it is optimal to post nearer the midprice to increase the chances of being filled (i.e. selling one more unit of the asset) because the price is not expected to move too much before expiry and the entire position will be unwound at the midprice – making a profit on the last unit of the asset that was sold.

It is also interesting to see that the optimal strategy induces mean reversion in inventories. For example, if $q_t = 2$ then the sell depth is lower than the buy

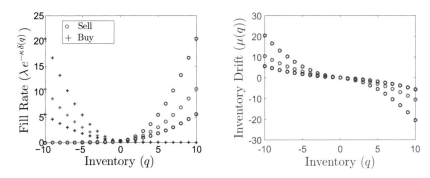

**Figure 10.2** Long-term inventory level. Model parameters are: $\lambda^{\pm} = 1$, $\kappa^{\pm} = 100$, $\bar{q} = -\underline{q} = 10$, $\alpha = 0.0001$, $\sigma = 0.01$, $S_0 = 100$, and $\phi = \{2 \times 10^{-3}, 10^{-3}, 5 \times 10^{-4}\}$.

depth, $\delta^+ < \delta^-$, so that it is more likely for the strategy to sell, than to buy, one unit of the asset. This asymmetry in the optimal depths is what induces mean reversion to zero in the inventory. Moreover, in panel (b) the strategy's running inventory penalty is much higher and it is clear that the higher $\phi$ is, the quicker inventories will revert to zero.

We also see that the strategy $\delta^{*\pm}(t,q)$ induces mean reversion in inventories, by observing that the expected drift in inventories is given by the difference in the arrival rates of filled orders. Thus, given the pair of optimal strategies $\delta^{+,*}(t,q), \delta^{-,*}(t,q)$, the expected drift in inventories is given by

$$\mu(t,q) \triangleq \lim_{s \downarrow t} \frac{1}{s-t} \mathbb{E}\left[Q_s - Q_t \mid Q_{t^-} = q\right]$$
$$= \lambda^- e^{-\kappa^- \delta^{-,*}(t,q)} - \lambda^+ e^{-\kappa^+ \delta^{+,*}(t,q)} .$$

(10.12)

Note that the drift $\mu(t,q)$ depends on time. For instance, it is clear that for the same level of inventory the speed will be different depending on how near or far the strategy is from the terminal date, because at time $T$ the strategy tries to unwind all outstanding inventory.

Figure 10.2 shows the optimal level of inventory to which the strategy reverts, where we assume that we are far away from $T - t \to \infty$ in (10.12). The model parameters are $\lambda^{\pm} = 1$, $\kappa^{\pm} = 100$, $\bar{q} = -\underline{q} = 10$, $\alpha = 0.0001$, $\sigma = 0.01$, $S_0 = 100$ and we vary the running penalty $\phi = \{2 \times 10^{-3}, 10^{-3}, 5 \times 10^{-4}\}$. Note that for the set of parameters we are using here it suffices to be a few seconds away from the terminal date so that the optimal postings are not affected by the proximity to $T$. In the left panel of the figure we plot the fill rate probabilities of both sides of the LOB which are given by $\lambda^{\pm} e^{-\kappa^{\pm} \delta^{\pm}(t,q),*}$. Blue circles and blue crosses are the fill rate probabilities for the sell side and buy side of the LOB respectively when $\phi = 2 \times 10^{-4}$.

Figure 10.3 shows the inventory and price path for one simulation of the strategy. The model parameters are $\lambda^{\pm} = 1$, $\kappa^{\pm} = 100$, $\bar{q} = -\underline{q} = 10$, $\phi = 2 \times 10^{-4}$,

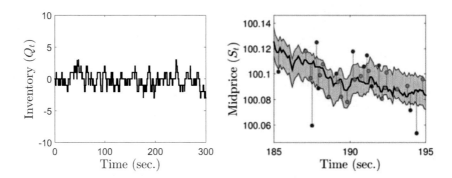

**Figure 10.3** Inventory and midprice path. Model parameters are: $\lambda^{\pm} = 1$, $\kappa^{\pm} = 100$, $\bar{q} = -\underline{q} = 10$, $\phi = 2 \times 10^{-4}$, $\alpha = 0.0001$, $\sigma = 0.01$, $S_0 = 100$.

$\alpha = 0.0001$, $\sigma = 0.01$, $S_0 = 100$. In the left panel we see how the inventory is mean reverting to zero and for this particular path we see that although the maximum and minimum amount of inventory that the strategy is allowed to hold is 10, it never goes beyond five units of the asset short or long.

The right panel of Figure 10.3 shows a window of the midprice path along with MM's buy and sell LOs. Solid circles in the figure show the incoming MO's which are filled by the MM's resting LOs (a red circle is a sell MO filled by the MM's buy LO and a blue circle is a buy MO filled by the MM's sell LO) and grey circles represent MOs that were filled by other market participants. The distance between the midprice and the MOs that arrive shows how far the MOs are walking into the LOB. At the beginning of the window, the agent's inventory is zero and we observe that the strategy acquires two units (one at $185.3s$ and another at $187.5s$) before the first sell order (at $187.8s$) is filled and then closed out an instant later (at $187.9s$). After the first filled buy order the strategy remains asymmetric and the agent posts closer to the midprice on the sell side of the book, compared to the buy side of the book, to rid herself of her inventory. At $189s$, $190.2s$, $190.8s$, $191.1s$ and $191.9s$, a sequence of sell orders is filled and the agent holds a short position of 2 assets after the last sale at $191.9s$. Her strategy is therefore to post closer to the midprice on the buy side of the book to increase her chance of unwinding her position. These shifts in her posts, which induce the unwinding of any inventory she acquires (long or short), continues until the end of the trading horizon.

Now we turn to discussing the financial performance of the strategy. The left panel of Figure 10.4 shows the profit and loss (P&L) of the optimal strategy and the right panel shows the lifetime inventory for different running penalty parameters $\phi = \{10^{-5}, 5 \times 10^{-5}, 10^{-3}, 10^{-2}\}$. We observe that when $\phi$ increases the histogram of P&L shifts to the left because the strategy does not allow inventory positions to stray away from zero, and hence expected profits decrease. The lifetime inventory histogram shows how much time the strategy holds an inventory of $n$. For example, when $\phi = 10^{-2}$ we know that the strategy heavily penalises

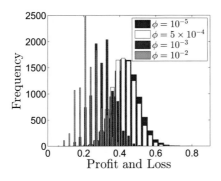 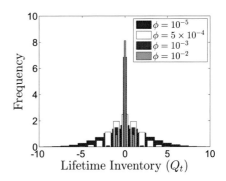

**Figure 10.4** P&L and Life Inventory of the optimal strategy for 10,000 simulations. The remaining model parameters are: $\lambda^{\pm} = 1$, $\kappa^{\pm} = 100$, $\bar{q} = -\underline{q} = 10$, $\alpha = 0.0001$, $\sigma = 0.01$, and $S_0 = 100$.

deviations of running inventory from zero, so the strategy spends most of the time at inventory levels of $-1$, $0$, $1$. As the running inventory penalty becomes smaller, the strategy spends more time at levels away from zero.

### 10.2.1    Market Making with no Inventory Restrictions

If we assume that the MM does not penalise running inventories, does not pick up a terminal inventory penalty, that is $\phi = \alpha = 0$, and there are no constraints on the amount of inventory the strategy may accumulate, i.e $|\underline{q}|, \bar{q} \to \infty$, then the MM's strategy simplifies to

$$\delta^{+,*}(t, q) = \frac{1}{\kappa^+}, \quad \text{and} \quad \delta^{-,*}(t, q) = \frac{1}{\kappa^-}. \tag{10.13}$$

This optimal strategy tells the MM to post in the LOB so that the probability of the LOs being filled is maximised. To see this we observe that if there are no penalties for liquidating terminal inventory, by assuming $\alpha = 0$ the terminal inventory is unwound at the midprice, and there is no running penalty for inventories straying away from zero, then we make the ansatz

$$H(t, x, q, S) = x + q \, S + h(t). \tag{10.14}$$

This is similar to the one proposed above, see (10.6), but here $h(t)$ does not depend on $q$ because the MM does not pose any restrictions on inventory throughout the life of the strategy and can liquidate terminal inventory at the midprice. Thus, substituting the ansatz into the DPE

$$0 = \partial_t h + \lambda^+ \sup_{\delta^+} \left\{ e^{-\kappa^+ \delta^+} \delta^+ \right\} + \lambda^- \sup_{\delta^-} \left\{ e^{-\kappa^- \delta^-} \delta^- \right\} \tag{10.15}$$

with terminal condition $h(T) = 0$, delivers the result (10.13).

Furthermore, we can show that

$$h(t) = e^{-1} \left( \frac{\lambda^+}{\kappa^+} + \frac{\lambda^-}{\kappa^-} \right) (T - t).$$

This result is simple to interpret. An MM who does not penalise inventories and who unwinds terminal inventory at the midprice, will make markets by maximising the probability of her LOs being filled at every instant in time regardless of the inventory position or how close the terminal date is. Therefore, the MM's problem reduces to choosing $\delta^\pm$ to maximise the expected depth conditional on an MO hitting or lifting the appropriate side of the LOB, i.e. to maximise $\delta^\pm e^{-\kappa^\pm \delta^\pm}$. The first order condition of this optimisation problem is

$$e^{-\kappa^\pm \delta^\pm} - \kappa^\pm \delta^\pm e^{-\kappa^\pm \delta^\pm} = 0, \tag{10.16}$$

so the optimal depths are as in (10.13).

### 10.2.2    Market Making At-The-Touch

In very liquid markets, most orders do not walk the book and instead tend to only lift or hit LOs posted at-the-touch. To capture this market feature, in this section we investigate the agent's optimal postings at-the-touch, i.e. at the best bid and best offer. Throughout we assume that the spread is constant and equal to $\Delta$. Next, let $\ell_t^\pm \in \{0, 1\}$ denote whether the agent is posted on the sell side $(+)$ or buy side $(-)$ of the LOB. In this way, the agent may be posted on both sides of the book, only the sell side, only the buy side, or not posted at all. Her performance criteria is

$$H^\ell(t, x, S, q) = \mathbb{E}_{t,x,S,q} \left[ X_T^\ell + Q_T^\ell \left( S_T - \left( \tfrac{\Delta}{2} + \varphi Q_T^\ell \right) \right) - \phi \int_t^T \left( Q_u^\ell \right)^2 du \right],$$

where her cash process $X_t^\ell$ now satisfies the SDE

$$dX_t^\ell = \left( S_t + \tfrac{\Delta}{2} \right) dN_t^{+,\ell} - \left( S_t - \tfrac{\Delta}{2} \right) dN_t^{-,\ell},$$

where $N_t^{\pm,\ell}$ denote the counting process for filled LOs. We also further assume that, if she is posted in the LOB, when a matching MO arrives her LO is filled with probability one. In this case, $N_t^{\pm,\ell}$ are controlled doubly stochastic Poisson processes with intensity $\ell_t^\pm \lambda^\pm$. Finally, at the terminal date any open inventory position is liquidated using an MO and the price obtained per share is the best bid $(Q_T > 0)$ or offer $(Q_T < 0)$ and picks up a penalty $\varphi Q_T^2$, with $\varphi \geq 0$, which includes market impact (walking the LOB) and liquidity taking fees.

As before, the set $\mathcal{A}$ of admissible strategies are $\mathcal{F}$-predictable such that the agent is not posted on the buy (sell) side if her inventory is equal to the upper (lower) inventory constraints $\bar{q}$ $(\underline{q})$ and her value function is denoted by

$$H(t, x, S, q) = \sup_{\ell \in \mathcal{A}} H^\ell(t, x, S, q).$$

## The Resulting DPE

Applying the DPP, we find the agent's value function $H$ should satisfy the DPE

$$0 = \left(\partial_t + \tfrac{1}{2}\sigma^2 \partial_{SS}\right) H - \phi q^2$$
$$+ \lambda^+ \max_{\ell^+ \in \{0,1\}} \left\{ \left( H\left(t, x + \left(S + \tfrac{\Delta}{2}\right) \ell^+, S, q - \ell^+\right) - H\right)\right\} \mathbb{1}_{q > \underline{q}}$$
$$+ \lambda^- \max_{\ell^- \in \{0,1\}} \left\{ \left( H\left(t, x - \left(S - \tfrac{\Delta}{2}\right) \ell^-, S, q + \ell^-\right) - H\right)\right\} \mathbb{1}_{q < \bar{q}},$$

subject to the terminal condition

$$H(T, x, S, q) = x + q\left(S - \left(\tfrac{\Delta}{2} + \varphi q\right)\right).$$

The various terms in the DPE carry the following interpretations:

- the first line in the DPE represents the diffusive component of the midprice and the running inventory penalisation,
- the maximisation terms represent the agent's control to post or not on the sell or buy side of the LOB,
- the maximisation term in the second line represents the change in value function, if the agent is posted, due to the arrival of an MO which lifts the agent's offer,
- the third line is for the other side of the book.

The terminal condition once again suggests the ansatz which splits out the accumulated cash, the book value of the shares marked-to-market at the midprice, and the added value from optimally making markets throughout the remaining life of the strategy:

$$H(t, x, S, q) = x + q\,S + h(t, q),$$

and on substituting this ansatz into the above DPE we find that $h$ satisfies

$$0 = \partial_t h - \phi q^2$$
$$+ \lambda^+ \max_{\ell^+ \in \{0,1\}} \left\{ \left( \ell^+ \tfrac{\Delta}{2} + \left[h(t, q - \ell^+) - h(t, q)\right]\right)\right\} \mathbb{1}_{q > \underline{q}}$$
$$+ \lambda^- \max_{\ell^- \in \{0,1\}} \left\{ \left( \ell^- \tfrac{\Delta}{2} + \left[h(t, q + \ell^-) - h(t, q)\right]\right)\right\} \mathbb{1}_{q < \bar{q}},$$

subject to the terminal condition

$$h(T, q) = -q\left(\tfrac{\Delta}{2} + \varphi q\right).$$

The form of the optimising terms allows us to characterise the optimal postings in a compact form. When $\ell = 0$ both terms that are being maximised are zero, hence, the optimal postings of the agent can be characterised succinctly as

$$\boxed{\begin{aligned} \ell^{+,*}(t, q) &= \mathbb{1}_{\left\{ \frac{\Delta}{2} + [h(t, q-1) - h(t,q)] > 0\right\} \cap \{q > \underline{q}\}}, \\ \ell^{-,*}(t, q) &= \mathbb{1}_{\left\{ \frac{\Delta}{2} + [h(t, q+1) - h(t,q)] > 0\right\} \cap \{q < \bar{q}\}}. \end{aligned}}$$

$$(10.17)$$

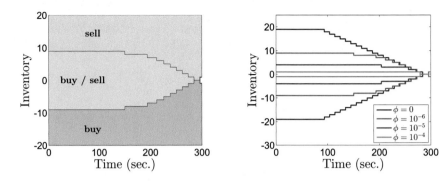

**Figure 10.5** The optimal strategy for the agent who posts only at-the-touch.

The interpretation of this result is that the agent posts an LO on the appropriate side of the LOB by ensuring that she only posts if the arrival of an MO, which hit/lifts her LO, produces a change in her value function larger than $-\frac{\Delta}{2}$.

### Strategy Features
Next, we illustrate some typical features of the optimal solution to gain some insight into the optimal strategy. For this purpose we use the following set of model parameters:

$$T = 300 \text{ sec}, \quad \bar{q} = -\underline{q} = 20, \quad \lambda^{\pm} = \frac{50}{300},$$

$$\Delta = 0.01, \quad \phi = 0.01, \quad \sigma = 0.001.$$

Note that the rate of arrival of market orders is chosen so that on average the agent's upper/lower inventory bounds are no more than 20% of the market.

Figure 10.5 shows how the agent's optimal posting varies with time and the running penalty. The left panel explicitly shows that the agent posts only sell LOs whenever her inventory is very high, and only buy LOs whenever her inventory is very low. In the central region, she posts both buy and sell LOs. In this manner, the agent's inventory is constrained to remain within one unit of the green region – once her inventory escapes she posts only on one side of the book thus pushing inventory back into the green region. We therefore see that despite the agent allowing herself to hold up to 20 units of the asset, long or short, the running penalty constrains her strategy. Furthermore, note that as the running penalty increases, the region over which the agent constrains her inventory shrinks, and eventually reaches the point at which she only takes on one single unit of the asset (long or short) and then immediately liquidates it.

Later, in subsection 10.4.2 we see how the agent modifies her strategy to account for adverse selection effects.

### 10.2.3     Market Making Optimising Volume

In the previous sections, when the agent posts an LO, she is assumed to be placing a single order. This single order can be thought of as the typical order size of, say, 100 shares. The agent may, however, wish to optimise the posted volume. In this case, the agent's performance criteria is taken to be

$$H^{\ell}(t, x, S, q) = \mathbb{E}_{t,x,S,q}\left[X_T^{\ell} + Q_T^{\ell}\left(S_T - \left(\tfrac{\Delta}{2} + \varphi Q_T^{\ell}\right)\right) - \phi \int_t^T \left(Q_u^{\ell}\right)^2 du\right],$$

and her cash process $X_t^{\ell}$ now satisfies the SDE

$$dX_t^{\ell} = \left(S_t + \tfrac{\Delta}{2}\right) \ell_t^{+} \, dN_t^{+,\ell} - \left(S_t - \tfrac{\Delta}{2}\right) \ell_t^{-} \, dN_t^{-,\ell},$$

where $\ell_t^{\pm}$ are $\mathcal{F}$-predictable such that

$$\ell_t^{+} \in \left\{0, 1, 2, , \dots, q_{t-} - \underline{q}\right\} \quad \text{and} \quad \ell_t^{-} \in \left\{0, 1, 2, , \dots, \bar{q} - q_{t-}\right\},$$

and $N_t^{\pm,\ell}$ denote the counting processes for her filled LOs – not accounting for the volume traded (i.e. it only counts whether an MO arrived and filled her posted LO). The restrictions on the volume ensure that the agent never posts a volume which, if filled, would send her inventory outside of her allowed trading bounds. We further assume that if she is posted in the LOB when a matching MO arrives, her LO is filled with probability $\rho(\ell)$ where $\ell$ is posted volume. For example, $\rho(\ell) = e^{-\kappa\ell}$ would represent an exponential fill probability. In this case, $N_t^{\pm,\ell}$ are controlled doubly stochastic Poisson processes with intensity $\rho(\ell_t^{\pm}) \lambda^{\pm} \mathbb{1}_{\ell_t^{\pm} > 0}$. In this formulation, we have further assumed that if the agent makes a post of a given volume, the entire volume is matched or none at all is matched. The approach here can be generalised to account for partial fills of postings, but this is left as an exercise for the reader.

As before, the set $\mathcal{A}$ of admissible strategies also restricts her so that the strategy does not post on the buy (sell) side of the LOB if her inventory is equal to the upper (lower) inventory constraints $\bar{q}$ ($\underline{q}$). Her value function is denoted by

$$H(t, x, S, q) = \sup_{\ell \in \mathcal{A}} H^{\ell}(t, x, S, q).$$

**The Resulting DPE**
This analysis is similar to that of the previous section, except now the set of strategies allows the agent to post multiple volumes. In this case, applying the DPP we find the agent's value function $H$ should satisfy the DPE

$$0 = \left(\partial_t + \tfrac{1}{2}\sigma^2 \partial_{SS}\right)H - \phi q^2$$
$$+ \lambda^{+} \max_{\ell^{+} \in \{0,1,\dots,q-\underline{q}\}} \left\{\rho(\ell^{+})\left(H\left(t, x + \left(S + \tfrac{\Delta}{2}\right)\ell^{+}, S, q - \ell^{+}\right) - H\right)\right\} \mathbb{1}_{q > \underline{q}}$$
$$+ \lambda^{-} \max_{\ell^{-} \in \{0,1,\dots,\bar{q}-q\}} \left\{\rho(\ell^{-})\left(H\left(t, x - \left(S - \tfrac{\Delta}{2}\right)\ell^{-}, S, q + \ell^{-}\right) - H\right)\right\} \mathbb{1}_{q < \bar{q}},$$

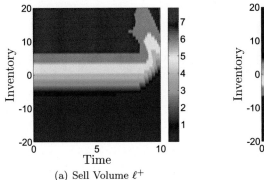 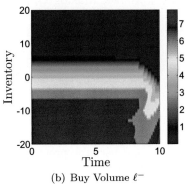

(a) Sell Volume $\ell^+$                    (b) Buy Volume $\ell^-$

**Figure 10.6** The optimal volume postings for the agent who posts only at-the-touch.

subject to the terminal condition

$$H(T, x, S, q) = x + q \left( S - \left( \tfrac{\Delta}{2} + \varphi q \right) \right) .$$

With the exception that the maxima are computed over the set of allowed volumes and cash jumps accordingly, this non-linear PDE is identical to the one derived in the previous section when the agent did not optimise over the volume of the LOs. The terminal condition suggests the usual ansatz $H(t, x, S, q) = x + q S + h(t, q)$ which splits out the accumulated cash, book value of the inventory marked-to-market using the midprice, and the added value from optimally making markets throughout the remaining life of the strategy. On substituting into the DPE we find that $h$ satisfies

$$0 = \partial_t h - \phi q^2$$
$$+ \lambda^+ \max_{\ell^+ \in \{0,1,\ldots,q-\underline{q}\}} \left\{ \rho(\ell^+) \left( \ell^+ \tfrac{\Delta}{2} + \left[ h(t, q - \ell^+) - h(t, q) \right] \right) \right\} \mathbb{1}_{q > \underline{q}}$$
$$+ \lambda^- \max_{\ell^- \in \{0,1,\ldots,q-\overline{q}\}} \left\{ \rho(\ell^-) \left( \ell^- \tfrac{\Delta}{2} + \left[ h(t, q + \ell^-) - h(t, q) \right] \right) \right\} \mathbb{1}_{q < \overline{q}},$$

subject to the terminal condition

$$h(T, q) = -q \left( \tfrac{\Delta}{2} + \varphi q \right) .$$

**Strategy Features**
Figure 10.6 shows the agent's optimal volume postings at-the-touch using the following model parameters:

$$T = 10 \text{ sec}, \quad \overline{q} = -\underline{q} = 20, \quad \lambda^{\pm} = 5,$$
$$\Delta = 0.01, \quad \varphi = 0.01, \quad \sigma = 0.001,$$
$$\rho(\ell) = e^{-0.01\,\ell}, \quad \text{and} \quad \phi = 10^{-3} .$$

Notice that the posted sell volume increases as inventory increases, while the posted buy volume increases as inventory decreases. Furthermore, there are large

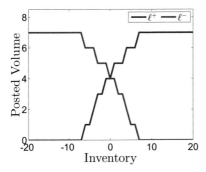 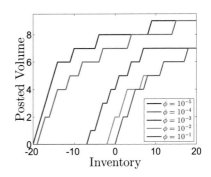

**Figure 10.7** The optimal volume postings (at $t = 0$) for the agent who posts only at-the-touch. The left panel has $\phi = 10^{-3}$ and the right panel shows only $\ell^+$.

regions where the agent is posted only on one side of the LOB. Figure 10.7 shows the asymptotic optimal volume postings at the start of trading. The left panel shows both the buy and sell postings when $\phi = 10^{-3}$, while the right panel shows how the postings vary when $\phi$ varies for the sell side of the book. For $\phi = 10^{-3}$, the agent stops posting limit sell orders when her inventory is at or below $-7$ and stops posting limit buy orders when her inventory is at or above $+7$. As the right panel in Figure 10.7 shows, the critical level below which the agent stops posting moves towards zero inventory as the inventory penalisation parameter $\phi$ increases. With $\phi = 0.1$, the agent will withdraw from the market and simply not post any limit orders.

Interestingly, the agent's posted volume is typically not sufficient to draw her inventory back to zero. For example, when $\phi = 10^{-3}$ and her inventory is 10, she will post only a sell LO for 7 units, which once filled, will draw her inventory down to 3. At this point, she will post a sell LO for 5 units and a buy LO for 3 units, neither of which if filled pulls her inventory to zero.

## 10.3 Utility Maximising Market Maker

In the previous section, the agent was indifferent to uncertainty in the cash value of her sales and instead maximised expected profit from making markets subject to inventory controls. Some agents, however, may instead wish to penalise uncertainty in their sales directly. Here, we show that if the agent uses exponential utility as a performance measure, her strategy will be identical (up to a constant and a re-scaling of parameters) to the one implied by the running penalty studied in Section 10.2.

To this end, suppose the agent sets preferences based on expected utility of terminal cash with exponential utility $u(x) = -e^{-\gamma x}$. In this case, her performance criteria is

$$G^\delta(t, x, S, q) = \mathbb{E}_{t,x,S,q}\left[-\exp\left\{-\gamma\left(X_T^\delta + Q_T^\delta\left(S_T - \alpha Q_T^\delta\right)\right)\right\}\right].$$

Using the standard approach, her value function,

$$G(t, x, S, q) = \sup_{\delta \in \mathcal{A}} G^\delta(t, x, S, q),$$

will satisfy the DPE

$$
\begin{cases}
\begin{aligned}
\left(\partial_t + \tfrac{1}{2}\sigma^2 \partial_{SS}\right) G & \\
+ \lambda^+ \sup_{\delta+} \left\{ e^{-\kappa^+ \delta^+} \left[G(t, x + (S + \delta^+), S, q - 1) - G\right] \mathbb{1}_{q > \underline{q}} \right\} & \\
+ \lambda^- \sup_{\delta-} \left\{ e^{-\kappa^- \delta^-} \left[G(t, x - (S - \delta^-), S, q + 1) - G\right] \mathbb{1}_{q < \overline{q}} \right\} & \\
= 0, & \\
G(t, x, S, 0) = -e^{-\gamma x}, & \\
G(T, x, S, q) = -e^{-\gamma(x + q(S - \alpha q))}. &
\end{aligned}
\end{cases}
$$

We leave it as an exercise for the reader to show that ansatz

$$G(t, x, S, q) = -e^{-\gamma(x + q\,S + g(t,q))}$$

leads to the following equation for $g(t, q)$:

$$
\partial_t g - \tfrac{1}{2}\sigma^2\,\gamma\,q^2 + \sup_{\delta+} \lambda^+\, e^{-\kappa^+ \delta^+} \frac{1 - e^{-\gamma\left(\delta^+ + g(t, q-1) - g(t, q)\right)}}{\gamma}
$$
$$
+ \sup_{\delta-} \lambda^-\, e^{-\kappa^- \delta^-} \frac{1 - e^{-\gamma\left(\delta^- + g(t, q+1) - g(t, q)\right)}}{\gamma} = 0,
\tag{10.18}
$$

with terminal and boundary conditions

$$g(t, 0) = 0, \quad \text{and} \quad g(T, q) = -\alpha\,q^2.$$

From the first order condition, we find that the optimal depths, in feedback control form, at which the agent posts are

$$
\boxed{
\begin{aligned}
\delta^{*,+} &= \frac{1}{\gamma}\log\left(1 + \frac{\gamma}{\kappa^+}\right) + g(t, q) - g(t, q - 1), & q > \underline{q}, \\
\delta^{*,-} &= \frac{1}{\gamma}\log\left(1 + \frac{\gamma}{\kappa^-}\right) + g(t, q) - g(t, q + 1), & q < \overline{q}.
\end{aligned}
}
\tag{10.19}
$$

This form is very similar to, but slightly differs from, the optimal depth in the previous section provided in (10.8). The $g$ function may differ from $h$ and the base line level $(\kappa^\pm)^{-1}$ is modified to $(\hat{\kappa}^\pm)^{-1} = \frac{1}{\gamma}\log\left(1 + \frac{\gamma}{\kappa^\pm}\right)$. This modification can be seen as a risk aversion bias. Indeed, in the limit of zero risk-aversion

$$\hat{\kappa}^\pm \xrightarrow{\gamma \downarrow 0} \kappa^\pm,$$

and the result from the previous section is recovered.

Substituting this feedback form into (10.18), we now find the non-linear ODE

$$\partial_t g - \tfrac{1}{2}\sigma^2\,\gamma\,q^2 + \frac{\hat{\lambda}^+}{\kappa^+}\,e^{-\kappa^+(g(t,q)-g(t,q-1))} + \frac{\hat{\lambda}^-}{\kappa^-}\,e^{-\kappa^-(g(t,q)-g(t,q+1))} = 0\,,$$

where

$$\hat{\lambda}^\pm = \left(\kappa^\pm/(\kappa^\pm+\gamma)\right)^{1+\kappa^\pm/\gamma}\lambda^\pm\,.$$

In the limit of zero risk-aversion $\hat{\lambda}^\pm \xrightarrow{\gamma\downarrow 0} e^{-1}\lambda^\pm = \tilde{\lambda}^\pm$ and once again we recover the parameter that appears in (10.9). The above ODE is in fact identical in structure to (10.9). Hence, matching parameters by setting

$$\phi = \frac{1}{2}\sigma^2\gamma\,,\qquad \lambda_0 = e^{+1}\hat{\lambda}\,,\qquad \text{and}\qquad \kappa_0 = \kappa\,,$$

where the subscript 0 denotes the parameters to use in the base model from the previous section, we see that $h(t,q)$ and $g(t,q)$ will coincide and the optimal strategies satisfy the relation

$$\boxed{\delta^{\pm,*} = \delta_0^{\pm,*} + \left((\hat{\kappa}^\pm)^{-1} - (\kappa_0^\pm)^{-1}\right)\,.} \tag{10.20}$$

In other words, with a re-scaling of model parameters, the optimal strategy for the utility maximising agent is the same, up to a constant shift, as that of the agent who only penalises running inventory.

In addition to the relationship between the optimal strategies, the value functions can be written in terms of one another. Since $h(t,q) = g(t,q)$, we have

$$H(t,x,S,q) = -\frac{1}{\gamma}\log\left(-G(t,x,S,q)\right)\,,$$

or writing the value function in the original control form, we obtain

$$\boxed{\begin{aligned} &\sup_{\delta\in\mathcal{A}} \mathbb{E}_{t,x,S,q}\left[X_\tau^\delta + Q_\tau^\delta(S_\tau - \alpha\,Q_\tau) - \phi\int_0^\tau (Q_s^\delta)^2\,ds\right]\\ &= -\frac{1}{\gamma}\log\left(-\sup_{\delta\in\mathcal{A}} \mathbb{E}_{t,x,S,q}\left[-\exp\left\{-\gamma\left(X_\tau^\delta + Q_\tau^\delta(S_\tau - \alpha\,Q_\tau)\right)\right\}\right]\right)\,. \end{aligned}} \tag{10.21}$$

This relationship between the value functions is in fact part of a more general result that relates optimisation problems to exponential utility, and optimisation problems to penalties (see the further readings section).

## 10.4  Market Making with Adverse Selection

The market place is populated by traders that come to the market for different reasons and with varying degrees of information. One of the most important risks faced by MMs is adverse selection risk. As discussed in Chapter 2 (see for example Sections 2.2 and 2.3), when trading with informed market participants the MM is exposed to having a sell limit order (LO) filled right before prices go up, or a buy LO filled before prices go down. In this section we extend the market making problem developed in Section 10.2, and present two ways in which midprice dynamics incorporate adverse selection effects. In the first model we assume that the midprice undergoes jumps in the direction of incoming market orders (MOs). In the second specification the midprice's drift has a short-term-alpha component which is affected by the arrival of MOs.

## 10.4.1 Impact of Market Orders on Midprice

Here we assume that the midprice dynamics follows

$$dS_t = \sigma dW_t + \epsilon^+ dM_t^+ - \epsilon^- dM_t^-, \tag{10.22}$$

where $M_t^+$ and $M_t^-$ are Poisson processes, with intensities $\lambda^+$ and $\lambda^-$ respectively, which count the number of buy $(+)$ and sell $(-)$ MOs. Every time an MO arrives, the midprice will undergo a jump of size $\epsilon^\pm$ which are i.i.d. and whose distribution functions are $F^\pm$ with finite first moment denoted by $\varepsilon^\pm = \mathbb{E}[\epsilon^\pm]$.

Intuitively, here we can view the dynamics of the midprice as the sum of two components. The first component, the Brownian motion on the right-hand side of (10.22), captures the changes in the midprice that are due to information flows that reach all or some market participants who subsequently update their quotes. The other component, the jump process with increments $\epsilon^+ dM_t^+ - \epsilon^- dM_t^-$, represents the changes in the midprice caused by the arrival of MOs that have a permanent price impact. MOs may come at times when there is enough liquidity in the market – hence prices remain unchanged or change by a negligible amount; or they may arrive at times when liquidity is thin or the orders are sent by traders with superior information, and these trades have a permanent impact on prices. The impact of trading on the midprice may also be viewed as the action of informed traders. If an informed trader purchases (sells) shares, he will only do so if the asset price is known to be going up (down). The resulting increase (decrease) of the midprice following informed trading can be approximated by an immediate, and permanent, price impact as we model here.

The rest of the MM's setup is as in Section 10.2, with the only difference that here the midprice follows (10.22). For convenience we repeat the MM's performance criteria:

$$H^\delta(t, x, S, q) = \mathbb{E}_{t,x,S,q}\left[X_T^\delta + Q_T^\delta(S_T - \alpha Q_T^\delta) - \phi \int_t^T (Q_u^\delta)^2 \, du\right],$$

so her value function is

$$H(t, x, S, q) = \sup_{\delta^\pm \in \mathcal{A}} H^\delta(t, x, S, q).$$

Thus the MM's value function satisfies the DPE

$$0 = \partial_t H + \tfrac{1}{2}\sigma^2 \partial_{SS} H - \phi q^2$$
$$+ \lambda^+ \sup_{\delta^+} \left\{ e^{-\kappa^+ \delta^+} \mathbb{1}_{q > \underline{q}} \, \mathbb{E}\left[ H(t, x + (S + \delta^+), S + \epsilon^+, q - 1) - H \right] \right.$$
$$\left. + \left( 1 - e^{-\kappa^+ \delta^+} \mathbb{1}_{q > \underline{q}} \right) \mathbb{E}\left[ H(t, x, S + \epsilon^+, q) - H \right] \right\} \quad (10.23)$$
$$+ \lambda^- \sup_{\delta^-} \left\{ e^{-\kappa^- \delta^-} \mathbb{1}_{q < \bar{q}} \, \mathbb{E}\left[ H(t, x - (S - \delta^-), S - \epsilon^-, q + 1) - H \right] \right.$$
$$\left. + \left( 1 - e^{-\kappa^- \delta^-} \mathbb{1}_{q < \bar{q}} \right) \mathbb{E}\left[ H(t, x, S - \epsilon^-, q) - H \right] \right\} ,$$

subject to the terminal condition

$$H(T, x, S, q) = x + q \left( S - \alpha q \right), \quad (10.24)$$

and where the expectation is over the random variables $\epsilon^\pm$ (not over $x$, $S$, or $q$) and $\mathbb{1}$ is the indicator function.

Intuitively, the various terms in the HJB equation have the same interpretation as the case above in Section 10.2, with a difference in how the value function changes when the midprice jumps upon the arrival of an MO. To see this, note that the sup over $\delta^+$ contains the terms due to the arrival of a market buy order (which is filled by a limit sell order). The first term represents the expected change in the value function $H$ due to the arrival of an MO which fills the LO and the midprice $S_t$ jumps up by the random amount $\epsilon^+$; and the second term represents the arrival of an MO which does not reach the LO's price level (but still causes a random jump in the midprice). Similarly, the sup over $\delta^-$ contains the analogous terms for the market sell orders which are filled by limit buy orders.

To solve the DPE we make the ansatz

$$H(t, x, S, q) = x + q S + h(t, q), \quad (10.25)$$

and substituting it into the DPE we obtain

$$\phi q^2 = \partial_t h + \lambda^+ \sup_{\delta^+} \left\{ e^{-\kappa^+ \delta^+} \left( \delta^+ - \varepsilon^+ + h_{q-1} - h_q \right) \right\} \mathbb{1}_{q > \underline{q}}$$
$$+ \lambda^- \sup_{\delta^-} \left\{ e^{-\kappa^- \delta^-} \left( \delta^- - \varepsilon^- + h_{q+1} - h_q \right) \right\} \mathbb{1}_{q < \bar{q}} \quad (10.26)$$
$$+ \left( \varepsilon^+ \lambda^+ - \varepsilon^- \lambda^- \right) q ,$$

subject to $h(T, q) = -\alpha q^2$ which allows us to solve for the optimal controls in feedback form:

$$\delta^{+,*}(t, q) = \frac{1}{\kappa^+} + \varepsilon^+ - h(t, q - 1) + h(t, q), \quad q \neq \underline{q}, \quad (10.27a)$$

$$\delta^{-,*}(t, q) = \frac{1}{\kappa^-} + \varepsilon^- - h(t, q + 1) + h(t, q), \quad q \neq \bar{q}. \quad (10.27b)$$

The interpretation of the optimal controls in feedback form is very similar to what was discussed above. The main difference is that here MOs impact the midprice and this affects the optimal controls in two ways: one is explicitly shown in (10.27) in the form of $\varepsilon^\pm$, and the other is encoded in the solution of $h(t, q)$.

It is clear that the MM incorporates the impact of MOs by including the expectation of the jump in prices, conditional on an MO arriving, by posting LOs which are $\varepsilon^\pm = \mathbb{E}[\epsilon^\pm]$ further away from the midprice. In this way the MM trader recovers, on average, the losses she incurs due to adverse selection.

Moreover, note that the effects of the jumps in the midprice also feed into the solution of $h(t, q)$ because the optimal strategy must take into account the future arrival of MOs, as these move the prices. Thus, it is important to note that the optimal controls derived here are not the controls as given in (10.8) plus the recovery of the average losses $\varepsilon^\pm$ adverse selection costs. This becomes clear when looking at the solution of the DPE which we discuss in the next subsection.

### Solving the DPE

If $\kappa^+ = \kappa^- = \kappa$, then write

$$h(t, q) = \frac{1}{\kappa} \log \omega(t, q) ,$$

and stack $\omega(t, q)$ into a vector

$$\boldsymbol{\omega}(t) = \left[ \omega(t, \bar{q}), \, \omega(t, \bar{q} - 1), \dots, \, \omega(t, \underline{q}) \right]' .$$

Furthermore, let $\mathbf{A}$ denote the $(\bar{q} - \underline{q} + 1)$-square matrix whose rows are labelled from $\bar{q}$ to $\underline{q}$ and whose entries are given by

$$\mathbf{A}_{i,q} = \begin{cases} q\kappa \left( \varepsilon^+ \lambda^+ - \varepsilon^- \lambda^- \right) - \phi \kappa \, q^2 , & i = q , \\ \tilde{\lambda}^+ , & i = q - 1 , \\ \tilde{\lambda}^- , & i = q + 1 , \\ 0, & \text{otherwise} , \end{cases} \tag{10.28}$$

where $\tilde{\lambda}^\pm = \lambda^\pm e^{-1 - \kappa \varepsilon^\pm}$. Then, on substituting $h$ in terms of $\omega$ in (10.26), we find that

$$\partial_t \boldsymbol{\omega}(t) + \mathbf{A} \, \boldsymbol{\omega}(t) = \mathbf{0} .$$

This matrix ODE can be easily solved to find

$$\boxed{\boldsymbol{\omega}(t) = e^{\mathbf{A}(T-t)} \mathbf{z} ,} \tag{10.29}$$

where $\mathbf{z}$ is a $(\bar{q} - \underline{q} + 1)$-dim vector where each component is $z_j = e^{-\alpha \kappa j^2}$, $j = \bar{q}, \dots, \underline{q}$. Note that this solution is similar to the one derived above in Section 10.2 but here we have the impact of the MOs on the midprice dynamics.

As a direct consequence of assuming that the shape of the LOB is symmetric, as well as assuming that the rate and impact on midprice of arrival of MOs is the same ($\kappa^\pm = \kappa$, $\lambda^\pm = \lambda$, $\varepsilon^\pm = \varepsilon$), the MM's optimal depths on the buy side with $q$ shares equals the optimal depth on the sell side with $-q$ shares, $\delta^{\pm*}(t, q) = \delta^{\mp*}(t, -q)$.

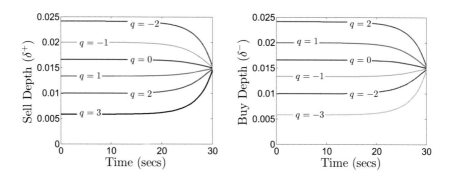

**Figure 10.8** The optimal depths as a function of time for various inventory levels and $T = 30$. The remaining model parameters are: $\lambda^{\pm} = 1$, $\kappa^{\pm} = 100$, $\overline{q} = -\underline{q} = 3$, $\phi = 0.02$, $\alpha = 0.0001$, $\sigma = 0.01$, $S_0 = 100$, $\varepsilon = 0.005$.

## The Behaviour of the Strategy

In this section we illustrate several aspects of the behaviour of the optimal strategy as a function of $q$, $\phi$, $\alpha$, $\lambda^{\pm}$, $\overline{q}$, and $-\underline{q}$. In Figure 10.8 we show the optimal sell and buy depths when $T = 30$, $\lambda^{\pm} = 1$, $\kappa^{\pm} = 100$, $\overline{q} = -\underline{q} = 3$, $\phi = 0.02$, $\alpha = 0.0001$, $\sigma = 0.01$, $S_0 = 100$, and $\varepsilon = 0.005$. The figure shows that when $q = 0$ the optimal buy and sell depths are the same, but when the inventory is $q \neq 0$ the optimal depths are asymmetric (sell depth is different from buy depth) and the fill rates are tilted to induce mean reversion in inventories. For instance, if $q = 2$ the optimal sell depth is lower than the optimal buy depth so that it is more likely for the strategy to sell one unit of the asset than to acquire a unit.

The top panel in Figure 10.9 illustrates further how the fill rates are tilted to induce mean reversion to the optimal level of inventory for different levels of the running inventory parameter $\phi = \{0.2, 0.1, 0.05\}$. The other model parameters are: $\lambda^{+} = 2$, $\lambda^{-} = 1$, $\varepsilon = 0.005$, $\kappa^{\pm} = 100$, $\overline{q} = -\underline{q} = 10$, $\alpha = 0.0001$, $\sigma = 0.01$, $S_0 = 100$. The two bottom panels show that the optimal level of inventory for this choice of parameters is to hold a positive amount of shares – this optimal point is located where the inventory drift is zero because this is the level at which the strategy 'pulls' inventories.

We discuss in detail the trajectory of inventories when $\phi = 0.05$ which is depicted by the red circles. For example, if $\phi = 0.05$ and the current level of inventory is $q = 4$, the strategy posts asymmetrically so that the buy LO is closer to the midprice than the sell LO. In this way it is more likely for inventory to increase (positive inventory drift). Similarly, if the current level of inventory is $q = 6$ the optimal strategy is to post sell LOs closer to the midprice than the buy LOs so that it is more likely that inventory will be reduced (negative inventory drift). The optimal level of inventory is when $q = 5$ where LOs are symmetrically posted around the midprice (buy and sell fill rates are the same). If we follow the same line of reasoning, we see that for $\phi = 0.1$, depicted by green circles, the

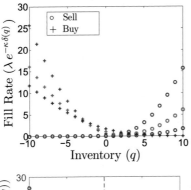

**Figure 10.9** Long-term inventory level. Model parameters are: $\lambda^+ = 2$, $\lambda^- = 1$, $\varepsilon = 0.005$, $\kappa^{\pm} = 100$, $\bar{q} = -\underline{q} = 10$, $\phi = \{0.2, 0.1, 0.05\}$, $\alpha = 0.0001$, $\sigma = 0.01$, $S_0 = 100$.

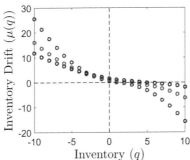

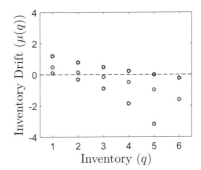

optimal level of inventory is between 2 and 3 units of the asset; and for $\phi = 0.2$, depicted by blue circles, the optimal level of inventory is approximately 1 unit of the asset.

To further understand these results, it is important to note that the optimal level of inventory is positive because the intensity of the arrival of buy MOs is $\lambda^+ = 2$ whilst the intensity of the arrival of sell MOs is $\lambda^- = 1$. Thus, on average, the midprice is drifting up because every time an MO arrives the midprice undergoes a jump (the distribution of the jumps up and down is the same), but since buy MOs arrive twice as often as sell MOs, the midprice is drifting upward – see the midprice dynamics (10.22). Therefore, it is optimal for the strategy to post LOs so that the fill rates are tilted in favour of holding positive inventory because it appreciates on average due to the midprice's upward trend.

## 10.4.2     Short-Term-Alpha and Adverse Selection

In this section we assume that the midprice of the asset follows

$$dS_t = (v + \alpha_t)\, dt + \sigma\, dW_t\,, \tag{10.30}$$

where the drift is given by a long-term component $v$ and by a short-term component $\alpha_t$ which is a predictable zero-mean reverting process. Here the long- and short-term components are important when devising market making strategies. For example, if the agent is an MM who trades at time scales where she does not 'see' the short-term component $\alpha_t$, then her strategy will not only be sub-optimal, but will lose money to better informed traders – traders who are better informed will pick-off the LOs posted by the less informed MM. On the other hand, if the MM has the ability to observe $\alpha_t$ then she will ensure that on average her strategy does not lose money to other traders, and will also use this knowledge to execute more speculative trades when $\alpha_t$ is different from zero as we shall show below.

One can specify the dynamics of the predictable drift $\alpha_t$ in many ways, depending on the factors that affect the short-term drift of the midprice. Here we assume that the MM is operating at high-frequency and short-term-alpha is driven by order flow. Thus, we model $\alpha_t$ as a zero-mean-reverting process which jumps by a random amount at the arrival times of MOs. The short-term drift jumps up when buy MOs arrive and jumps down when sell MOs arrive. As such, $\alpha_t$ satisfies

$$d\alpha_t = -\zeta\,\alpha_t\,dt + \eta\,dW_t^\alpha + \epsilon_{1+M_{t^-}^+}^+\,dM_t^+ - \epsilon_{1+M_{t^-}^-}^-\,dM_t^-\,, \tag{10.31}$$

where $\{\epsilon_1^\pm, \epsilon_2^\pm, \dots\}$ are i.i.d. random variables (independent of all processes) representing the size of the sell/buy MO impact on the drift of the midprice. Moreover, $W_t^\alpha$ denotes a Brownian motion independent of all other processes, $\zeta$, $\eta$ are positive constants, and the MOs arrive at an independent constant rate of $\lambda^\pm$.

Now we pose the market making problem where the MM posts only at-the-touch, as we did in Section 10.2.2. However, here the agent's strategy accounts for the influence of short-term-alpha. To this end, let $\ell_t^\pm \in \{0,1\}$ denote whether she is posted on the sell side $(+)$ or buy side $(-)$ of the LOB. Her performance criteria is as usual

$$H^\ell(t,x,S,\alpha,q) = \mathbb{E}_{t,x,S,\alpha,q}\left[X_T^\ell + Q_T^\ell\left(S_T - \left(\tfrac{\Delta}{2} + \varphi Q_T^\ell\right)\right) - \phi\int_t^T \left(Q_u^\ell\right)^2 du\right],$$

and her cash process $X_t^\ell$ satisfies the SDE

$$dX_t^\ell = \left(S_t + \tfrac{\Delta}{2}\right)dN_t^{+,\ell} - \left(S_t - \tfrac{\Delta}{2}\right)dN_t^{-,\ell}\,,$$

where $N_t^{\pm,\ell}$ denote the counting processes for her filled LOs. We further assume that, if she is posted in the LOB, when a matching MO arrives her LO is filled with probability one. In this case, $N_t^{\pm,\ell}$ are controlled doubly stochastic Poisson processes with intensity $\ell_t^\pm \lambda^\pm$. (In Exercise E.10.2 we ask the reader to generalise the problem to account for a fill probability less than 1.)

As before, the set $\mathcal{A}$ of admissible strategies are $\mathcal{F}$-predictable such that the agent is not posted on the buy (sell) side if her inventory is equal to the upper (lower) inventory constraints $\overline{q}$ $(\underline{q})$ and her value function is denoted by

$$H(t,x,S,\alpha,q) = \sup_{\ell\in\mathcal{A}} H^\ell(t,x,S,\alpha,q)\,.$$

### The Resulting DPE

Applying the DPP we find that the agent's value function $H$ should satisfy the DPE

$$0 = \left( \partial_t + \alpha\, \partial_S + \tfrac{1}{2}\sigma^2 \partial_{SS} - \zeta\, \alpha \partial_\alpha + \tfrac{1}{2}\eta^2 \partial_{\alpha\alpha} \right) H - \phi\, q^2$$

$$+ \lambda^+ \max_{\ell^+ \in \{0,1\}} \left\{ \mathbb{1}_{q > \underline{q}}\, \mathbb{E} \left[ H \left( t, x + (S + \tfrac{\Delta}{2}\, \ell^+)\, \ell^+, S, \alpha + \epsilon^+, q - \ell^+ \right) - H \right] \right.$$

$$\left. + \left( 1 - \ell^+\, \mathbb{1}_{q > \underline{q}} \right) \mathbb{E} \left[ H \left( t, x, S, \alpha + \epsilon^+, q \right) - H \right] \right\}$$

$$+ \lambda^- \max_{\ell^- \in \{0,1\}} \left\{ \mathbb{1}_{q < \bar{q}}\, \mathbb{E} \left[ H \left( t, x - (S - \tfrac{\Delta}{2}\, \ell^-)\, \ell^-, S, \alpha - \epsilon^-, q + \ell^- \right) - H \right] \right.$$

$$\left. + \left( 1 - \ell^-\, \mathbb{1}_{q < \bar{q}} \right) \mathbb{E} \left[ H \left( t, x, S, \alpha - \epsilon^-, q + \ell^- \right) - H \right] \right\},$$

subject to the terminal condition

$$H(T, x, S, \alpha, q) = x + q \left( S - \left( \tfrac{\Delta}{2} + \varphi\, q \right) \right).$$

Here, the expectations are over the random jump sizes $\epsilon^\pm$. The various terms in the DPE carry the following interpretations:

- the first line in the DPE represents the drift and diffusive components of the midprice and the short-term-alpha, as well as the alpha's mean-reverting feature,

- the maximisation terms represent the agent's control whether to post an LO at-the-touch,

- the second line represents the change in value function, if the agent is posted, due to the arrival of an MO which lifts the agent's offer and simultaneously induces a jump in the short-term-alpha,

- the third line represents the change in the value function when an MO arrives, but the agent is not posted – in which case only the short-term-alpha jumps,

- the fourth and fifth lines are for the other side of the book.

The terminal condition once again suggests the ansatz which splits out the accumulated cash, the book value of the shares which are marked-to-market at the midprice, and the added value from making markets optimally:

$$H(t, x, S, \alpha, q) = x + q\, S + h(t, \alpha, q).$$

In this context, $h$ depends on time, inventory, and the short-term-alpha. Substituting the ansatz into the above DPE we find that $h$ satisfies

$$0 = \left( \partial_t - \zeta\alpha\partial_\alpha + \tfrac{1}{2}\eta^2\partial_{\alpha\alpha} \right)h + \alpha q - \phi q^2$$

$$+ \lambda^+ \max_{\ell^+\in\{0,1\}} \left\{ \mathbb{1}_{q>\underline{q}}\, \mathbb{E}\left[ \ell^+ \tfrac{\Delta}{2} + h(t,\alpha+\epsilon^+,q-\ell^+) - h(t,\alpha+\epsilon^+,q) \right] \right\}$$

$$+ \lambda^- \max_{\ell^-\in\{0,1\}} \left\{ \mathbb{1}_{q<\overline{q}}\, \mathbb{E}\left[ \ell^- \tfrac{\Delta}{2} + h(t,\alpha-\epsilon^-,q+\ell^-) - h(t,\alpha-\epsilon^-,q) \right] \right\}$$

$$+ \lambda^+ \, \mathbb{E}\left[ h(t,\alpha+\epsilon^+,q) - h(t,\alpha,q) \right]$$

$$+ \lambda^- \, \mathbb{E}\left[ h(t,\alpha-\epsilon^-,q) - h(t,\alpha,q) \right] ,$$

subject to the terminal condition

$$h(T,\alpha,q) = -q\left( \tfrac{\Delta}{2} + \varphi q \right) .$$

The term $\alpha q$ which appears in the first line of the above equation is responsible for making the solution to this problem explicitly dependent on $\alpha$. If it were absent, then the optimal postings and $h$ function would be independent of $\alpha$, since the terminal conditions do not depend on $\alpha$ and there would be no source terms in $\alpha$. However, it is precisely this dependence on $\alpha$ which renders the strategy interesting and allows it to adapt to the adverse selection induced by the arrival of order flow. Finally, the expectation operator $\mathbb{E}$ is with respect to the random jump size $\epsilon$.

The form of the optimising terms allows us to characterise the optimal postings in a compact form. When $\ell = 0$ both terms that are being maximised are zero, hence, the optimal postings of the agent can be characterised succinctly as

$$\boxed{\begin{aligned} \ell^{+,*}(t,q) &= \mathbb{1}_{\left\{ \frac{\Delta}{2}+\mathbb{E}[h(t,\alpha+\epsilon^+,q-1)-h(t,\alpha+\epsilon^+,q)]>0 \right\}\cap\left\{ q>\underline{q} \right\}} , \\ \ell^{-,*}(t,q) &= \mathbb{1}_{\left\{ \frac{\Delta}{2}+\mathbb{E}[h(t,\alpha-\epsilon^-,q+1)-h(t,\alpha-\epsilon^-,q)]>0 \right\}\cap\left\{ q<\overline{q} \right\}} . \end{aligned}} \qquad (10.32)$$

These postings are the analog of the optimal postings in (10.17) from subsection 10.2.2 where we investigated how the agent trades when posting only at-the-touch. The key difference here is that the agent knows that when an MO arrives, $\alpha$ jumps up/down and therefore she compares the expected change in the value functions evaluated at $\alpha \pm \epsilon^\pm$, rather than at $\alpha$, with the half-spread.

### Features of the Strategy

For the purpose of focusing solely on the effect of short-term-alpha, we set the running penalty $\phi = 0$, and the remaining model parameters are

$$T = 60 \text{ sec}, \quad \overline{q} = -\underline{q} = 20, \quad \lambda^\pm = 0.8333, \quad \Delta = 0.01, \quad \varphi = 0.01,$$

$$\eta = 0.001, \quad \zeta = 0.5, \quad \mathbb{E}[\epsilon] = 0.005 .$$

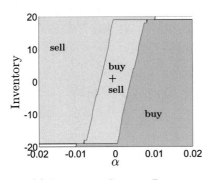

**Figure 10.10** The optimal postings when accounting for short-term-alpha.

(a) Asymptotic Strategy Posts

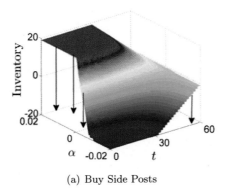

(a) Buy Side Posts

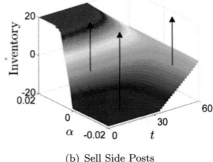

(b) Sell Side Posts

The choice of $\lambda^{\pm}$ ensures that the agent has a maximum inventory equal to 20% of the expected number of trades. With these parameters, Figure 10.10 shows how the optimal strategy behaves as a function of time and short-term-alpha. The agent posts limit buys whenever her inventory is below the surface in the left panel, and she posts limit sells whenever her inventory is above the surface in the right panel. There are a number of notable features here.

(i) Due to the symmetry of rates of arrival of MOs, the surfaces are mirror reflections of one another.

(ii) As maturity approaches her strategy becomes essentially independent of short-term-alpha, and instead induces her to sell when her inventory is positive and buy when inventory is negative. Therefore, the optimal strategy attempts to close the trading period with zero inventory.

(iii) The optimal strategies become independent of time far from maturity.

(iv) Far from maturity, the agent tends to post symmetrically when short-term-alpha and inventory are close to zero. As $\alpha$ increases, she is willing to take on inventory, but keeps posting on both sides, until $\alpha$ becomes quite large, then she posts only buy LOs. The opposite holds when $\alpha$ decreases.

Next, Figure 10.11 shows a sample path of the agent's posts together with the short-term-alpha. In the left panel, the green lines demonstrate when (and at

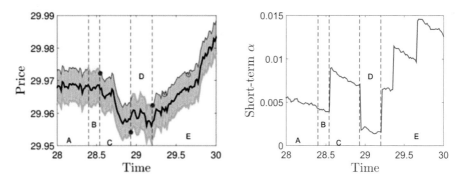

**Figure 10.11** Sample path of the optimal strategy. Green lines show when and at what price the agent is posted. Solid red circles indicate MOs that arrive and hit/lift the posted bid/offers. Open red circles indicate MOs that arrive but do not hit/lift the agent's posts. Shaded region is the bid-ask spread.

what price) she is posted, and the solid red circles indicate arrival of MOs that fill the agent's posts. In this sample path, her postings change a total of five times and her inventory begins at zero ($Q = 0$). In regime $A$ she is posted only on the buy side since $\alpha$ is large enough to suggest that purchasing the asset is worthwhile. As time evolves and she enters regime $B$, $\alpha$ decays and she begins to post symmetrically. A buy MO arrives and lifts her offer (so that $Q_t = -1$), short-term-alpha immediately jumps upwards and she removes her sell LO in regime $C$. A sell MO eventually arrives and hits her bid (so that $Q_t = 0$) and immediately induces a downward jump in $\alpha$. Since $\alpha$ in regime $D$ is relatively small, and her inventory is zero, she posts symmetrically once again. Eventually a buy MO once again lifts her offer (so that $Q_t = -1$) and induces an upward jump in $\alpha$. In regime $E$ she now only posts on the buy side. A sequence of buy MOs arrive in this interval inducing more upward jumps in short-term-alpha; however, since she has no LO sell posted, her inventory remains one short and she remains posted only on the buy side.

## 10.5     Bibliography and Selected Readings

Ho & Stoll (1981), Avellaneda & Stoikov (2008), Stoikov & Sağlam (2009), Laruelle, Lehalle & Pagès (2013), Cartea & Jaimungal (2013), Cartea, Donnelly & Jaimungal (2013), Guéant, Lehalle & Fernandez-Tapia (2013), Jaimungal & Kinzebulatov (2013), Cartea, Jaimungal & Ricci (2014), Kharroubi & Pham (2010), Guilbaud & Pham (2013), Bouchard, Dang & Lehalle (2011), Carmona & Webster (2012), Carmona & Webster (2013).

## 10.6    Exercises

E.10.1 Consider the framework developed in subsection 10.2.2, where the MM posts only at-the-touch, but assume that when an MO arrives, and the agent is posted on the matching side of the LOB, her order is filled with probability $\rho < 1$. Derive the DPE and compute the optimal strategy in feedback form. Also, implement the resulting non-linear coupled system of ODEs and show how the strategy is altered by the fill probability.

E.10.2 Consider the framework developed in subsection 10.4.2, where the MM is subject to adverse selection from active traders, but assume that when an MO arrives, and the agent is posted on the matching side of the LOB, her order is filled with probability $\rho < 1$. Derive the DPE and compute the optimal strategy in feedback form. Also, implement the resulting non-linear coupled system of ODEs and show how the strategy is altered by the fill probability.

# 11 Pairs Trading and Statistical Arbitrage Strategies

## 11.1 Introduction

The success of many trading algorithms depends on the quality of the predictions of stock price movements. Predictions of the price of a single stock are generally less accurate than predictions of a portfolio of stocks. A classical strategy which makes the most of the predictability of the joint, rather than the individual, behaviour of two assets is **pairs trading** where a portfolio consisting of a linear combination of two assets is traded. At the heart of the strategy is how the two assets co-move – some of these statistical issues were discussed early in Section 3.7. As an example, take two assets whose spread, that is the difference between their prices, exhibits a marked pattern and deviations from it are temporary. Then, pairs trading algorithms profit from betting on the empirical fact that spread deviations tend to return to their historical or predictable level. Thus, pairs trading fall under the class of strategies sometimes labeled as **statistical arbitrage** (or StatArb for short). They are not true arbitrages (which are strategies that produce returns in excess of the risk-free rate with zero risk), but rather are strategies which bet off of the typical behaviour of asset prices, and hence are not risk-free.

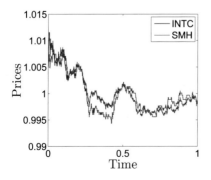
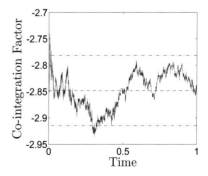

**Figure 11.1** INTC and SMH on November 1, 2013 for the whole day of trading: (left panel) midprice relative to mean midprice; (right panel) co-integration factor. The dashed line indicates the mean-reverting level, the dash-dotted lines indicate the 2 standard deviation bands.

In Figure 11.1 we show an example with Intel Inc. (INTC) and the Market Vectors Semiconductor ETF (SMH) for November 1, 2013. The left panel shows the midprice paths of INTC and SMH (scaled by the mean midprices), and it is clear that the two assets tend to move together and in the same direction. Thus, a portfolio consisting of long one asset and short the other will exhibit a less volatile and more predictable behaviour than that of the individual assets. In this case, since the assets tend to co-move in the same direction, a simple pairs trading strategy is to buy the portfolio if its value is less than a threshold and sell it if its value is greater than the threshold. This strategy will deliver profits as long as the value of the portfolio fluctuates about and reverts to the threshold.

A more sophisticated approach is to look at the co-integration factor of the prices of the two stocks. The right panel of Figure 11.1 shows the path of a co-integration factor $\varepsilon_t = A\,S_t^{(INTC)} + B\,S_t^{(SMH)}$, where $A$ and $B$ are estimated from the data that day to be $A \sim 0.95$ and $B \sim -0.63$, see Section 3.7. Thus, if the mean-reverting behaviour we have observed is persistent, then we expect the value of a portfolio long 0.95 shares in INTC and short 0.63 shares in SMH to hover around the mean of the co-integration factor which is zero. And how can we profit from the mean-reverting to zero value of this portfolio? The answer is a pairs trading strategy which consists in going long the portfolio when it is 'cheap' and then closing the position when the portfolio's value increases, or going short the portfolio when it is 'dear' and closing the position when the portfolio's value decreases.

In this chapter we present different trading algorithms based on co-integration in the stock price level or in the drift component of a collection of assets. In Section 11.2 we show naive approaches which place ad hoc bands around the mean-reverting level of the co-integration factor so that the strategy enters a position, long or short the portfolio when either band is hit, and then another pair of ad hoc bands to unwind the position. In Section 11.3 we develop more sophisticated approaches which determine the optimal bands to enter and close a position, and in Section 11.4 the drifts of a collection of assets are co-integrated.

## 11.2    Ad Hoc Bands

A simple strategy to profit from the co-integration factor's mean-reversion, as seen in Figure 11.1, is to place bands which are one standard deviation above and below the mean-reverting level, which is zero, and buy one unit of the portfolio if the lower band is hit or sell one unit of the portfolio if the upper band is hit. Once the strategy has entered into a position, either long or short, the next step is to close it. To close the position the strategy waits for the value of the portfolio to be within a small interval, say 1/10 standard deviation of the mean-reverting level of the co-integration factor, and at that point the agent liquidates the position.

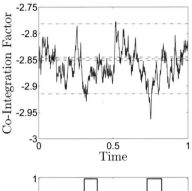

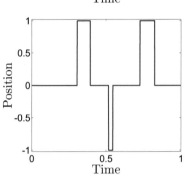

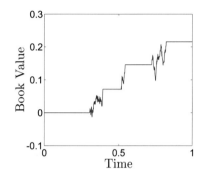

**Figure 11.2** A sample path of the co-integration factor, the trading position, and the book value of the trade, using the two standard deviation banded strategy.

Figure 11.2 shows three pictures: a simulated sample path of the co-integration factor, the path of the inventory of the strategy which opens and closes positions a few times during the trading horizon, and finally, the accumulated cash and marked-to-market value of the strategy. The strategy starts by waiting for the path of the co-integration factor to breach one of the outer bands. At around $t \sim 0.3$ the path hits the lower outer band so the agent longs the portfolio in anticipation of its value appreciating. Then for a short period of time the agent holds on to the portfolio whose value fluctuates one-to-one with changes in the co-integration factor. The book value of the strategy is given by

$$BV_t = X_t + \beta_t \left( A\, S_t^{(INTC)} + B\, S_t^{(SMH)} \right),$$

where $X_t$ is the strategy's accumulated cash position, and $\beta_t$ denotes the units of the portfolio held by the strategy, in this case assuming that $\beta_t \in \{-1, 0, 1\}$.

The next step is to wait until the co-integration factor hits the inner band to close the position. Here this occurs at around $t \sim 0.4$ and the strategy goes back to holding zero units of the portfolio and locks in a profit equal to the difference between the outer and inner bands. Next, the strategy waits for a little while and enters into a short position at around $t \sim 0.51$ and liquidates at around $t \sim 0.55$. Finally, the strategy enters into a long position around $t \sim 0.73$ and liquidates at around $t \sim 0.83$. In all, for this simulated path, the strategy makes a profit of three times the outer-inner band spread.

In the scenario in Figure 11.2, the agent ends the trading horizon with zero

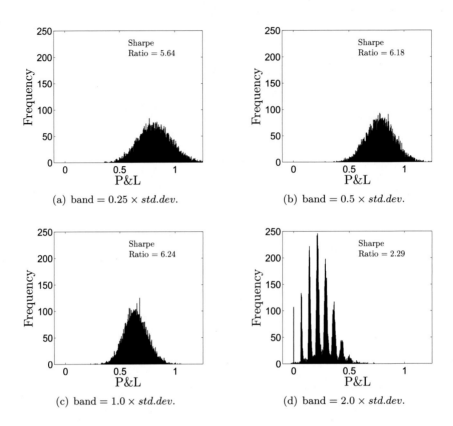

**Figure 11.3** P&L histograms from 10,000 scenarios using the naive strategy with various trigger bands.

inventory. This is not guaranteed by the strategy, and in fact the strategy may have entered a long/short position which never reverted back to the inner band by the end of the trading horizon. This would induce potential losses into the strategy. The wider the trigger bands, the more likely it is to end with inventory. Also, while wider bands have larger profits when the position closes out, the con-integration factor makes fewer outer-inner band transitions when the band size increases.

Figure 11.3 shows the profit and loss (P&L) histogram from generating 10,000 scenarios from the estimated model, computing the co-integration factor and placing trades as described above. The figure shows the effect that the band size has on the P&L as well as the Sharpe ratio (i.e. the mean P&L divided by the standard deviation of the P&L). Notice that the Sharpe ratio first increases as the band widens, but then starts decreasing. Also notice that when the band size is largest at $2 \times std.dev.$, the distribution is multimodal. In fact, on close examination, all of the distributions are multimodal. The reason is because the profit from closing out a long/short position equals approximately the band size (since you enter into a long/short position once the factor hits the band and close it out near the mean). Hence, the P&L is concentrated near integer multiples of the band size and the weight on a given multiple equals the probability of making that many round trip trades during the trading horizon. The reason the P&L is not concentrated solely on the band is because towards the end of the trading horizon, the co-integration factor may not return to the equilibrium value prior to trading end. Hence, the trader must close out a position that might make less profit than the band size, or in fact may take a loss if the co-integration factor moves away from the equilibrium prior to ending the trading horizon.

## 11.3   Optimal Band Selection

In the previous section, we introduced a very simple but naive strategy for entering and exiting a long/short position in the co-integration portfolio. Here, we determine the optimal strategy to enter and exit by posing the problem as an optimal stopping problem. In this context, the trader will make a single round trip trade and there is no terminal time horizon.

First, assume that there is a portfolio with $A$ shares in one asset and $B$ shares in another asset so that the portfolio dynamics are given by $\varepsilon_t$, which is the co-integration factor, and we assume that

$$d\varepsilon_t = \kappa \left(\theta - \varepsilon_t\right) dt + \sigma \, dW_t \,,$$

with $W_t$ a standard Brownian motion. The coefficient $\kappa$ is the rate of mean-reversion, $\theta$ is the level that the process mean-reverts to, and $\sigma$ is the volatility of the process. To formulate the problem, we first solve for the optimal time at which to exit a long position in the portfolio and then use this as the input to determine when the agent should optimally enter a long position in the portfolio. The agent's performance criteria for exiting the long position is given by

$$H_+^{(\tau)}(t, \varepsilon) = \mathbb{E}_{t,\varepsilon} \left[e^{-\rho\,(\tau-t)}(\varepsilon_\tau - c)\right] \,,$$

where $c$ is a transaction cost for closing out the portfolio, $\rho > 0$ is the agent's discount factor (akin to an urgency parameter since increasing $\rho$ will push the exit boundary in towards the long-run level), $\mathbb{E}_{t,\varepsilon}[\cdot]$ denotes expectation conditional on $\varepsilon_t = \varepsilon$, and her corresponding value function is

$$H_+(t, \varepsilon) = \sup_\tau H_+^{(\tau)}(t, \varepsilon) \,.$$

The value function seeks for the optimal stopping time which maximises the performance criteria, because once the agent is long the portfolio the objective is to unwind the position when its value has increased.

Next, the agent's performance criteria for entering the long position is given by

$$G_+^{(\tau)}(t, \varepsilon) = \mathbb{E}_{t,\varepsilon} \left[e^{-\rho(\tau-t)} \left(H_+(\tau, \varepsilon_\tau) - \varepsilon_\tau - c\right)\right] \,,$$

and her corresponding value function is

$$G_+(t, \varepsilon) = \sup_\tau G_+^{(\tau)}(t, \varepsilon).$$

The intuition here is that the agent pays $\varepsilon + c$ for the portfolio, but receives the exit option which has value $H_+(\varepsilon, \tau)$.

Now, due to the stationary properties of the OU process, the performance criteria and, therefore, the value functions do not depend on time. In what follows we suppress the time dependence. The dynamic programming principle (DPP) implies that the value functions $H$ and $G$ should satisfy the coupled system of variational inequalities (VIs)

$$\max \{(\mathcal{L} - \rho) H_+(\varepsilon) \ ; \ (\varepsilon - c) - H_+(\varepsilon)\} = 0 \,,$$
$$\max \{(\mathcal{L} - \rho) G_+(\varepsilon) \ ; \ (H_+(\varepsilon) - \varepsilon - c) - G_+(t, \varepsilon)\} = 0 \,,$$

where $\mathcal{L}$ is the infinitesimal generator of the co-integration process, i.e.

$$\mathcal{L} = \kappa (\theta - \varepsilon) \partial_\varepsilon + \tfrac{1}{2} \sigma^2 \partial_{\varepsilon\varepsilon} \,.$$

## 11.3.1 The Optimal Exit Problem

The VI for $H$ is very similar to the value of a perpetual call option, and can be obtained by finding the fundamental solutions of the ODE

$$(\mathcal{L} - \rho) F(\varepsilon) = 0 \,, \tag{11.1}$$

which we denote $F_\pm(\varepsilon)$, and write $H_+(\varepsilon) = A \, F_+(\varepsilon) + B \, F_-(\varepsilon)$ in the continuation region $(\varepsilon < \varepsilon^*)$ and $H_+(\varepsilon) = (\varepsilon - c)$ in the exercise region $\varepsilon > \varepsilon^*$. We then need to impose the value matching and smooth pasting conditions

$$H_+(\varepsilon^*) = (\varepsilon^* - c) \,, \qquad \text{and} \qquad \partial_\varepsilon H_+(\varepsilon^*) = 1 \,,$$

to solve for the optimal point $\varepsilon^*$ where the agent closes out the position.

To this end, one can check that

$$F_+(\varepsilon) = \int_0^\infty u^{\frac{\rho}{\kappa} - 1} e^{-\sqrt{\frac{2\kappa}{\sigma^2}} (\theta - \varepsilon) u - \frac{1}{2} u^2} \, du \,,$$

$$F_-(\varepsilon) = \int_0^\infty u^{\frac{\rho}{\kappa} - 1} e^{+\sqrt{\frac{2\kappa}{\sigma^2}} (\theta - \varepsilon) u - \frac{1}{2} u^2} \, du \,.$$

are solutions of (11.1).

Moreover, it is easy to see (by differentiating under the integral) that $F_+'(\varepsilon) > 0$, $F_+''(\varepsilon) > 0$, $F_-'(\varepsilon) < 0$, and $F_-''(\varepsilon) > 0$, so that $F_+$ is strictly positive, increasing and convex, while $F_-$ is strictly positive, decreasing and convex. (As a side note, these integrals can be written in terms of Whittaker or confluent hypergeometric functions.)

The value function $H$ must vanish as $\varepsilon \to -\infty$ (since the time to exiting the position will tend to infinity and the discount factor will render such strategies worthless), hence we must have

$$H_+(\varepsilon) = A \, F_+(\varepsilon) \, \mathbb{1}_{\varepsilon < \varepsilon^*} + (\varepsilon - c) \, \mathbb{1}_{\varepsilon \geq \varepsilon^*} \,,$$

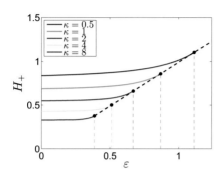 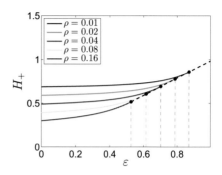

**Figure 11.4** The optimal exit trigger levels (given by the black circles) and corresponding value function $H_+$.

for some constant $A$, i.e, $B = 0$ otherwise $H_+$ would blow up. Applying the value matching and smooth pasting conditions, we have

$$A F_+(\varepsilon^*) = (\varepsilon^* - c), \qquad \text{and} \qquad A F'_+(\varepsilon^*) = 1.$$

Taking the ratio of these equations and re-arranging, the optimal level at which to close out the position is the unique solution to the non-linear equation

$$\boxed{(\varepsilon^* - c) F'_+(\varepsilon^*) = F_+(\varepsilon^*),} \qquad (11.2)$$

and further we have $A = \frac{\varepsilon^* - c}{F_+(\varepsilon^*)}$. The value function $H_+$ can then be written as

$$H_+(\varepsilon) = \frac{F_+(\varepsilon)}{F_+(\varepsilon^*)} (\varepsilon^* - c) \mathbb{1}_{\varepsilon < \varepsilon^*} + (\varepsilon - c) \mathbb{1}_{\varepsilon \geq \varepsilon^*}.$$

Figure 11.4 shows the value function $H$ in the continuation regions (the solid lines) and exercise regions (the dashed lines) as the mean-reversion rate $\kappa$ and discount rate $\rho$ vary. We also set $\theta = 0$, $\sigma = 0.5$ and $c = 0.01$. As the mean-reversion rate increases, the optimal trigger levels decrease since the co-integration factor is drawn more strongly to the mean-reversion level. Similarly, as the discount rate $\rho$ increases, the trigger levels decrease, to draw the stopping time nearer since future gains are discounted more.

## 11.3.2 The Optimal Entry Problem

Armed with the optimal exit strategy, we can solve for the optimal entry problem. This also amounts to solving for the price of a perpetual American-style option, albeit now with the exercise value of $H_+(\varepsilon) - \varepsilon - c$ rather than a simple call payoff. Now, we anticipate that the value function $G_+$ should be decreasing, rather than increasing in $\varepsilon$. The reason is simple: suppose that $\varepsilon < \theta$, i.e. the co-integration portfolio value is currently less than its long-run level; then as $\varepsilon$ increases, if the agent enters into a long position, she will extract less value from

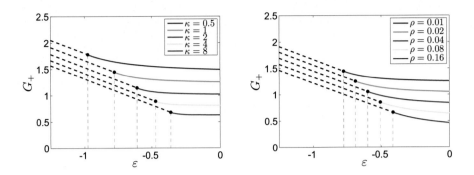

**Figure 11.5** The optimal entry trigger level (given by the black circles) and corresponding value function $G_+$.

it once she exercises at the optimal exit point $\varepsilon^*$. This value should reduce to zero as $\varepsilon$ tends to infinity, hence, we can write

$$G_+(\varepsilon) = D\,F_-(\varepsilon)\,\mathbb{1}_{\varepsilon > \varepsilon_*} + (H_+(\varepsilon) - \varepsilon - c)\,\mathbb{1}_{\varepsilon \le \varepsilon_*}$$

for some constant $D$. The value matching and smooth pasting conditions are now

$$D\,F_-(\varepsilon_*) = H_+(\varepsilon_*) - \varepsilon_* - c, \qquad \text{and} \qquad D\,F'_-(\varepsilon_*) = H'_+(\varepsilon_*) - 1,$$

or solving for $C$ and re-arranging, we have

$$\boxed{(H_+(\varepsilon_*) - \varepsilon_* - c)\,F'_-(\varepsilon_*) = (H'_+(\varepsilon_*) - 1)\,F_-(\varepsilon_*).} \tag{11.3}$$

Naturally, we anticipate that $\varepsilon_* < \varepsilon^*$. Putting these results together, the value function $G$ can be written as

$$G_+(\varepsilon) = \frac{F_-(\varepsilon)}{F_-(\varepsilon_*)}\,(H_+(\varepsilon_*) - \varepsilon_* - c)\,\mathbb{1}_{\varepsilon > \varepsilon_*} + (H_+(\varepsilon) - \varepsilon - c)\,\mathbb{1}_{\varepsilon \le \varepsilon_*}.$$

Figure 11.5 shows the value function $G_+$ in the continuation regions (the solid lines) and exercise regions (the dashed lines) as the mean-reversion rate $\kappa$ and discount rate $\rho$ vary. We also set $\theta = 0$, $\sigma = 0.5$ and $c = 0.01$. As the mean-reversion rate increases, the optimal trigger levels increase (and move towards the mean-reversion level) since the co-integration factor is drawn more strongly to the mean-reversion level. Similarly, as the discount rate $\rho$ increases, the trigger levels increase, to draw the stopping time nearer since future gains are discounted more.

Comparing Figure 11.5 to 11.4, we observe that even with the same parameters, the optimal entry and exit prices are not symmetric around the mean-reversion level. This is at first a somewhat surprising result. However, since the entry and exit times are ordered, the discount factor plays a role in biasing the entry point to occur closer to the mean-reversion level, relative to the exit point. For example, with $\kappa = 4$, we see the entry price is about $-0.47$, while the exit price is about 0.51. With $\kappa = 0.5$, the asymmetry is even larger as the entry price is about $-0.97$ while the exit price is about 1.1.

## 11.3.3    Double-Sided Optimal Entry-Exit

In the previous sections we studied how an agent would behave if she wishes to enter into and then exit from a long position in a co-integration portfolio. This ignores the possibility that the agent may instead wish to enter into a short position. Here we incorporate a double-sided strategy which considers the optimal time in which to enter either a long or short position and then optimally exit the position.

Let the performance criteria for exiting from a long or short position be given by

$$H_+^{(\tau)}(t,\varepsilon) = \mathbb{E}_{t,\varepsilon}\left[e^{-\rho(\tau-t)}\left(\varepsilon_\tau - c\right)\right],$$

$$H_-^{(\tau)}(t,\varepsilon) = \mathbb{E}_{t,\varepsilon}\left[e^{-\rho(\tau-t)}\left(-\varepsilon_\tau - c\right)\right],$$

with the corresponding value functions

$$H_+(t,\varepsilon) = \sup_\tau H_+^{(\tau)}(t,\varepsilon),$$

$$H_-(t,\varepsilon) = \sup_\tau H_-^{(\tau)}(t,\varepsilon).$$

Naturally, $H^+$ coincides with $H_+$ from the previous section. We therefore only need to focus on computing $H_-$.

As before, we see that $H_-$ is independent of time and should satisfy the VI

$$\max\left\{(\mathcal{L} - \rho)H_-(\varepsilon)\ ;\ (-\varepsilon - c) - H_-(\varepsilon)\right\} = 0.$$

Clearly, since this is the value of exiting from a short position, the agent's value function must be decreasing in $\varepsilon$ and must vanish as $\varepsilon \to \infty$. Hence, we must have

$$H_-(\varepsilon) = A\,F_-(\varepsilon)\,\mathbb{1}_{\varepsilon > \varepsilon_-^*} - (\varepsilon + c)\,\mathbb{1}_{\varepsilon \le \varepsilon_-^*},$$

where $\varepsilon_-^*$ is the trigger level at which the agent will close out the position. Applying the value matching and smooth pasting conditions we have

$$A\,F_-(\varepsilon_-^*) = -(\varepsilon_-^* + c), \qquad \text{and} \qquad A\,F_-'(\varepsilon_-^*) = -1,$$

and taking the ratio of these equations and re-arranging, the optimal level at which to close out the position is the unique solution to the non-linear equation

$$\boxed{(\varepsilon_-^* + c)\,F_-'(\varepsilon_-^*) = F_-(\varepsilon_-^*)\,.} \tag{11.4}$$

In all, the value functions $H_\pm$ can be written as

$$H_+(\varepsilon) = \frac{F_+(\varepsilon)}{F_+(\varepsilon_-^*)} \left(\varepsilon_+^* - c\right) \mathbb{1}_{\varepsilon < \varepsilon_+^*} + (\varepsilon - c) \mathbb{1}_{\varepsilon \geq \varepsilon_-^*},$$

$$H_-(\varepsilon) = -\frac{F_-(\varepsilon)}{F_-(\varepsilon_-^*)} \left(\varepsilon_-^* + c\right) \mathbb{1}_{\varepsilon > \varepsilon_-^*} - (\varepsilon + c) \mathbb{1}_{\varepsilon \leq \varepsilon_+^*}.$$

Next, the agent's performance criteria $G^{(\tau)}(t, \varepsilon)$ for the entry problem can be written as

$$G^{(\tau)}(t, \varepsilon) = \mathbb{E}_{t,\varepsilon} \Big[ \; e^{-\rho(\tau_+ - t)} \left(H_+(\tau_+, \varepsilon_{\tau_+}) - \varepsilon_{\tau_+} - c\right) \mathbb{1}_{\min(\tau_+, \tau_-) = \tau_+}$$

$$+ e^{-\rho(\tau_- - t)} \left(H_-(\tau_-, \varepsilon_{\tau_-}) + \varepsilon_{\tau_-} - c\right) \mathbb{1}_{\min(\tau_+, \tau_-) = \tau_-} \Big].$$

The corresponding VI is

$$\max \Big\{ (\mathcal{L} - \rho) \, G(\varepsilon) \; ;$$

$$\left(H_+(\varepsilon) - \varepsilon - c\right) - G(t, \varepsilon) \; ; \; \left(H_-(\varepsilon) + \varepsilon - c\right) - G(t, \varepsilon) \Big\} = 0.$$

Now there will be two trigger points $\varepsilon_{\pm *}$ corresponding to entering the long or short position. In this case, in the continuation region, the function $G$ will be a linear combination of both $F_\pm$, so that

$$G(\varepsilon) = (A \, F_+(\varepsilon) + B \, F_-(\varepsilon)) \, \mathbb{1}_{\varepsilon \in (\varepsilon_{*+}, \varepsilon_{*-})}$$

$$+ \left(H_+(\varepsilon) - \varepsilon - c\right) \mathbb{1}_{\varepsilon \leq \varepsilon_{*+}} + \left(H_-(\varepsilon) + \varepsilon - c\right) \mathbb{1}_{\varepsilon \geq \varepsilon_{*-}}.$$

The constant coefficients $A$ and $B$ will be determined via value matching and smooth pasting at both trigger points. As such we have,

$$A \, F_+(\varepsilon_{*+}) + B \, F_-(\varepsilon_{*+}) = H_+(\varepsilon_{*+}) - \varepsilon_{*+} - c,$$

$$A \, F'_+(\varepsilon_{*+}) + B \, F'_-(\varepsilon_{*+}) = H'_+(\varepsilon_{*+}) - 1,$$

$$A \, F_+(\varepsilon_{*-}) + B \, F_-(\varepsilon_{*-}) = H_-(\varepsilon_{*-}) + \varepsilon_{*-} - c,$$

$$A \, F'_+(\varepsilon_{*-}) + B \, F'_-(\varepsilon_{*-}) = H'_-(\varepsilon_{*-}) + 1.$$

We can solve for $A$ and $B$ in terms of $\varepsilon_{*\pm}$ from the value matching conditions (first and third equations):

$$A = \frac{\left(H_+(\varepsilon_{*+}) - \varepsilon_{*+} - c\right) F_-(\varepsilon_{*-}) - \left(H_-(\varepsilon_{*-}) + \varepsilon_{*-} - c\right) F_-(\varepsilon_{*+})}{F_+(\varepsilon_{*+}) F_-(\varepsilon_{*-}) - F_+(\varepsilon_{*-}) F_-(\varepsilon_{*+})},$$

$$B = \frac{\left(H_+(\varepsilon_{*+}) - \varepsilon_{*+} - c\right) F_+(\varepsilon_{*-}) - \left(H_-(\varepsilon_{*-}) + \varepsilon_{*-} - c\right) F_+(\varepsilon_{*+})}{F_-(\varepsilon_{*+}) F_+(\varepsilon_{*-}) - F_-(\varepsilon_{*-}) F_+(\varepsilon_{*+})},$$

and substitute these expressions back in to determine $\varepsilon_{*\pm}$ as roots of the smooth pasting conditions (second and fourth equations). To solve for the trigger points, it is reasonable to use initial starting points $\varepsilon_{*\pm}^0$ for $\varepsilon_{*\pm}$ implied from the one-sided trade results of the last section, i.e $\varepsilon_{*+}^0 \sim \varepsilon_*$ and $\varepsilon_{*-}^0 \sim -\varepsilon_*$.

In Figure 11.6 we show the resulting value functions for the case when

$$\rho = 0.01, \quad \sigma = 0.5, \quad \theta = 1, \quad \text{and} \quad c = 0.01,$$

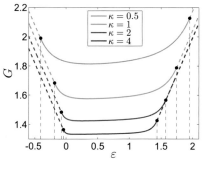

**Figure 11.6** The optimal entry trigger level and corresponding value function for the double entry-exit problem.

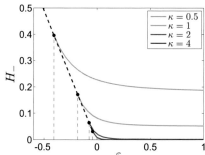

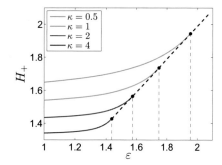

and we allow the mean-reversion rate $\kappa$ to vary. There are a few interesting points to notice. First, $H_+$ and $H_-$ behave quite differently. Second, the optimal exit points are not symmetric around $\theta$. Third, the optimal entry points are in fact equal to the corresponding optimal exit point from the opposite portfolio: e.g., the optimal entry point for the long portfolio equals the optimal exit point from the short position. The table below illustrates this point more clearly.

| $\kappa$ | exit triggers | | entry triggers | |
|---|---|---|---|---|
| | $\varepsilon_+^*$ (long) | $\varepsilon_-^*$ (short) | $\varepsilon_{*+}$ (long) | $\varepsilon_{*-}$ (short) |
| 0.5 | 1.9537 | -0.4060 | -0.4060 | 1.9537 |
| 1.0 | 1.7460 | -0.1815 | -0.1815 | 1.7460 |
| 2.0 | 1.5744 | -0.0740 | -0.0740 | 1.5744 |
| 4.0 | 1.4367 | -0.0410 | -0.0410 | 1.4367 |

## 11.4 Co-integrated Log Prices with Short-Term-Alpha

Another approach to devise trading algorithms that take advantage of structural dependencies between assets, is one where the drifts of a collection of assets are co-integrated, rather than the prices themselves.

### 11.4.1    Model Setup

Suppose that we have a collection of risky assets whose vector of prices $\boldsymbol{Y} = \left(Y_t^1, \ldots, Y_t^n\right)_{0 \leq t \leq T}$ satisfy the coupled system of SDEs

$$\frac{dY_t^k}{Y_t^k} = \delta_k \, \alpha_t \, dt + \sum_{i=1}^n \sigma_{ki} \, dW_t^i, \tag{11.5}$$

where

$$\alpha_t = a_0 + \sum_{i=1}^n a_i \, \log Y_t^i,$$

and $\boldsymbol{W} = \left(W_t^1, \ldots, W_t^n\right)_{0 \leq t \leq T}$ is a vector of independent Brownian motions. Focusing on the diffusion term above, it is not difficult to see that the instantaneous covariance, loosely interpreted as $\mathbb{C}\left[\frac{dY_t^i}{Y_t^i}, \frac{dY_t^j}{Y_t^j} \middle| \mathcal{F}_t\right]$, between assets $i$ and $j$ is given by

$$\Omega_{ij} = \sum_{k=1}^n \sigma_{ik} \sigma_{kj}.$$

Thus, the matrix $\boldsymbol{\sigma}$ whose elements are $\sigma_{ij}$ is the Cholesky decomposition of the instantaneous variance-covariance matrix $\boldsymbol{\Omega}$ so that in matrix notation $\boldsymbol{\Omega} = \boldsymbol{\sigma}\boldsymbol{\sigma}'$. We assume that there are no redundant degrees of freedom here, so that $\boldsymbol{\sigma}$ is invertible. Furthermore, we show below that $\alpha_t$ acts as a co-integration factor. First, note that when $\alpha_t = 0$ all assets are simply correlated geometric Brownian motions with zero drift, and are hence martingales. In general, however, $\alpha_t$ will be non-zero representing short-term deviations from martingale behaviour, and may also be considered as a 'short-term-alpha' component – see subsection 10.4.2.

We justify calling $\alpha_t$ a co-integration factor by demonstrating that it is indeed a mean-reverting process. First, note that the log-prices satisfy the SDEs

$$d \log Y_t^k = \left(\delta_k \, \alpha_t - \tfrac{1}{2}\Omega_{kk}\right) dt + \sum_{i=1}^n \sigma_{ki} \, dW_t^i.$$

Next we compute the differential of $\alpha_t$ as follows:

$$d\alpha_t = d\left(a_0 + \sum_{k=1}^n a_k \, \log Y_t^k\right)$$

$$= \sum_{k=1}^n a_k \left(\delta_k \, \alpha_t - \tfrac{1}{2}\Omega_{kk}\right) dt + \sum_{k=1}^n a_k \sum_{i=1}^n \sigma_{ki} \, dW_t^i,$$

and therefore,

$$d\alpha_t = \kappa \left(\theta - \alpha_t\right) dt + \boldsymbol{a}' \boldsymbol{\sigma} \, d\boldsymbol{W}_t. \tag{11.6}$$

Here, $a'$ represents the transpose of the column vector $a$ where

$$a = (a_1, \ldots, a_n)',$$

$$\kappa = -\sum_{k=1}^{n} a_k \delta_k = -\delta' a,$$

$$\theta = \frac{\sum_{k=1}^{n} a_k \Omega_{kk}}{2 \sum_{k=1}^{n} a_k \delta_k} = \frac{1}{2} \frac{\text{Tr}(A\Omega)}{\delta' a}.$$

Here, $\text{Tr}(\cdot)$ denotes the trace of the matrix in braces (i.e. the sum of its diagonal elements) and $A = \text{diag}(a)$ is a diagonal matrix whose diagonal entries are $a$. To ensure that the model does indeed describe a mean-reverting process, as opposed to a mean-repelling one, we assume that $\delta' a < 0$.

From the SDE above, we can see that $\alpha$ mean-reverts at a rate which depends on the various strengths of the impact that $\alpha$ has on each asset (through $\delta$) as well as the strength of the log-asset price's contribution to $\alpha$ itself (through $a$). The mean-reversion level $\theta$ depends on the ratio of the volatility relative to the impact each component has on the drift of the assets.

Another alternative representation of the model stems from inserting the expression for $\alpha_t$ directly into the SDE for $\log Y_t$. In this case, we have

$$d \log Y_t^k = \delta_k \left( a_0 - \tfrac{1}{2}\Omega_{kk} + \sum_{i=1}^{n} a_i \log Y_t^i \right) dt + \sum_{i=1}^{n} \sigma_{ki} \, dW_t^i,$$

so that if we let $Z_t = \log Y_k$, where the log is interpreted componentwise, then

$$dZ_t = (c - B Z_t) \, dt + \sigma \, dW_t,$$

with

$$B = - \begin{pmatrix} \delta_1 a_1 & \delta_1 a_2 & \cdots & \delta_1 a_n \\ \vdots & \vdots & & \vdots \\ \delta_n a_1 & \delta_n a_2 & \cdots & \delta_n a_n \end{pmatrix}, \quad c = \begin{pmatrix} a_0 - \Omega_{11} \\ \vdots \\ a_0 - \Omega_{nn} \end{pmatrix}.$$

In this representation, we directly see that the log-prices are a vector-autoregressive model (VAR). Although this representation is in some sense more compact, we will find in the next section that the short-term-alpha version of the model is a preferable representation for solving the agent's control problem. Note that here, $B$ is singular and contains exactly one positive eigenvalue due to the singular co-integration factor that we incorporate into the model. If we have $m \leq n$ co-integration factors then there will be in general $m$ positive eigenvalues.

In contrast to our earlier work on price impact models of trading (see Chapters 6 and 7), here we assume there is no impact from trading, and instead optimise the agent's utility of expected wealth. Thus, let $\pi = (\pi_t^0, \pi_t^1, \ldots, \pi_t^n)_{0 \leq t \leq T}$ denote the **dollar value invested** in the riskless ($\pi_t^0$) and risky assets ($\pi_t^1, \ldots, \pi_t^n$), and let $X^\pi = (X_t^\pi)_{0 \leq t \leq T}$ denote the agent's (controlled) wealth process. With this convention, the number of units $m_t^k$ the agent holds in asset $k$ is $m_t^k = \pi_t^k / Y_t^k$.

Hence, the wealth process can be written as

$$X_t^{\pi} = \sum_{k=0}^{n} \pi_t^k = \sum_{k=0}^{n} m_t^k Y_t^k ,$$

so that

$$dX_t^{\pi} = \sum_{k=0}^{n} d(m_t^k Y_t^k) = \sum_{k=0}^{n} m_t^k dY_t^k = \sum_{k=1}^{n} \pi_t^k \frac{dY_t^k}{Y_t^k}$$

$$= \sum_{k=1}^{n} \pi_t^k \left\{ \delta_k \alpha_t \, dt + \sum_{i=1}^{n} \sigma_{ki} \, dW_t^i \right\} = (\pi_t' \delta) \alpha_t \, dt + \pi_t' \sigma dW_t ,$$

where the second equality follows from the usual self-financing constraint – which can be interpreted as the change in the wealth process due to the change in each asset's value, assuming the positions are held fixed over a small interval of time. The third equality is obtained by assuming that the time horizon is short enough that interest rates are zero (so that $dY_t^0 = 0$), but long enough that the geometric model we employ is required. (It is not too difficult to include a deterministic discount factor and the interested reader is urged to try this out.) The last equality is a rewrite of the equations using vector/matrix notation.

To set up the dynamic programming equations below we require the quadratic variation and cross-variations of the processes $X$ and $\boldsymbol{y}$. First, it is easy to see that

$$d[Y^i, Y^j]_t = \sum_{k,l=1}^{n} Y_t^i Y_t^j \sigma_{ki} \sigma_{lj} d[W^i, W^j]_t = \sum_{k=1}^{n} Y_t^i Y_t^j \sigma_{ki} \sigma_{lk} dt = (\boldsymbol{y}_t' \boldsymbol{\Omega} \, \boldsymbol{y}_t)_{ij} \, dt ,$$

where $(\boldsymbol{y}_t' \boldsymbol{\Omega} \, \boldsymbol{y}_t)_{ij}$ represents the $ij$th element of the matrix in the braces. Next,

$$d[X^{\pi}, X^{\pi}]_t = \sum_{i,j,k,l=1}^{n} \pi_t^k \pi_t^l \sigma_{ki} \sigma_{lj} d[W^i, W^j]_t = \sum_{i,k,l=1}^{n} \pi_t^k \pi_t^l \sigma_{ki} \sigma_{li} \, dt = \pi_t' \boldsymbol{\Omega} \pi_t \, dt .$$

Finally,

$$d[X^{\pi}, Y^k]_t = \sum_{i,j,l=1}^{n} \pi_t^l \sigma_{li} \sigma_{kj} Y_t^k \, d[W^i, W^j]_t = \sum_{j,l=1}^{n} \pi_t^l \sigma_{lj} \sigma_{kj} Y_t^k \, dt = (\pi_t' \boldsymbol{\Omega} \, \boldsymbol{y}_t)_k ,$$

where $(\pi_t' \boldsymbol{\Omega} \, \boldsymbol{y}_t)_k$ represents the $k$th element of the vector in the braces.

## 11.4.2    The Agent's Optimisation Problem

Next, the agent will optimise her dollar value in assets directly, rather than the rate of trading, and has exponential utility $u(x) = -e^{-\gamma x}$. Her performance criteria is given by

$$H^{\pi}(t, x, \boldsymbol{y}) = \mathbb{E}_{t,x,\boldsymbol{y}} \left[ -\exp\left(-\gamma X_T^{\pi}\right) \right] ,$$

where $\mathbb{E}_{t,x,y}[\cdot]$ represents expectation conditional on $X_t^{\pi} = x$ and $y_t = y$. The agent's value function is as usual

$$H(t,x,y) = \sup_{\pi \in \mathcal{A}} H^{\pi}(t,x,y),$$

where the set of admissible strategies $\mathcal{A}$ are those for which

$$\mathbb{E}\left[\sum_{k=0}^{n} \int_0^T (\pi_u^k)^2 \, du\right] < \infty.$$

Alternatively, we can enforce the condition that each component of $\pi$ is bounded.

Applying the dynamic programming principle leads to the DPE which the value function should satisfy:

$$\partial_t H + \alpha \, \boldsymbol{\delta}' \boldsymbol{\mathcal{D}}_y H + \tfrac{1}{2} \boldsymbol{\mathcal{D}}_{yy}^{\Omega} H$$
$$+ \sup_{\pi} \left\{ (\pi' \boldsymbol{\delta}) \, \alpha \, \partial_x H + \tfrac{1}{2} (\pi_t' \boldsymbol{\Omega} \pi) \, \partial_{xx} H + \pi' \boldsymbol{\Omega} \, \boldsymbol{\mathcal{D}}_{xy} H \right\} = 0 \,,$$

subject to $H(T, x, y) = -e^{-\gamma x}$, and where $\alpha = a_0 + \boldsymbol{a}' \log y$ (the log being interpreted componentwise) represents the state of the co-integration process, and the following linear differential operators were introduced to reduce clutter:

$$\boldsymbol{\mathcal{D}}_y H = (y^1 \partial_{y^1} H, \dots, y^n \partial_{y^n} H)' \,,$$
$$\boldsymbol{\mathcal{D}}_{yy}^{\Omega} = \textstyle\sum_{i,j=1}^{n} y^i \Omega_{ij} y^j \partial_{y^i y^j} H \,,$$
$$\boldsymbol{\mathcal{D}}_{xy} = (y^1 \partial_{xy^1} H, \dots, y^n \partial_{xy^n} H)' \,.$$

Next, let us obtain the optimal investment in feedback form. For this we focus on the supremum term and perform a matrix completion of the square. Specifically, assuming that $\partial_{xx} H \neq 0$, we write

$$\mathcal{M} = (\pi' \boldsymbol{\delta}) \, \alpha \, \partial_x H + (\pi_t' \boldsymbol{\Omega} \pi) \, \partial_{xx} H + \pi' \boldsymbol{\Omega} \, \boldsymbol{\mathcal{D}}_{xy} H$$
$$= \tfrac{1}{2} \partial_{xx} H \left\{ (\pi_t' \boldsymbol{\Omega} \pi) + 2\pi' \frac{(\boldsymbol{\delta} \, \alpha \, \partial_x H + \boldsymbol{\Omega} \, \boldsymbol{\mathcal{D}}_{xy} H)}{\partial_{xx} H} \right\}$$
$$= \tfrac{1}{2} \partial_{xx} H \left\{ (\pi_t + \boldsymbol{\Omega}^{-1} \boldsymbol{\mathcal{L}} H)' \boldsymbol{\Omega} (\pi_t + \boldsymbol{\Omega}^{-1} \boldsymbol{\mathcal{L}} H) - \frac{\boldsymbol{\mathcal{L}}' H \boldsymbol{\Omega}^{-1} \boldsymbol{\mathcal{L}} H}{(\partial_{xx} H)^2} \right\} \,,$$

where the vector-valued linear operator $\boldsymbol{\mathcal{L}}$ acts on $H$ as follows:

$$\boldsymbol{\mathcal{L}} H = \boldsymbol{\delta} \, \alpha \, \partial_x H + \boldsymbol{\Omega} \, \boldsymbol{\mathcal{D}}_{xy} H \,,$$

and we have used the fact that $\boldsymbol{\Omega}$ is symmetric (since it is a variance-covariance matrix) so that $(\boldsymbol{\Omega}^{-1})' = (\boldsymbol{\Omega}')^{-1} = \boldsymbol{\Omega}^{-1}$. From the above expression we can immediately identify the optimal investment in feedback control form as

$$\boxed{\pi^* = -\frac{\boldsymbol{\Omega}^{-1} \boldsymbol{\delta} \, \alpha \, \partial_x H + \boldsymbol{\mathcal{D}}_{xy} H}{\partial_{xx} H} \,.} \tag{11.7}$$

From the above matrix completion of the square, we also have the maximum term

$$\mathcal{M} = -\tfrac{1}{2} \frac{\mathcal{L}'H\,\Omega^{-1}\mathcal{L}H}{\partial_{xx}H},$$

so that upon reinsertion into the DPE, we obtain the following non-linear PDE for the value function:

$$\boxed{\partial_t H + \alpha\,\delta'\mathcal{D}_y H + \tfrac{1}{2}\mathcal{D}^{\Omega}_{yy}H - \frac{\mathcal{L}'H\,\Omega^{-1}\mathcal{L}H}{2\,\partial_{xx}H} = 0\,.}$$    (11.8)

## 11.4.3    Solving the DPE

In the classical Merton problem, where the asset prices are geometric Brownian motions, the value function for exponential utility has the form $-e^{-\gamma(x+h(t))}$, where $h$ is a deterministic function of time. Here, due to the presence of the co-integration factor we expect instead that the value function depends also on a combination of the price state variables which equals the co-integration factor. That is we expect to be to able to write

$$H(t,x,\boldsymbol{y}) = -\exp\left\{-\gamma\left(x + h(t,a_0) + \sum_{i=1}^{n} a_i \log y^i\right)\right\}$$

for some function $h(t,\alpha)$, with $\alpha = a_0 + \sum_{i=1}^{n} a_i \log y^i$, and subject to the terminal condition $h(T,\alpha) = 0$. Note that differentiation with respect to $y^k$ acts simply on $h$, specifically

$$\partial_{y^k} h(t,\alpha) = \frac{a_k}{y^k}\,\partial_{\alpha} h(t,\alpha)\,,$$

so that

$$\partial_{y^k} e^{-\gamma h(t,\alpha)} = -\gamma\,\frac{a_k}{y^k}\,\partial_{\alpha} h(t,\alpha)\,e^{-\gamma h(t,\alpha)}\,,$$

$$\partial_{y^j y^k} h(t,\alpha) = -\gamma\,\partial_{y^j}\left(\frac{a_k}{y^k}\,\partial_{\alpha} h(t,\alpha)\,e^{-\gamma h(t,\alpha)}\right)\,.$$

Hence, for $j \neq k$,

$$\partial_{y^j y^k} e^{-\gamma h(t,\alpha)} = -\gamma\,\frac{a_k\,a_j}{y^k\,y^j}\,\partial_{\alpha}\left(\partial_{\alpha} h(t,\alpha)\,e^{-\gamma h(t,\alpha)}\right)$$

$$= -\gamma\,\frac{a_k\,a_j}{y^k\,y^j}\,\left(\partial_{\alpha\alpha} h(t,\alpha) - \gamma\,(\partial_{\alpha} h(t,\alpha))^2\right)\,e^{-\gamma h(t,\alpha)}\,,$$

while for $j = k$,

$$\partial_{y^j y^j} e^{-\gamma h(t,\alpha)}$$

$$= -\gamma\left(-\frac{a_k}{(y^k)^2}\,\partial_{\alpha} h(t,\alpha)\,e^{-\gamma h(t,\alpha)} + \frac{a_k^2}{(y^k)^2}\,\partial_{\alpha}\left(\partial_{\alpha} h(t,\alpha)\,e^{-\gamma h(t,\alpha)}\right)\right)$$

$$= -\gamma\,\frac{a_k}{(y^k)^2}\,\left(-\partial_{\alpha} h(t,\alpha) + a_k\,(\partial_{\alpha\alpha} h(t,\alpha) - \gamma\,(\partial_{\alpha} h(t,\alpha))^2)\right)\,e^{-\gamma h(t,\alpha)}\,.$$

Putting these two results together we can write

$$\partial_{y^j y^k} e^{-\gamma h(t,\alpha)} = -\gamma \frac{a_k a_j}{y^k y^j} \left( \partial_{\alpha\alpha} h(t,\alpha) - \gamma \left( \partial_\alpha h(t,\alpha) \right)^2 \right) e^{-\gamma h(t,\alpha)}$$
$$+ \delta_{jk} \gamma \frac{a_k}{(y^k)^2} \partial_\alpha h(t,\alpha) e^{-\gamma h(t,\alpha)},$$

where $\delta_{jk}$ is the Kronecker delta which equals 1 if $j = k$ and 0 otherwise.

Armed with these results, the various linear differential operators that appear in the non-linear PDE (11.8) can be written as follows:

$$\mathcal{D}_y H = (-\gamma H) \, a \, \partial_\alpha h,$$
$$\mathcal{D}_{yy}^\Omega = (-\gamma H) \sum_{i,j=1}^n \Omega_{ij} a_i a_j \left( \partial_{\alpha\alpha} h - \gamma (\partial_\alpha h)^2 \right) + (\gamma H) \sum_{j=1}^n a_j \Omega_{jj} \partial_\alpha h$$
$$= -(\gamma H) (a'\Omega a) \left( \partial_{\alpha\alpha} h - \gamma (\partial_\alpha h)^2 \right) + (\gamma H) \operatorname{Tr}(A\Omega) \partial_\alpha h,$$

and recall that $\operatorname{Tr}(\cdot)$ denotes the trace of the matrix (i.e. the sum along its diagonal elements), and $A = \operatorname{diag}(a)$ is a diagonal matrix with the elements of the vector $a$ along the diagonal. Furthermore,

$$\mathcal{D}_{xy} H = (\gamma^2 H) \, a \, \partial_\alpha h,$$
$$\mathcal{L} H = (-\gamma H) \, \delta \, \alpha + (\gamma^2 H) \, \Omega a \, \partial_\alpha h.$$

Inserting these expression into the PDE (11.8), allows us to write

$$0 = (-\gamma H)\partial_t h + (-\gamma H) (\delta' a) \, \alpha \, \partial_\alpha h$$
$$+ \tfrac{1}{2} \left( (-\gamma H) (a'\Omega a) \left( \partial_{\alpha\alpha} h - \gamma (\partial_\alpha h)^2 \right) + (\gamma H) \operatorname{Tr}(A\Omega) \partial_\alpha h \right)$$
$$- \tfrac{1}{2} \frac{\left( (-\gamma H) \, \delta\alpha + (\gamma^2 H) \, \Omega a \, \partial_\alpha h \right)' \Omega^{-1} \left( (-\gamma H) \, \delta\alpha + (\gamma^2 H) \, \Omega a \, \partial_\alpha h \right)}{\gamma^2 H}.$$

At this point, there are three important simplifications. First, cancelling $-\gamma H$ in all terms, we find that

$$0 = \partial_t h + (\delta' a) \, \alpha \, \partial_\alpha h$$
$$+ \tfrac{1}{2} \left( (a'\Omega a) \left( \partial_{\alpha\alpha} h - \gamma (\partial_\alpha h)^2 \right) - \operatorname{Tr}(A\Omega) \partial_\alpha h \right) \tag{11.9}$$
$$+ \frac{1}{2\gamma} (\delta\alpha - \gamma \, \Omega a \, \partial_\alpha h)' \, \Omega^{-1} (\delta\alpha - \gamma \, \Omega a \, \partial_\alpha h).$$

Next, expanding the third line, we have

$$(\delta\alpha - \gamma \, \Omega a \, \partial_\alpha h)' \, \Omega^{-1} (\delta\alpha - \gamma \, \Omega \, a \, \partial_\alpha h)$$
$$= \delta' \, \Omega^{-1} \, \delta \, \alpha^2 - 2\gamma \, \delta' \, a \, \alpha \, \partial_\alpha h + \gamma^2 \, a' \, \Omega a \, (\partial_\alpha h)^2,$$

so that upon substituting into the previous expression there are two important cancellations:

(i) the non-linear terms containing $(\partial_\alpha h)^2$ from the expansion above and the second line in (11.9) cancel one another;

(ii) the terms $\delta' \, a \, \alpha \, \partial_\alpha h$ from the first line in (11.9) and the above expansion also cancel.

Putting these observations together we then find that $h$ satisfies a very simple linear PDE:

$$\partial_t h - \tfrac{1}{2}\operatorname{Tr}(\boldsymbol{A}\,\boldsymbol{\Omega})\,\partial_\alpha h + \tfrac{1}{2}(\boldsymbol{a}'\boldsymbol{\Omega}\boldsymbol{a})\,\partial_{\alpha\alpha}h + \frac{\boldsymbol{\delta}'\boldsymbol{\Omega}\boldsymbol{\delta}}{2\gamma}\,\alpha^2 = 0\,, \qquad (11.10)$$

subject to $h(T,\alpha) = 0$.

Returning to the optimal investment (11.7) in the assets, the ansatz for the value function $H = -\exp\{-\gamma(x + h(t,\alpha))\}$ allows us to write

$$\boldsymbol{\pi}^* = -\frac{-\gamma\,H\,(\boldsymbol{\Omega}^{-1}\boldsymbol{\delta})\,\alpha + (\gamma^2\,H)\,\boldsymbol{a}\,\partial_\alpha h}{\gamma^2\,H}$$

$$= -\frac{-(\boldsymbol{\Omega}^{-1}\boldsymbol{\delta})\,\alpha + \gamma\,\boldsymbol{a}\,\partial_\alpha h}{\gamma}$$

$$= \tfrac{1}{\gamma}\,(\boldsymbol{\Omega}^{-1}\boldsymbol{\delta})\,\alpha - \boldsymbol{a}\,\partial_\alpha h\,. \qquad (11.11)$$

Let us recall that the optimal investment in the classical Merton problem is $\tfrac{1}{\gamma}\boldsymbol{\Omega}^{-1}(\boldsymbol{\nu} - r)$ where $r$ is the risk-free rate and $\boldsymbol{\nu}$ is (in the classical Merton problem) the drift of the GBMs that drives the asset prices. The first term of the above expression is quite similar, since here $r = 0$, and the drift of the assets are $\boldsymbol{\delta}\,\alpha_t$ – here, however, the drift is stochastic and hence the Merton solution cannot be applied directly. The optimal investment scales with the co-integration factor. Moreover, the optimal investment is perturbed from the Merton solution by the second term. In the coming sections we will see precisely what this contribution is to the investment policy.

First, from (11.10), we can see that the solution to $h$ must be quadratic in $\alpha$. Indeed in the next section we construct an explicit probabilistic representation for $h(t,\alpha)$ and also characterise the value function in terms of a risk-neutral probability measure which makes the quadratic dependence explicit. Given that $h$ is quadratic in $\alpha$, we also see that the optimal dollar invested in each asset $\boldsymbol{\pi}^*$ will be linear in $\alpha$. Thus, the agent reacts to the co-integration factor in at most a linear fashion, but the size of the investment will vary with time.

### A Probabilistic Interpretation

This surprisingly simple PDE (11.10) has an interesting probabilistic interpretation. First, consider the probability measure change induced by a vector of market price of risk $\boldsymbol{\lambda}_t$, which induces a new measure $\mathbb{P}^*$ through the Radon-Nikodym derivative

$$\frac{d\mathbb{P}^*}{d\mathbb{P}} = \exp\left\{-\tfrac{1}{2}\int_0^T \|\boldsymbol{\lambda}_s\|^2\,ds - \int_0^T \boldsymbol{\lambda}_s'\,d\boldsymbol{W}_s\right\}\,.$$

Girsanov's Theorem implies that the stochastic processes

$$\boldsymbol{W}_t^* = -\int_0^t \boldsymbol{\lambda}_s\,ds + \boldsymbol{W}_t$$

are independent $\mathbb{P}^*$-Brownian motions. Let us choose $\boldsymbol{\lambda}_t$ such that the drift of the traded assets $\boldsymbol{Y}_t$ are martingales, i.e. let us find the measure transformation which produces the risk-neutral measure. To this end, rewrite the SDE as

$$dY_t^k = \delta_k \, Y_t^k \, \alpha_t \, dt + \sum_{i=1}^n \sigma_{ki} \, Y_t^k \, dW_t^i$$
$$= \left(\delta_k \, \alpha_t + \sum_{i=1}^n \sigma_{ki} \lambda_t^i\right) Y_t^k \, dt + \sum_{i=1}^n \sigma_{ki} \, Y_t^k \, dW_t^{*i} \,.$$

The martingale condition requires that

$$\sum_{i=1}^n \sigma_{ki} \, \lambda_t^i = -\delta_k \, \alpha_t, \quad k = 1, \ldots, n, \quad \Longleftrightarrow \quad \boldsymbol{\sigma} \, \boldsymbol{\lambda}_t = -\boldsymbol{\delta} \, \alpha_t \,.$$

Therefore, since $\boldsymbol{\sigma}$ is invertible by assumption, we have

$$\boxed{\boldsymbol{\lambda}_t = -\boldsymbol{\sigma}^{-1} \, \boldsymbol{\delta} \, \alpha_t \,.} \tag{11.12}$$

At this point, we have found the probability measure $\mathbb{P}^*$ which renders the traded assets martingales. That is, we have found the risk-neutral measure. Next, we can ask what the dynamics of the co-integration factor are in terms of the risk-neutral Brownian motions. Hence, from (11.6) we have

$$d\alpha_t = \left(-\tfrac{1}{2} \operatorname{Tr}(\boldsymbol{A}\,\boldsymbol{\Omega}) + (\boldsymbol{a}'\,\boldsymbol{\delta})\alpha_t\right) dt + \boldsymbol{a}'\boldsymbol{\sigma} d\boldsymbol{W}_t$$
$$= \left(-\tfrac{1}{2} \operatorname{Tr}(\boldsymbol{A}\,\boldsymbol{\Omega}) + (\boldsymbol{a}'\,\boldsymbol{\delta})\alpha_t\right) dt + \boldsymbol{a}'\boldsymbol{\sigma} \left(\boldsymbol{\lambda}_t \, dt + d\boldsymbol{W}_t^*\right)$$
$$= \left(-\tfrac{1}{2} \operatorname{Tr}(\boldsymbol{A}\,\boldsymbol{\Omega}) + (\boldsymbol{a}'\,\boldsymbol{\delta})\alpha_t + \boldsymbol{a}'\,\boldsymbol{\sigma}\,\boldsymbol{\lambda}_t\right) dt + \boldsymbol{a}'\boldsymbol{\sigma} \, d\boldsymbol{W}_t^* \,,$$

so that

$$\boxed{d\alpha_t = -\tfrac{1}{2} \operatorname{Tr}(\boldsymbol{A}\,\boldsymbol{\Omega}) \, dt + \boldsymbol{a}'\boldsymbol{\sigma} \, d\boldsymbol{W}_t^* \,.} \tag{11.13}$$

The last equality follows from (11.12). Surprisingly, although $\alpha_t$ is mean-reverting in the real-world $\mathbb{P}$-measure, it is a Brownian motion under the risk-neutral $\mathbb{P}^*$-measure.

Let us now return to the PDE (11.10). It can be re-written in the form

$$(\partial_t + \mathcal{L}^{*,\alpha}) h + \frac{\boldsymbol{\delta}'\boldsymbol{\Omega}\boldsymbol{\delta}}{2\gamma} \alpha^2 = 0 \,,$$

subject to $h(T, \alpha) = 0$, where $\mathcal{L}^{*,\alpha}$ is the $\mathbb{P}^*$-infinitesimal generator of $\alpha_t$ and therefore, applying a Feynman-Kac formula we see that $h(t, \alpha)$ can be expressed as the following expectation:

$$\boxed{h(t, \alpha) = \mathbb{E}_{t,\alpha}^* \left[\frac{\boldsymbol{\delta}'\boldsymbol{\Omega}\boldsymbol{\delta}}{2\gamma} \int_t^T \alpha_s^2 \, ds\right] \,,} \tag{11.14}$$

where $\mathbb{E}_{t,\alpha}^*$ denotes $\mathbb{P}^*$-expectation given that $\alpha_t = \alpha$. Putting this expression for $h$ back into the value function we then have the following relationship:

$$\boxed{\begin{aligned} \sup_{\pi \in \mathcal{A}} &\mathbb{E}_{t,x,y} \left[- \exp\left(-\gamma \, X_T^\pi\right)\right] \\ &= -\exp\left(-\gamma x - \tfrac{1}{2} \boldsymbol{\delta}' \, \boldsymbol{\Omega} \, \boldsymbol{\delta} \, \mathbb{E}_{t,x,y}^* \left[\int_t^T \alpha_s^2 \, ds\right]\right) \,. \end{aligned}} \tag{11.15}$$

The future expectation of integrated $\alpha_t^2$ determines the incremental value of trading on this co-integrated collection of assets. The strength of the contribution increases with volatility (through $\mathbf{\Omega}$) as well as through the strength that $\alpha$ has on each asset (through $\boldsymbol{\delta}$).

## Explicit Construction of the Optimal Investment Strategy

Based on the representation from the previous section, we can construct an explicit formula for $h(t, \alpha)$. In particular, we have that

$$
h(t, \alpha) = \mathbb{E}^*_{t,\alpha} \left[ \tfrac{\delta'\mathbf{\Omega}\delta}{2\gamma} \int_t^T \alpha_s^2 \, ds \right]
$$

$$
= \tfrac{\delta'\mathbf{\Omega}\delta}{2\gamma} \mathbb{E}^* \left[ \int_t^T \left( \alpha - \tfrac{1}{2}\mathrm{Tr}(\mathbf{A}\mathbf{\Omega})(s-t) + \sqrt{\mathbf{a}'\mathbf{\Omega}\mathbf{a}} \left( B_s^* - B_t^* \right) \right)^2 ds \right],
$$

where we have introduced the process $B_t^* = \frac{1}{\sqrt{\mathbf{a}'\mathbf{\Omega}\mathbf{a}}} \mathbf{a}'\boldsymbol{\sigma} W_t^*$ which is a standard $\mathbb{P}^*$-Brownian motion. Hence,

$$
\begin{aligned}
h(t, \alpha) = \tfrac{\delta'\mathbf{\Omega}\delta}{2\gamma} \mathbb{E}^* \Big[ \int_t^T \Big\{ & \left( \alpha - \tfrac{1}{2}\mathrm{Tr}(\mathbf{A}\mathbf{\Omega})(s-t) \right)^2 \\
& + 2 \left( \alpha - \tfrac{1}{2}\mathrm{Tr}(\mathbf{A}\mathbf{\Omega})(s-t) \right) \sqrt{\mathbf{a}'\mathbf{\Omega}\mathbf{a}} \left( B_s^* - B_t^* \right) \\
& + (\mathbf{a}'\mathbf{\Omega}\mathbf{a}) \left( B_s^* - B_t^* \right)^2 \Big\} \, ds \Big]
\end{aligned}
$$

$$
= \tfrac{\delta'\mathbf{\Omega}\delta}{2\gamma} \left\{ -\tfrac{2}{3\,\mathrm{Tr}(\mathbf{A}\mathbf{\Omega})} \left[ \left( \alpha - \tfrac{1}{2}\mathrm{Tr}(\mathbf{A}\mathbf{\Omega})\tau \right)^3 - \alpha^3 \right] + \tfrac{1}{2}(\mathbf{a}'\mathbf{\Omega}\mathbf{a})\tau^2 \right\},
$$

where $\tau = T - t$. The fourth equality follows by using Fubini to interchange the integral and expectation, in which case the second term in the third equation vanishes identically, and in the third term of the third equation we use $\mathbb{E}^*[(B_s^* - B_t^*)^2] = (s - t)$.

In all, we see that $h(t, \alpha)$ is quadratic in $\alpha$. Consequently, the optimal investment from (11.11) takes on the explicit form

$$
\pi^* = \tfrac{1}{\gamma} \left\{ (\mathbf{\Omega}^{-1}\boldsymbol{\delta})\,\alpha + \tfrac{\delta'\mathbf{\Omega}\delta}{\mathrm{Tr}(\mathbf{A}\mathbf{\Omega})} \left[ \left( \alpha - \tfrac{1}{2}\mathrm{Tr}(\mathbf{A}\mathbf{\Omega})\tau \right)^2 - \alpha^2 \right] \mathbf{a} \right\}
$$

$$
= \tfrac{1}{\gamma} \left\{ (\mathbf{\Omega}^{-1}\boldsymbol{\delta})\,\alpha + \tfrac{\delta'\mathbf{\Omega}\delta}{\mathrm{Tr}(\mathbf{A}\mathbf{\Omega})} \left[ \tfrac{1}{2}\mathrm{Tr}(\mathbf{A}\mathbf{\Omega})\,\tau\,\alpha + \tfrac{1}{4}(\mathrm{Tr}(\mathbf{A}\mathbf{\Omega}))^2\,\tau^2 \right] \mathbf{a} \right\},
$$

so that

$$
\boxed{\pi^* = \tfrac{1}{\gamma} \left\{ (\mathbf{\Omega}^{-1}\boldsymbol{\delta})\,\alpha + (\boldsymbol{\delta}'\mathbf{\Omega}\boldsymbol{\delta}) \left[ \tfrac{1}{2}\tau\,\alpha + \tfrac{1}{4}\mathrm{Tr}(\mathbf{A}\mathbf{\Omega})\,\tau^2 \right] \mathbf{a} \right\}.}
\tag{11.16}
$$

As a reminder, the first term is what you would expect from the classical Merton problem since $\boldsymbol{\delta}\,\alpha$ are the drifts of the assets. The second term proportional to $\mathbf{a}$ represents the correction due to co-integration. As the above expression shows, the perturbation around the 'Merton' portfolio decays as the terminal date approaches. That is, the agent ignores the short-term-alpha effect when trading is coming to a close.

### 11.4.4    Numerical Experiments

In this section we showcase how the strategy behaves in a three-asset case. For this purpose the following modelling parameters are chosen:

$$
\boldsymbol{\delta} = (1\ 1\ 0)', \qquad \mathbf{a} = (-1\ 0\ 1)', \qquad a_0 = 0,
$$

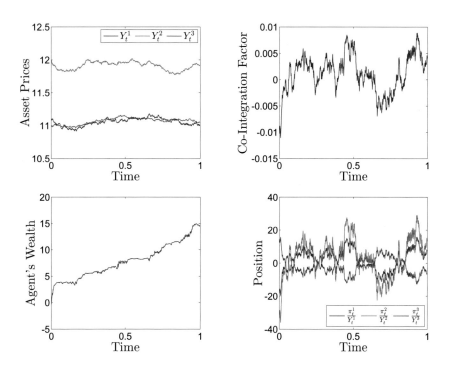

**Figure 11.7** A single sample path of the co-integrated asset prices and the resulting optimal strategy.

and

$$\sigma = \begin{pmatrix} 0.2000 & 0 & 0 \\ 0.0375 & 0.1452 & 0 \\ 0.0250 & 0.0039 & 0.0967 \end{pmatrix} \quad \Rightarrow \quad \Omega = \begin{pmatrix} 0.04 & 0.0075 & 0.005 \\ 0.0075 & 0.0225 & 0.0015 \\ 0.005 & 0.0015 & 0.01 \end{pmatrix}.$$

The volatilities of the three assets can be read off of the diagonal of $\Omega$ to be $\{2\%, 1.5\%, 1\%\}$, and the corresponding correlation matrix $\rho$ is

$$\rho = \begin{pmatrix} 1.00 & 0.25 & 0.25 \\ 0.25 & 1.00 & 0.10 \\ 0.25 & 0.10 & 1.00 \end{pmatrix}.$$

Finally, we start the asset prices at the following levels:

$$Y_t^1 = 11.10, \qquad Y_t^2 = 12.00, \qquad Y_t^3 = 11.00.$$

The particular choice of the co-integration vector $\boldsymbol{a}$ implies that only the first and third assets feed into the co-integration factor. Yet, the strength term $\boldsymbol{\delta}$ implies that only the first and second asset's drift are affected by that factor.

In Figure 11.7 we show a sample path of the asset prices, co-integration factor, optimal strategy, and agent's wealth process. In Figure 11.8 we show the histogram of the P&L.

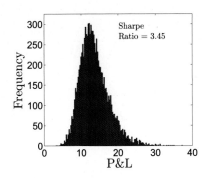

**Figure 11.8** Histogram of the P&L of the optimal pairs trading strategy.

## 11.5    Bibliography and Selected Readings

Elliott, Van Der Hoek & Malcolm (2005), Tourin & Yan (2013), Leung & Li (2014), Leung & Ludkovski (2011).

# 12 Order Imbalance

## 12.1 Introduction

In many of the previous chapters, the agent made trading decisions based on three key ingredients: (i) the midprice, (ii) the arrival of incoming market orders (MOs), and (iii) the agent's own inventory. In some cases, these state variables were supplemented by observables such as order flow (see, e.g., Section 7.3 and Chapter 9), short-term-alpha in 10.4.2, and co-integration of prices in Chapter 11, among others. In this chapter, we investigate the role that another important state variable plays: the quoted volume order imbalance (or simply order imbalance). This is a measure of the buy versus sell pressure on an asset and, as we will see, it contains predictive power on both the arrival rates of MOs, and the direction and size of future price movements. Hence, it is an important factor to include when designing trading algorithms.

The chapter is organised as follows: in Section 12.2 we define order imbalance, and show using NASDAQ data how it typically evolves at an intraday level. The section also introduces three Markov Chain models for order imbalance, arrival of MOs and price jumps, and develops maximum likelihood estimators of the model parameters. Section 12.3 provides a brief discussion of the daily features exhibited by order imbalance using functional data analysis. Finally, Section 12.4 provides an analysis of the optimal liquidation problem, using limit orders (LOs) only, in the presence of order imbalance.

## 12.2 Intraday Features

We define limit order imbalance $\rho_t$ at time $t$ as the ratio of the quoted volume imbalance to the total quoted volume, i.e.

$$\rho_t = \frac{V_t^b - V_t^a}{V_t^b + V_t^a},$$

where $V_t^b$ denotes the volume of LOs posted on the bid side of the LOB and $V_t^a$ denotes the volume of LOs posted on the ask side of the LOB. For simplicity, from now on we refer to the LO imbalance as order imbalance. The volumes may be computed by looking only at-the-touch, the best $n$-levels of the LOB,

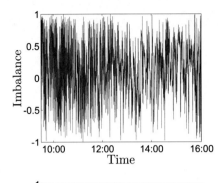

**Figure 12.1** Order imbalance for ORCL on Nov 1, 2013. In the bottom panels, blue (red) dots show times and imbalance for buy (sell) MOs.

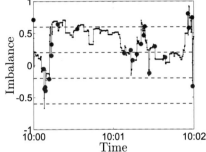

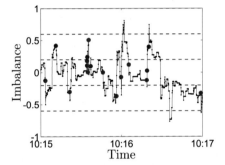

or volume that is posted within $n$ ticks of the midprice. All of these are viable measures, and it is up to the agent to decide which is best in a given situation. Some studies suggest that the best trade-off between predictive power versus model complexity is strongest using only the touch. For simplicity we use only this information for our estimations, although nothing in the model dictates this restriction.

Figure 12.1 shows order imbalance for Oracle Corporation (ORCL) computed from the Nov 1, 2013 event data (NASDAQ exchange) sampled every millisecond and averaged over 100ms, and as mentioned above, using volume posted only at-the-touch. The bottom left and right panels show the imbalance for two-minute periods starting at 10:00am and 10:15am together with 5 regimes of imbalance chosen to be equally spaced along the points $\rho_t \in \{-1, -0.6, -0.2, 0.2, 0.6, 1\}$. The top panel shows the order imbalance for the entire day and illustrates the significant fluctuations in order imbalance throughout the day. When imbalance is placed into bins, the fluctuation rates are somewhat mitigated. Figure 12.2 shows some properties of MO arrivals. The left panel shows the percentage of MOs which are buys/sells/total when order imbalance is in a particular regime – so that, e.g., when buy MOs arrive, 40.1 percent of the time order imbalance is in regime 4. The right panel shows the arrival rate (per second) of MOs by normalising the number of MOs that arrive in a regime according to the time that order imbalance spends in that regime. It appears that the arrival of buy MOs is biased towards times when order imbalance is high, and similarly sell

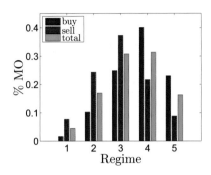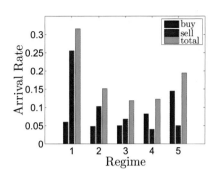

**Figure 12.2** MO arrival for ORCL on Nov 1, 2013 conditional on imbalance regimes.

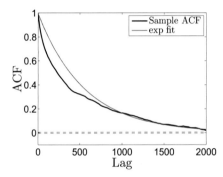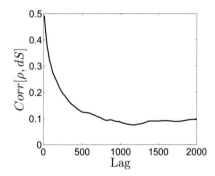

**Figure 12.3** Auto-correlation of imbalance (left panel) and correlation of order imbalance with price changes (right panel) for ORCL on Nov 1, 2013.

MOs arrive more frequently when order imbalance is negative. The total arrival rates exhibit a U-shaped pattern as a function of order imbalance, indicating that MOs tend to arrive more frequently when the LOB is bid-heavy ($\rho$ close to 1) or ask-heavy ($\rho$ close to $-1$). We explore these features more deeply later in this chapter.

Figure 12.3 shows the auto-correlation function (ACF) over 2,000 lags which equals 200 secs, as well as the correlation between imbalance and the price change (conditional on a price change occurring) over the next $n$-intervals. As the plot shows, there is a significant amount of auto-correlation in order imbalance. Moreover, order imbalance is positively correlated with price changes. This correlation naturally becomes less significant as the lag increases.

## 12.2.1  A Markov Chain Model

We provide a simple Markov chain model for order imbalance and show how to calibrate it to market data. Let $Z_t \in \{1, \ldots, K\}$ denote the order imbalance regime observed at time $t$, here taken to be discrete $t \in \{1, 2, \ldots, T\}$. We assume that order imbalance is described by a Markov chain with transition matrix $\boldsymbol{A}$.

## Maximum Likelihood Estimator

Let $z_1$, $z_2$, ..., $z_T$ denote a sequence of observations from the Markov chain $Z$. Below we show that the maximum likelihood estimator (MLE) of the elements of the transition matrix $\boldsymbol{A}$ is given by

$$\hat{A}_{ij} = \frac{n_{ij}}{\sum_{j=1}^{K} n_{ij}}, \tag{12.1}$$

where $n_{ij} = \sum_{t=2}^{T} \mathbb{1}_{\{Z_{t-1}=i,\, Z_t=j\}}$, i.e. $n_{ij}$ is the number of observed transitions from regime $i$ to regime $j$.

To demonstrate this result, we first write the likelihood $L$ of the sequence of observations as

$$L = A_{Z_1,Z_2} \times A_{Z_2,Z_3} \times \cdots \times A_{Z_{T-1},Z_T} = \prod_{i,j=1}^{K} A_{ij}^{n_{ij}}.$$

The second equality follows from collecting like terms together. Next, note that the transition matrix is constrained so that $A_{ij} \geq 0$ and the sum along each row equals 1, i.e. $\sum_{j=1}^{K} A_{ij} = 1$ for $i = 1, \ldots, K$. Next, we wish to maximise $L$, or equivalently, the log-likelihood ($\log L$) subject to this summation constraint (the positivity constraint will be automatic). To this end, we introduce the Lagrange multipliers, $\gamma_1, \ldots, \gamma_K$ and aim to maximise

$$f(\boldsymbol{A}, \boldsymbol{\gamma}) = \log L + \sum_{i=1}^{K} \gamma_i \left( \sum_{j=1}^{K} A_{ij} - 1 \right)$$

$$= \sum_{i,j=1}^{K} n_{ij} \log A_{ij} + \sum_{i=1}^{K} \gamma_i \left( \sum_{j=1}^{K} A_{ij} - 1 \right).$$

The first order conditions, $\partial_{A_{ij}} f(\boldsymbol{A}, \boldsymbol{\gamma}) = 0$ and $\partial_{\gamma_i} f(\boldsymbol{A}, \boldsymbol{\gamma}) = 0$, imply that the MLE estimator $\hat{\boldsymbol{A}}$ of $\boldsymbol{A}$ satisfies

$$0 = \frac{n_{ij}}{\hat{A}_{ij}} + \gamma_i, \tag{12.2a}$$

$$0 = \sum_{j=1}^{K} \hat{A}_{ij} - 1. \tag{12.2b}$$

From (12.2a) we can write the Lagrange multiplier in terms of $n_{ij}$ and $\hat{A}_{ij}$ as

$$\gamma_i = -\frac{n_{ij}}{\hat{A}_{ij}}. \tag{12.3}$$

Next, multiplying (12.2a) by $\hat{A}_{ij}$ and then summing over $j$ from 1 to $K$ implies that

$$0 = \sum_{j=1}^{K} n_{ij} + \gamma_i \sum_{j=1}^{K} \hat{A}_{ij} = \sum_{j=1}^{K} n_{ij} + \gamma_i = \sum_{j=1}^{K} n_{ij} - \frac{n_{ij}}{\hat{A}_{ij}},$$

where in the second equality we use the constraint (12.2b) to write $\sum_{j=1}^{K} \hat{A}_{ij} = 1$ and in the third equality we use (12.3) to eliminate the Lagrange multiplier. Solving for $\hat{A}_{ij}$ provides us with the result in (12.1).

### Estimation from Data

For the model estimation from data we use ORCL on Nov 1, 2013 with order balance measured every millisecond (volume at-the-touch) and averaged over the last 100ms. We then bin the order imbalance into 5 equally spaced regimes, and find the following estimator for the transition matrix:

$$\hat{A} = \begin{pmatrix} 0.946 & 0.050 & 0.003 & 0.000 & 0.000 \\ 0.006 & 0.973 & 0.020 & 0.001 & 0.000 \\ 0.000 & 0.009 & 0.979 & 0.012 & 0.000 \\ 0.000 & 0.000 & 0.013 & 0.980 & 0.008 \\ 0.000 & 0.000 & 0.001 & 0.023 & 0.976 \end{pmatrix}. \tag{12.4}$$

The estimated transition rates indicate that the chain tends to move only between its neighbours. Moreover, for this specific asset and day, the slightly bid-heavy regime 4 appears to be the "stickiest" since the probability of remaining there is highest. There also appears to be transition pressure towards cycling back and forth between the neutral (3) and slightly bid-heavy (4) regimes – because $\hat{A}_{34} > \hat{A}_{32}$ and $\hat{A}_{43} > \hat{A}_{45}$.

### From Discrete- to Continuous-Time Markov Models

In the previous section, we focused on a discrete-time Markov model for order imbalance. When developing stochastic models for algorithmic trading, however, it proves more useful to utilise continuous-time Markov models because we are using the tools of continuous-time stochastic control. Indeed, there is a simple transformation that transforms the discrete-time model into a continuous-time one. First, recall that a continuous-time Markov model $Z = \{Z_t\}_{0 \le t \le T}$ with $Z_t \in \{1, \dots, K\}$, has a generator matrix $B$ which produces transition probabilities

$$\mathbb{P}\{Z_t = j \mid Z_s = i\} = [\exp\{B\,(t - s)\}]_{ij}, \qquad t \ge s,$$

where the exponential here is a matrix exponential and $[\cdot]_{ij}$ denotes the $ij^{th}$ element of the matrix in the braces. Also recall that the generator matrix must satisfy the conditions

$$B_{ij} \ge 0, \quad \forall\, i \ne j, \qquad \text{and} \qquad \sum_{i=1}^{K} B_{ij} = 0,$$

so that the diagonal elements equal negative the sum of the off diagonal elements. The absolute value of the diagonal elements represent the rate of flow out of that regime.

The continuous-time model can then be estimated from the discrete-time one by matching the estimated transition probability in the discrete setting from the previous section. Specifically, we set

$$\exp\{\hat{B}\,\Delta T\} = \hat{A} \quad \Rightarrow \quad \hat{B} = \tfrac{1}{\Delta T}\log\hat{A}\,.$$

In the above, the logarithm is to be interpreted as a matrix logarithm and $\Delta T$ is the time between observations in the discrete Markov chain. In the next section we show a more formal approach to the estimation which also takes into account the arrival of MOs.

For the 100ms transition rate matrix which we estimated at the start of the previous subsection (see (12.4)), the corresponding generator matrix is

$$\hat{B} = \begin{pmatrix} -0.553 & 0.521 & 0.030 & 0.002 & 0.000 \\ 0.068 & -0.279 & 0.205 & 0.005 & 0.001 \\ 0.001 & 0.089 & -0.219 & 0.128 & 0.001 \\ 0.000 & 0.002 & 0.128 & -0.209 & 0.078 \\ 0.000 & 0.001 & 0.008 & 0.235 & -0.244 \end{pmatrix}. \tag{12.5}$$

Focusing on the diagonal elements, which represent minus one times the sum of the rate of transition out of the corresponding regime, we see that the chain tends to flow out of the ask-heavy regimes towards the slightly bid-heavy regime. Once there, it tends to flow back to the neutral regime.

## 12.2.2    Jointly Modelling Market Orders

We extend the continuous-time Markov model of the previous section to include the modelling of MO arrivals. We also note that the continuous-time approximation above has one important flaw – the discrete-time model is estimated from average order imbalance over a 100ms window, therefore a continuous-time limit based on this estimate will not account properly for the interdependence of overlapping time windows.

To address these issues, we assume that, conditional on being within a given regime $k$, market buy and sell orders arrive independently at the arrival times of independent Poisson processes with rates $\lambda_k^+$ and $\lambda_k^-$, respectively. That is, the counting processes $M^\pm = \{M_t^\pm\}_{0\le t\le T}$ of market buy (sell) orders are doubly stochastic Poisson processes with activity rates $\Lambda_t^\pm = \lambda_{Z_t}^\pm$, where $\lambda_1^\pm, \lambda_2^\pm, \ldots, \lambda_K^\pm$ denote the activity rates in the various regimes. In addition, the order imbalance will be observed at every event, and the inter-arrival times of the events will play a role.

Referring to Figure 12.4, let $\tau_1, \tau_2, \ldots, \tau_N$ denote the switching times of the regimes, i.e. the times at which the continuous-time Markov chain $Z$ changes (with generator matrix denoted by $B$). We call the time interval $[\tau_r, \tau_{r+1})$ the

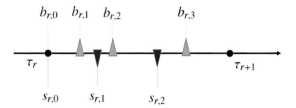

**Figure 12.4** Snapshot of the event timeline for computing the within epoch likelihood.

$r^{th}$ epoch ($r = 0, \ldots, N$) with $\tau_{N+1} = T$ the end of the observation time horizon. Next, let $\{b_{r,1}, b_{r,2}, \ldots, b_{r,m_r}\}$ and $\{s_{r,1}, s_{r,2}, \ldots, s_{r,n_r}\}$ denote the arrival times of buy and sell MOs, respectively, during the $r^{th}$ epoch. Then, since arrivals within an epoch of buys and sells are i.i.d., the likelihood within the $r^{th}$-epoch is

$$
L_r = \underbrace{\lambda^+_{Z_{\tau_r}} e^{-\lambda^+_{Z_{\tau_r}} (b_{r,1}-b_{r,0})} \times \cdots \times \lambda^+_{Z_{\tau_r}} e^{-\lambda^+_{Z_{\tau_r}} (b_{r,m_r}-b_{r,m_r-1})}}_{\textbf{buy arrivals}}
$$

$$
\times \underbrace{e^{-\lambda^+_{Z_{\tau_r}} (\tau_{r+1}-b_{r,m_r})}}_{\textbf{no more buys}}
$$

$$
\times \underbrace{\lambda^-_{Z_{\tau_r}} e^{-\lambda^-_{Z_{\tau_r}} (s_{r,1}-s_{0,1})} \times \cdots \times \lambda^-_{Z_{\tau_r}} e^{-\lambda^-_{Z_{\tau_r}} (s_{r,n_r}-s_{r,n_r-1})}}_{\textbf{sell arrivals}}
$$

$$
\times \underbrace{e^{-\lambda^-_{Z_{\tau_r}} (\tau_{r+1}-s_{r,n_r})}}_{\textbf{no more sells}}
$$

$$
= \left(\lambda^+_{Z_{\tau_r}}\right)^{M^+_r} e^{-\lambda^+_{Z_{\tau_r}} (\tau_{r+1}-\tau_r)} \times \left(\lambda^-_{Z_{\tau_r}}\right)^{M^-_r} e^{-\lambda^-_{Z_{\tau_r}} (\tau_{r+1}-\tau_r)} .
$$

Here, we set $b_{r,0} = s_{r,0} = \tau_r$ for notational convenience. In the first equality, the first (second) line represents the likelihood of the sequence of buy (sell) orders. In each line, the first to second last terms, of the form $\lambda^\pm_{Z_{\tau_r}} e^{-\lambda^\pm_{Z_{\tau_r}} (t_{r,l}-t_{r,l-1})}$, represent the survival since the last order arrival, and then the arrival of an MO at time $t_r$. The last term in each line represents the probability that no event arrives between the last buy (sell) MO and the time at which the Markov chain switches. The second equality is obtained by re-arranging the terms and letting $M^\pm_r$ denote the number of buy and sell MOs which arrive in the $r^{th}$-epoch.

The full likelihood is obtained by sewing together the within epoch likelihoods with the transition probability and the arrival of orders that occur after the last regime change, but before the sample ends. In all,

$$
L = L_0 \times [e^{B\tau_1}]_{Z_{\tau_0} Z_{\tau_1}} \times [B]_{Z_{\tau_0} Z_{\tau_1}} \times
$$

$$
\cdots \times L_{N-1} \times [e^{B(\tau_N - \tau_{N-1})}]_{Z_{\tau_{N-1}} Z_{\tau_N}} \times [B]_{Z_{\tau_{N-1}} Z_{\tau_N}} .
$$

Recall that $\tau_0 = 0$ and $\tau_{N+1} = T$ is the time of the last MO (buy or sell) after the last regime change. We in fact exclude the data from this last regime change to avoid some issues with censored data – which render the maximisation analytically intractable, although it is still amenable to numerical methods. Next, recall that a continuous-time Markov chain generator matrix has rows which sum to zero, and the non-diagonal elements are non-negative. Hence, we write the generator matrix $\boldsymbol{B}$ as

$$
\boldsymbol{B} = \begin{pmatrix}
-\Lambda_1 & \lambda_{12} & \lambda_{13} & \cdots & \lambda_{1K} \\
\lambda_{21} & -\Lambda_2 & \lambda_{23} & \cdots & \lambda_{2K} \\
\lambda_{31} & \lambda_{32} & -\Lambda_3 & \cdots & \lambda_{3K} \\
\vdots & \vdots & & \ddots & \vdots \\
\lambda_{K1} & \lambda_{K2} & \lambda_{K3} & \cdots & -\Lambda_K
\end{pmatrix} \, ,
$$

where $\Lambda_i := \sum_{j \neq i}^{K} \lambda_{ij}$ represents the total rate of outflow from regime $i$. The survival probability of the time $\tau_i$ at which the chain transitions out of regime $i$ is given by

$$
\mathbb{P}(\tau_i > t \mid Z_s = i) = e^{-\Lambda_i \, (t-s)} \, , \qquad s \leq t \, ,
$$

and conditional on a transition occurring, the chain will switch from regime $i$ to regime $j$ with probability

$$
[\boldsymbol{P}]_{ij} = \frac{\lambda_{ij}}{\Lambda_i} \, , \quad \text{for } i \neq j \, .
$$

Using this representation for the transition probability and expanding the likelihood above, we have

$$
L = \prod_{n=1}^{N} \left\{ \left( \lambda_{Z_{\tau_{n-1}}}^{+} \right)^{M_{n-1}^{+}} e^{-\lambda_{Z_{\tau_{n-1}}}^{+} \, (\tau_n - \tau_{n-1})} \left( \lambda_{Z_{\tau_{n-1}}}^{-} \right)^{M_{n-1}^{-}} e^{-\lambda_{Z_{\tau_{n-1}}}^{-} \, (\tau_n - \tau_{n-1})} \right.
$$

$$
\left. \times \, \Lambda_{Z_{\tau_{n-1}}} \, e^{-\Lambda_{Z_{\tau_{n-1}}} \, (\tau_n - \tau_{n-1})} \, [\boldsymbol{P}]_{Z_{\tau_{n-1}} Z_{\tau_n}} \right\} \, .
$$

Next, we re-arrange the terms by collecting like regimes and like transitions to obtain

$$
L = \left( \prod_{i=1}^{K} \left( \lambda_i^{+} \right)^{\hat{M}_i^{+}} e^{-\lambda_i^{+} \, \Delta\tau_i} \times \left( \lambda_i^{-} \right)^{\hat{M}_i^{-}} e^{-\lambda_i^{-} \, \Delta\tau_i} \right)
$$

$$
\times \left( \prod_{i,j=1}^{K} \left( [\boldsymbol{P}]_{ij} \right)^{n_{ij}} \right) \times \left( \prod_{i=1}^{K} \Lambda_i^{n_i} \, e^{-\Lambda_i \, \Delta\tau_i} \right) \, .
$$

Here, $\hat{M}_i^{\pm}$ denotes the number of market buy (sell) orders that occur while the chain is in regime $i$, $\Delta\tau_i$ denotes the total time the Markov chain spent in regime $i$, and $n_{ij}$ denotes the number of times the chain switches from $i$ to $j$. Since the times $\tau_1, \tau_2, \ldots, \tau_N$ are by definition the times at which the Markov chain switches regime, we must have $n_{ii} = 0$ and so $[\boldsymbol{P}]_{ii} = 0$.

Finally, we optimise the likelihood over the parameters $\lambda_i^{\pm}$, $\boldsymbol{P}$ and $\Lambda_i$. Each of these optimisations can be carried out independently, and follow along the lines outlined in the earlier subsection. Here, we simply record the final results for the MLE estimators of the model parameters:

$$\hat{\lambda}_i^{\pm} = \frac{\hat{M}_i^{\pm}}{\Delta \tau_i}, \qquad \hat{\Lambda}_i = \frac{\sum_{j \neq i} n_{ij}}{\Delta \tau_i}, \qquad \text{and} \qquad [\hat{P}]_{ij} = \frac{n_{ij}}{\sum_{j \neq i} n_{ij}}, \qquad (12.6)$$

from which we find that the MLE estimate of the off-diagonal elements of the generator matrix is

$$\hat{\lambda}_{ij} = \frac{n_{ij}}{\Delta \tau_i}.$$

The above results are similar to the simple Markov chain model, but now we also account for the regime specific arrival rate of market buy and sell orders.

Applying the MLE procedure to the ORCL data on Nov 1, 2013, using the imbalance at every event time, we obtain the following estimates (with time measured in seconds):

$$\hat{\lambda}^+ = \begin{pmatrix} 0.074 \\ 0.042 \\ 0.037 \\ 0.074 \\ 0.216 \end{pmatrix}, \qquad \hat{\lambda}^- = \begin{pmatrix} 0.856 \\ 0.123 \\ 0.048 \\ 0.027 \\ 0.025 \end{pmatrix},$$

$$\text{and} \qquad \hat{B} = \begin{pmatrix} -3.34 & 2.34 & 0.59 & 0.28 & 0.13 \\ 0.31 & -0.93 & 0.54 & 0.01 & 0.07 \\ 0.01 & 0.26 & -0.58 & 0.30 & 0.02 \\ 0.03 & 0.01 & 0.29 & -0.56 & 0.23 \\ 0.03 & 0.03 & 0.09 & 0.74 & -0.88 \end{pmatrix}. \qquad (12.7)$$

Comparing the estimate (12.7) with the estimate (12.5) we see some difference in the overall rates. This difference results from the fact that (12.5) is estimated using the average order imbalance over the previous 100ms, while (12.7) is estimated using the instantaneous order imbalance. Nonetheless, the chain has the same tendency to move to the slightly bid-heavy regime 4. The estimated arrival rates show a clear bias towards sell MOs on this day, with a total sell arrival rate of 1.079 per second versus a total buy arrival rate of 0.444 per second. Interestingly, although the overall day was a sell-heavy one, if we condition on being in a bid-heavy regime (4 or 5), the arrival rate of buy MOs is significantly larger than that of sell MOs. This observation indicates that the order imbalance is indeed a good predictor of order flow.

## 12.2.3    Modelling Price Jumps

Up to this point, we have been concerned with understanding how order imbalance influences the rate of arrival of MOs. An important and interesting related question is to determine the distribution of price changes conditional on the arrival of buy and sell MOs and the order imbalance regime prior to the arrival of that MO. To answer this question, we record the time of each MO and compute the midprice change 1s afterwards, conditional on the order imbalance regime when split into 5 equal bins with knots at $\{-1, -\frac{3}{5}, -\frac{1}{5}, +\frac{1}{5}, +\frac{3}{5}, +1\}$, as in the previous section. The corresponding states and imbalance $\rho$ is given by

$$
Z = \begin{cases}
-2, & \rho \in [-1, -\frac{3}{5}], & \text{sell-heavy}, \\
-1 & \rho \in [-\frac{3}{5}, -\frac{1}{5}), & \text{sell-bias}, \\
0, & \rho \in [-\frac{1}{5}, +\frac{1}{5}), & \text{neutral}, \\
1, & \rho \in [+\frac{1}{5}, +\frac{3}{5}), & \text{buy-bias}, \\
2, & \rho \in [+\frac{3}{5}, +1], & \text{buy-heavy}.
\end{cases}
$$

Table 12.1 shows the price change distribution conditional on the order imbalance regimes.

| | Buy Market Orders | | | | |
|---|---|---|---|---|---|
| | $Z=-2$ | $Z=-1$ | $Z=0$ | $Z=+1$ | $Z=+2$ |
| $\lambda^+$ | 0.074 | 0.042 | 0.037 | 0.075 | 0.216 |
| $\Delta S$ | | | | | |
| -300 | - | - | - | - | - |
| -0.02 | - | - | - | - | - |
| -0.01 | - | 0.05 | 0.01 | 0.01 | - |
| 0.00 | 1.00 | 0.86 | 0.77 | 0.78 | 0.70 |
| 0.01 | - | 0.09 | 0.21 | 0.20 | 0.28 |
| 0.02 | - | - | - | - | 0.02 |
| 0.03 | - | - | - | - | - |

| | Sell Market Orders | | | | |
|---|---|---|---|---|---|
| | $Z=-2$ | $Z=-1$ | $Z=0$ | $Z=+1$ | $Z=+2$ |
| $\lambda^-$ | 0.856 | 0.123 | 0.048 | 0.027 | 0.025 |
| $\Delta S$ | | | | | |
| -300 | - | - | - | 0.01 | 0.03 |
| -0.02 | 0.01 | - | 0.01 | 0.02 | 0.01 |
| -0.01 | 0.21 | 0.36 | 0.36 | 0.28 | 0.21 |
| 0.00 | 0.79 | 0.64 | 0.63 | 0.69 | 0.75 |
| 0.01 | - | - | - | - | - |
| 0.02 | - | - | - | - | - |
| 0.03 | - | - | - | - | - |

**Table 12.1** Price change distribution conditional on the order imbalance regimes for ORCL on Nov 1, 2013 (opening price is $33.72), ask-heavy $Z = -2$, ask-bias $Z = -1$, neutral $Z = 0$, bid-bias $Z = +1$, bid-heavy $Z = +2$.

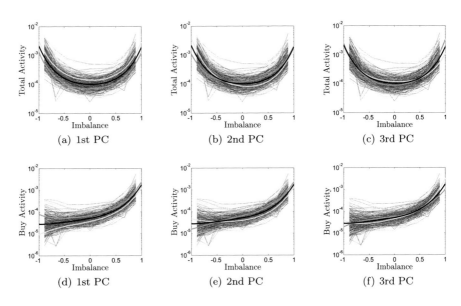

**Figure 12.5** Order imbalance for ORCL using 2013 data showing the first three functional principle components.

## 12.3　Daily Features

In the previous section, we focused on intraday features of order imbalance. In a trading environment, an understanding of daily features can assist in augmenting and tweaking the intraday model to reflect historical overall behaviour. To see these daily effects, Figure 12.5 shows order imbalance for Oracle Corporation (ORCL) from the entire 2013 event data. We place order imbalance into ten equal buckets from $-1$ to $+1$ and compute the rate of arrival of buy and sell MOs together, conditional on an order imbalance bucket. These estimates are shown in the top panels. In the bottom panels, we show the rate of arrival of only buy MOs. The thick black, green and yellow lines are the mean curves, obtained through a functional regression using Legendre polynomials, and the $\pm$ one standard deviation show the impact of the first three functional principle components. (We do not go into the details of how the functional principle components are obtained here.)

It is useful, however, to describe how the mean curves are obtained using functional data analysis (FDA) techniques. FDA takes the view that a sequence of observations is a realisation of a random draw from a function space, but observed discretely (in this case at the order imbalance buckets) with error. Each new collection of observations is a new random draw from this function space. In our present context, we view the trade activity $\mu(\rho)$ as being samples from a function space given by

$$\boldsymbol{\mu}(\rho) = \sum_{n=1}^{N} \boldsymbol{\alpha}_n \, P_n(\rho), \qquad \rho \in [0, 1] \,,$$

where $P_n(x)$ are the (normalised) Legendre polynomials of order 0 through $N$ and $\boldsymbol{\alpha}_n$ are viewed as random variables on a probability space $(\Omega, \mathbb{P}, \mathcal{F})$. The observed imbalance activity $\mu_t(\rho)$ on a given day is then viewed as a sample realisation of the random variables $\boldsymbol{\alpha}_n$. In this sense, we see that $\boldsymbol{\mu}$ is in fact a random field, since it is parameterised continuously by $\rho$, but it is projected onto a finite dimensional space through the $N$ random coefficients $\boldsymbol{\alpha}_n$.

At the end of each trading day, we regress the activity as a function of imbalance (using the middle of each bucket) onto the Legendre polynomials, and determine an estimate of the realisation $\hat{\boldsymbol{\alpha}}$ of $\{\boldsymbol{\alpha}_n : n = 1, \ldots, N\}$ on that day:

$$\hat{\boldsymbol{\alpha}}_t = \arg\min_{\boldsymbol{\alpha}} \sum_{m=1}^{M} \left( \left( \sum_{n=1}^{N} \alpha_n \, P_n(\rho_m) \right) - \mu_{t,m} \right)^2.$$

Here, $\mu_{t,m}$ denotes the sample activity on day $t$ in imbalance bin $m$, and $\rho_m$ denotes the imbalance in the middle of regime $m$. In this manner, we obtain a time series of coefficients $\boldsymbol{\alpha}_t$ covering the year.

The mean curve is obtained by computing the average of these daily estimates: $\bar{\alpha}_k = \frac{1}{T} \sum_{t=1}^{T} \hat{\alpha}_{k,t}$, substituting that average into (12.3) to obtain the mean curve

$$\bar{\mu}(\rho) = \sum_{n=1}^{N} \bar{\alpha}_k \, P_n(\rho), \qquad \rho \in [0, 1] \,.$$

There are many other useful objects one can compute using FDA, such as the functional principle components, the estimate of the distribution of shape of the curve for the remainder of the day given what has been observed so far, and so on. For discussions on this and other tools that FDA provides we refer the interested reader to consult Ramsay & Silverman (2010).

## 12.4    Optimal Liquidation

In this section we analyse how to incorporate order flow information in the optimal liquidation problem. Recall that in the optimal liquidation problem, the agent's goal is to liquidate all shares by the end of the trading horizon. In this section, we pose the problem for an agent who uses only LOs to make her decision and is allowed to post her order at an arbitrary depth $\delta$ in the LO book (LOB). In particular, we use the same approach as in Section 8.2 for posing the agent's optimisation problem.

As in Section 8.2, the agent posts LOs at a depth $\delta = (\delta_t)_{0 \le t \le T}$ from the midprice $S = (S_t)_{0 \le t \le T}$. As such, the agent's controlled cash process $X^{\delta} =$

$(X_t^\delta)_{0 \le t \le T}$ satisfies the SDE

$$dX_t^\delta = (S_t + \delta_t)\, dN_t^\delta\,,$$

where $N^\delta = (N_t^\delta)_{0 \le t \le T}$ is the controlled counting process which counts the agent's filled LOs posted at depth $\delta_t$ at time $t$ – assumed to be $\mathcal{F}$-predictable. To keep the framework simple, we assume the agent's orders are filled at a rate $\Lambda_t^{+,\delta} = e^{-\kappa\,\delta_t}\lambda_t^+$, where $\lambda^\pm = (\lambda_t^\pm)_{0 \le t \le T}$ represent the rate of arrival of buy (sell) MOs from other agents. In this manner, the deeper the agent posts in the book, the less likely it is that her order is filled. More specifically, conditional on an MO arriving, the probability that her order is filled when she is posted at depth $\delta_t$ is $e^{-\kappa\delta_t}$. We further let $M^\pm = (M_t^\pm)_{0 \le t \le T}$ denote the counting processes for other agent's buy (sell) MOs, with intensities $\lambda^\pm$.

Now we proceed by developing a model that incorporates the empirical observations described in the previous section, and in particular we jointly model the rate of arrival of MOs, price movements, and order imbalance. To this end, the midprice process should have jumps that are biased upwards when order imbalance is near $+1$, it should have jumps that are biased downwards when order imbalance is near $-1$, and it should have symmetric jumps when order imbalance is near zero.

Moreover, arrival rates of MOs should also be biased in a similar fashion. For this purpose, we will use the Markov chain $Z = (Z_t)_{0 \le t \le T}$ to represent the order imbalance regime, with $Z_t \in \{-1, 0, +1\}$ representing sell-heavy, neutral and buy-heavy regimes – one could incorporate more refined regimes, but the framework remains essentially the same. We let $G$ denote the generator of the order imbalance regime, and write $\lambda_t^\pm = \lambda^\pm(Z_t)$ with a slight abuse of notation. Moreover, let $\{\varepsilon_{0,k}^+, \varepsilon_{1,k}^+, \ldots\}$ denote i.i.d. random variables with distribution function $F_k^+$, and $\{\varepsilon_{0,k}^-, \varepsilon_{1,k}^-, \ldots\}$ denote i.i.d. random variables with distribution function $F_k^-$, for $k = -1, 0, 1$, also mutually independent of one another. These random variables will generate jumps in the midprice when an MO arrives and the order imbalance is in regime $k$ (see, e.g., Table 12.1).

Armed with the counting process $M^\pm$ for MOs, their intensities $\lambda^\pm$, the Markov chain driving order imbalance regimes $Z$, and the sequence of i.i.d. random variables for midprice jumps $\varepsilon_{(\cdot)}^\pm$, we can now state a candidate model for the midprice $S$ which is driven by order imbalance:

$$dS_t = \varepsilon_{M_{t-}^+,\, Z_{t-}}^+\, dM_t^+ - \varepsilon_{M_{t-}^-,\, Z_{t-}}^-\, dM_t^-\,.$$

The random variables $\varepsilon_{(\cdot)}^\pm$ are subordinated by the left-limit of the corresponding processes – this is a technical condition required to ensure that stochastic integrals with respect to the compensated counting processes are still martingales, and is the reason we indexed the random variables $\varepsilon_{(\cdot)}^\pm$ beginning from 0 rather than 1. Intuitively, the above model says that the midprice jumps the instant an MO arrives, and the rate of arrival of the orders and the distribution of the jump are regime dependent.

The model above is missing one more ingredient, which is to include the mid-price changes that we observe between MO arrivals. Thus, we modify the above to include exogenous jumps which result from, e.g., additions and cancellations in the LOB:

$$dS_t = \varepsilon_{M_{t-}^+, Z_{t-}} \, dM_t^+ - \varepsilon_{M_{t-}^-, Z_{t-}} \, dM_t^- + \eta_{J_{t-}^+, Z_{t-}} \, dJ_t^+ - \eta_{J_{t-}^-, Z_{t-}} \, dJ_t^-.$$

Here, $\{\eta_{0,k}^+, \eta_{1,k}^+, \dots\}$ denote i.i.d. random variables with distribution function $L_k^+$, and $\{\eta_{0,k}^-, \eta_{1,k}^-, \dots\}$ denote i.i.d. random variables with distribution function $L_k^-$, for $k = -1, 0, 1$, and all random variables are mutually independent. These random variables generate jumps in the midprice between the arrival of MOs, due to other agents posting and cancelling orders in the LOB, and these changes can in principle be dependent on the order imbalance regime. For example, when order imbalance is buy-heavy, agents may pull their orders from the sell side of the LOB and place them in the buy side, resulting in a general upward pressure on the midprice. Reshuffling of orders generally occurs at a higher frequency than the arrival of MOs themselves.

### 12.4.1    Optimisation Problem

Up to this point, we have specified the joint model for order imbalance, arrival of MOs, and midprice movements. Here, we pose and solve the agent's optimisation problem subject to this modelling assumption. First, the agent continues to trade until the stopping time

$$\tau = T \wedge \min\{t : Q_t^\delta = 0\},$$

i.e. the minimum of $T$ or the first time that the inventory hits zero, because then no more trading is necessary. The agent's performance criteria is essentially the same as in Chapter 6, and is given by

$$H^\delta(t, x, S, z, q) = \mathbb{E}_{t,x,S,z,q} \left[ X_\tau^\delta + Q_\tau^\delta \left( S_\tau - \alpha \, Q_\tau^\delta \right) - \phi \int_t^\tau \left( Q_s^\delta \right)^2 ds \right], \quad (12.8)$$

where the notation $\mathbb{E}_{t,x,S,z,q}[\cdot]$ represents expectation conditional on $X_{t-}^\delta = x$, $S_{t-} = S$, $Z_{t-} = z$ and $Q_{t-}^\delta = q$. As usual, her value function is the one which maximises this performance criteria, over all admissible strategies $\mathcal{A}$, taken to be the set of $\mathcal{F}$-predictable, bounded from below, processes, so that

$$H(t, x, S, z, q) = \sup_{\delta \in \mathcal{A}} H^\delta(t, x, S, z, q).$$

Applying the dynamic programming principle, we expect the value function to satisfy the dynamic programming equation (using $(t, \cdot)$ to denote $(t, x, S, z, q)$):

$$\phi q^2 = \partial_t H + \lambda^+(z) \sup_\delta \left\{ e^{-\kappa\delta} \, \mathbb{E}[H(t, x + (S + \delta), S + \varepsilon^+_{0,z}, z, q - 1) - H(t, \cdot)] \right.$$

$$+ (1 - e^{-\kappa\delta}) \, \mathbb{E}[H(t, x, S + \varepsilon^+_{0,z}, z, q) - H(t, \cdot)] \left. \right\}$$

$$+ \lambda^-(z) \, \mathbb{E}[H(t, x, S - \varepsilon^-_{0,z}, z, q) - H(t, \cdot)]$$

$$+ \eta^+(z) \, \mathbb{E}[H(t, x, S + \eta^+_{0,z}, z, q) - H(t, \cdot)]$$

$$+ \eta^-(z) \, \mathbb{E}[H(t, x, S - \eta^-_{0,z}, z, q) - H(t, \cdot)]$$

$$+ \sum_{k=-1,0,1} G_{z,k} \left[ H(t, x, S, k, q) - H(t, \cdot) \right],$$

where the expectations are over the random variables $\varepsilon^\pm_{0,z}$ and $\eta^\pm_{0,z}$, and the boundary and terminal conditions are

$$H(t, x, S, z, 0) = x, \qquad \text{and} \qquad H(T, x, S, z, q) = x + q\,(S - \alpha\,q).$$

The various terms in the equation have the interpretations given below.

(i) The left-hand side of the first line contains the running penalty the agent has from holding inventory different from zero.

(ii) The supremum takes into account the agent's ability to control the depth at which she posts her LOs.

(iii) The term $\lambda^+(z)\,e^{-\kappa\delta}$ represents the rate of arrival of MOs which fill the agent's posted LO at price $S + \delta$.

(iv) The expectation in the first line represents the expected change in the valuation when a buy MO arrives which fills the agent's post. The agent's wealth increases by $S + \delta$, her inventory decreases by 1 and the midprice jumps.

(v) The term $\lambda^+(z)\,(1 - e^{-\kappa\delta})$ represents the rate of arrival of buy MOs which do not fill the agents posted LO, but still induce a jump in midprice.

(vi) The expectation in the second line represents the expected change in the valuation when a buy MO arrives which does not fill the agent's post but causes a jump in midprice.

(vii) The third line represents the expected change in the value function when a sell MO arrives and the midprice jumps.

(viii) The fourth and fifth lines represent the expected change in the value function when the midprice jumps due to posts and cancellations in the LOB (i.e. between MO arrivals).

(ix) The last line represents the change in value function when the order imbalance switches regimes.

As seen several times, the terminal and boundary conditions suggest the ansatz

$$H(t, x, S, z, q) = x + q\,S + h(t, z, q), \quad h(T, z, q) = -\alpha\,q^2, \quad h(t, z, 0) = 0,$$

so that the term $x + q\,S$ is the book value of the agent's inventory and cash, while

$h$ represents the excess value that optimal trading generates. Upon substituting this ansatz we find that $h$ satisfies the coupled system of PDEs

$$
\begin{aligned}
0 = {} & \partial_t h(t, z, q) + \mu(z)\, q - \phi\, q^2 \\
& + \sup_\delta \left\{ \lambda^+(z)\, e^{-\kappa\delta}\, (\delta + h(t, z, q-1) - h(t, z, q)) \right\} \\
& + \sum_{k=-1,0,1} G_{z,k}\, [h(t, k, q) - h(t, z, q)],
\end{aligned}
\tag{12.9}
$$

where

$$
\mu(z) = \lambda^+(z)\, \mathbb{E}[\varepsilon_{0,z}^+] - \lambda^-(z)\, \mathbb{E}[\varepsilon_{0,z}^-] + \eta^+(z)\, \mathbb{E}[\eta_{0,z}^+] - \eta^-(z)\, \mathbb{E}[\eta_{0,z}^-],
$$

is the expected drift of the midprice while the order imbalance is in regime $z$. The first two lines of this equation are of the same form as that of the liquidation problem when there are no regime changes and the midprice is a drifted Brownian motion. The third line represents the jumps between the order imbalance regimes. It is somewhat surprising how this rather rich model reduces to something intuitively simple.

The first order conditions provides us with the optimal depth as

$$
\boxed{\delta^*(t, q, z) = \tfrac{1}{\kappa} - \Delta h(t, z, q),}
$$

where

$$
\Delta h(t, z, q) = h(t, z, q) - h(t, z, q-1),
$$

and upon substituting this feedback form into the previous equation, we find that $h$ satisfies the equation

$$
\begin{aligned}
0 = {} & \partial_t h(t, z, q) + \mu(z)\, q - \phi\, q^2 + \lambda^+(z)\, \tfrac{1}{\kappa}\, e^{-\kappa\left(\frac{1}{\kappa} - \Delta h(t, z, q)\right)} \\
& + \sum_{k=-1,0,1} G_{z,k}\, [h(t, k, q) - h(t, z, q)].
\end{aligned}
\tag{12.10}
$$

The above coupled system of non-linear ODEs does not appear to have a simple analytic solution and one must resort to a numerical scheme such as finite differences.

The above optimal depth may become negative. One interpretation of a negative depth is that it represents executing an MO. This interpretation is purely heuristic and is not accounted for in the model. To properly account for executing MOs, we would have to pose the problem as a sequence of stopping problems similar to what we have done in Section 8.4. An alternative is to pose the problem as a constrained optimisation problem and modify the admissible set $\mathcal{A}$ to non-negative $\mathcal{F}$-predictable processes. The resulting DPE will receive a modification on the optimisation set, so that (12.9) becomes

$$
\begin{aligned}
0 = {} & \partial_t h(t, z, q) + \mu(z)\, q - \phi\, q^2 \\
& + \sup_{\delta \geq 0} \left\{ \lambda^+(z)\, e^{-\kappa\delta}\, (\delta + h(t, z, q-1) - h(t, z, q)) \right\} \\
& + \sum_{k=-1,0,1} G_{z,k}\, [h(t, k, q) - h(t, z, q)].
\end{aligned}
$$

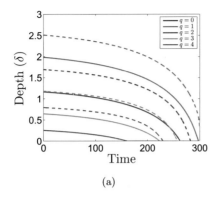 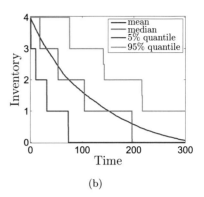

(a)                                    (b)

**Figure 12.6** (a) The optimal depth to post for the sell-heavy (solid lines) and buy-heavy (dashed lines) regimes as a function of time and inventory. (b) The mean, median, 5% and 95% quantiles of inventory through time from 10,000 simulations.

Hence, the optimal depth is modified to

$$\delta^*(t, q, z) = \max\left(\tfrac{1}{\kappa} - \Delta h(t, z, q) \; ; \; 0\right). \tag{12.11}$$

Upon substitution of this feedback control into the DPE, we have

$$0 = \partial_t h + \mu(z) q - \phi q^2$$
$$+ \lambda^+(z) \left\{ \Delta h(t, q, z) \, \mathbb{1}_{\Delta h(t,q,z) \geq \frac{1}{\kappa}} + \frac{e^{-\kappa(\frac{1}{\kappa} - \Delta h(t,z,q))}}{\kappa} \mathbb{1}_{\Delta h(t,q,z) < \frac{1}{\kappa}} \right\}$$
$$+ \sum_{k=-1,0,1} G_{z,k} \left[ h(t, k, q) - h(t, z, q) \right]. \tag{12.12}$$

## Simulations

In this section, we perform simulations using the estimated parameters provided in Table 12.1 and (12.7). For simplicity we do not model the movements of the LOB between MO events. Note that the invariant distribution of the Markov chain with generator matrix in (12.7) is

$$[0.0196 \; 0.1548 \; 0.3577 \; 0.3534 \; 0.1160],$$

and from Table 12.1 we find that the invariant arrival of buy MOs is 0.0727 per second. In the experiments below, we use $T = 300$sec and $\mathfrak{N} = 4$ so that the trader is liquidating approximately 20 percent of the buy MO flow – which has the potential to lift her posted sell LOs. Finally, we use the following remaining parameters:

$$S_0 = 33.61, \qquad \alpha = 0.01, \qquad \kappa = 100, \qquad \phi = 10^{-5}.$$

We numerically solve the constrained equation (12.12) for $h$ and then compute the optimal depth $\delta^*$ in (12.11). Figure 12.6(a) shows the optimal depth the agent

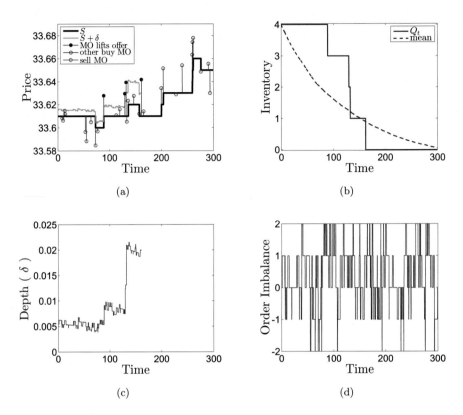

**Figure 12.7** Sample path of the optimal strategy showing price, market buy and sell arrivals, inventory path, optimal depth and order imbalance regime.

posts for the sell-heavy (solid lines) and buy-heavy (dashed lines) regimes as a function of time and inventory. In the buy-heavy regime, the agent posts deeper in the LOB in order to capture the expected upward tendency that prices have in this regime, and hence avoids being adversely selected. Similarly, in the sell-heavy regime, the agent posts closer to the midprice since she expects prices to have a downward pressure and would rather capture the current price than wait for prices to fall. Moreover, as the agent liquidates her position, she posts deeper in the LOB. Panel (b) of Figure 12.6 shows the mean, median, 5 and 95 percent quantiles of the inventory paths when the agent follows the optimal strategy.

To gain a better sense of how the strategy behaves, in Figure 12.7 we show a single sample path of the strategy. Panel (a) shows the midprice path (black line) and the optimal offer posting (green line) which reacts to the changes in the regimes shown in panel (d) as can be seen more clearly in the optimal depth shown in panel (c). In this case, the depth, and therefore the posting, reacts to the agent's remaining inventory which we have shown in panel (c) together with the mean inventory path. In panel (a) we also show the arrival of market sell orders (open red circles), which naturally are never matched with the agent's posted offers, and market buy orders (blue circles) which are filled if that specific order lifted the agent's offer.

## 12.5     Bibliography and Selected Readings

Bechler & Ludkovski (2014), Lipton, Pesavento & Sotiropoulos (2013), Cebiroglu & Horst (2015), Zheng, Moulines & Abergel (2012).

## 12.6     Exercises

E.12.1 Modify the setup developed in Section 12.4 for optimal liquidation using LOs only in the following cases:
  (a) the agent optimally executes MOs only,
  (b) the agent optimally executes MOs and optimally decides the depth at which to place LOs,
  (c) the agent optimally places LOs only at-the-touch,
  (d) the agent optimally places LOs only at-the-touch and optimally decides when to execute MOs.

E.12.2 Extend the analysis in Section 12.4 to the case when the agent is a market maker and decides whether to post buy and sell LOs.

# Appendix A Stochastic Calculus for Finance

This chapter provides a concise overview of stochastic calculus for diffusion and jump processes. There are many excellent textbooks (see, e.g., the bibliography and other readings at the end of this appendix) which cover these topics in an inordinate amount of detail. Here, however, we focus on the main tools and results that we require to pose and solve the algorithmic trading problems that appear in Part III of this textbook.

In all sections, we work on a **filtered probability space** (also called a **stochastic basis**) denoted by $(\Omega, \mathcal{F}, \{\mathcal{F}_t\}_{0 \le t \le T}, \mathbb{P})$ where as is standard:

- $\Omega$ denotes the space of all events,
- $\mathcal{F}$ denotes the set of all measurable events,
- $\{\mathcal{F}_t\}_{0 \le t \le T}$ (each contained in $\mathcal{F}$) denotes the filtration – a sequence of sigma-algebras which refine one another in the sense that $\mathcal{F}_s \subseteq \mathcal{F}_t$ for all $0 \le s < t \le T$.

We further assume that the *usual conditions* apply and that the filtered probability is completed:

- the probability space $(\Omega, \mathcal{F}, \mathbb{P})$ is complete – every subset of a measure zero set is measurable itself and (therefore) has zero measure; more precisely, for all $\omega \in \mathcal{F}$ s.t. $\mathbb{P}(\omega) = 0$ and for all $\omega' \subset \omega$, we have $\omega' \in \mathcal{F}$ and so $\mathbb{P}(\omega') = 0$,
- each $\mathcal{F}_t$ contains the zero measure sets of $\mathcal{F}$,
- the filtration is right-continuous, i.e. $\mathcal{F}_{t+} = \cap_{s>t} \mathcal{F}_s = \mathcal{F}_t$.

These technical conditions allow us to construct a *version* of a stochastic process which is cádlág, i.e. right continuous with left limit (RCLL).

## A.1    Diffusion Processes

In this section, we investigate the very important stochastic process known as a Brownian motion or Wiener process.

## A.1.1     Brownian Motion

DEFINITION A.1     A **standard Brownian motion** $W = (W_t)_{0 \le t \le T}$ is a stochastic process defined on the completed probability space $(\Omega, \mathcal{F}, \{\mathcal{F}_t\}_{0 \le t \le T}, \mathbb{P})$, where the filtration is the natural one generated by $W$, satisfying the following properties:

(i) $W_0 = 0$, almost surely,

(ii) $W$ has **independent increments**: for all $0 = t_0 \le t_1 \le t_2 \le \cdots \le t_{n-1} \le t_n = T$ the increments $W_{t_{k+1}} - W_{t_k}$ for $k = 0, \ldots, n-1$ are independent random variables,

(iii) $W$ has **stationary increments**: for all $0 \le t < t + h \le T$, the increment $W_{t+h} - W_t$ (which is a random variable) has distribution that is independent of $t$,

(iv) the random variable $W_t$ is **normally distributed** with mean 0 and variance $t$, and we will often write $W_t \sim \mathcal{N}(0, t)$,

(v) the function $t \mapsto W_t$ is **almost surely continuous**.

It is not clear whether specifying the marginal distributions of the Brownian motion, which is done in points (ii)–(iv) above, is consistent with the continuity condition in (v). This condition is necessary, because certainly if we modify the Brownian motion at say an exponential time $\tau$ and make it equal zero there, then the marginal properties remain the same, but it will fail to satisfy the continuity requirement. In fact, Paul Lévy proves (Jacod & Shiryaev (1987)) that indeed there is no contradiction and such processes do in fact exist.

## A.1.2     Stochastic Integrals

Stochastic integration with respect to Brownian motion can be most easily viewed through the lens of Itô integrals. Let $\Pi_1, \Pi_2, \ldots$ denote an infinite sequence of refining partitions of the interval $[0, T]$, so that each $\Pi_k$ represents an increasing sequence $0 = t_0^{(k)} < t_1^{(k)} < t_2^{(k)} < \cdots < t_{n^{(k)}}^{(k)} = T$, and the $L^\infty$ norm

$$||\Pi_k|| = \max_{m=1,\ldots,n^{(k)}} \left( t_m^{(k)} - t_{m-1}^{(k)} \right),$$

tends to zero as $k$ tends to infinity.

DEFINITION A.2     The **Itô integral** of an $\mathcal{F}_t$-adapted stochastic process $g = (g_t)_{0 \le t \le T}$ which is $\mathbb{P}$-square integrable, i.e. $\mathbb{E} \left[ \int_0^T g_s^2 \, ds \right] < +\infty$, often simply said to be square integrable, or in $L^2$, is defined as follows:

$$I_t = \int_0^t g_s \, dW_s = \lim_{||\Pi_k|| \to 0} \sum_{m=1}^{n_k} g_{t_{m-1}} \left( W_{t_m} - W_{t_{m-1}} \right). \tag{A.1}$$

In the definition above, the function $g$ is evaluated at the left-hand end point of the interval $[t_{m-1}, t_m)$. This has the financial interpretation of entering a position $g_{t_{m-1}}$ at the start of the interval, and the Brownian increment $(W_{t_m} - W_{t_{m-1}})$

represents the increase in the asset's value while keeping the position constant over that time interval. This is why Itô integrals are used ubiquitously in the mathematical finance literature.[1] Strictly, one should introduce the Itô integral for simple functions, i.e. functions which are piecewise defined: $H_t(\omega) = \sum_k h_k(\omega) \mathbb{1}_{t \in [t_k, t_{k+1})}$ for $h_k \in \mathcal{F}_{t_k}$, show that any $g \in L^2$ can be approximated, in $L^2$, by a sequence of such functions, and finally define the Itô integral of $g$ as the limiting value of the Itô integral of the sequence. See, e.g., Øksendal (2010). The Itô isometry is one of the tools which allows this programme to be developed.

THEOREM A.3    *Itô's isometry.* *If* $g \in L^2$, *that is,* $g$ *is* $\mathcal{F}_t$ *adapted and* $\mathbb{E}\left[\int_0^T g_s^2\, ds\right] < +\infty$, *then*

$$\mathbb{E}\left[\left(\int_0^T g_s\, dW_s\right)^2\right] = \mathbb{E}\left[\int_0^T g_s^2\, ds\right].$$

The stochastic integral (A.1) is called an Itô process, and is often written in differential form

$$dI_t = g_t\, dW_t,$$

despite the fact that a Brownian motion is not differentiable anywhere, so that the expression $dW_t$ without the integral is meaningless. It is, however, convenient to write it this way and renders computations more digestible. Itô processes can more generally be written as

$$I_t = \int_0^t \mu_s\, ds + \int_0^t \sigma_s\, dW_s,$$

where $\mu$ and $\sigma$ are $\mathcal{F}_t$-adapted and satisfy certain integrability requirements. The first term should be interpreted as a Riemann integral and the second term as an Itô integral. This is often written in shorthand as

$$dI_t = \mu_t\, dt + \sigma_t\, dW_t. \tag{A.2}$$

When $\mu_t = \mu(t, I_t)$ and $\sigma_t = \sigma(t, I_t)$, the above equation has the form of a differential equation, albeit one with diffusive noise, and these are referred to as stochastic differential equations (SDEs). Not all SDEs have solutions, and solutions come in two flavours, strong and weak solutions. A strong solution is a stochastic process which satisfies (A.2) and for which $I_t$ is given explicitly in terms of the version of Brownian motion $(W_t)_{0 \le t \le T}$ provided, i.e. it is adapted to the given filtration. A weak solution is one in which we seek a filtration $\mathcal{H}_t$ on which we have a Brownian motion $\tilde{W}_t$, and the solution $I_t$ satisfying (A.2) with $W$ replaced by $\tilde{W}$. We will be concerned with only strong solutions of SDEs.

Another interesting property of Itô integrals is that they are in fact martingales.

---

[1] There are other approaches to stochastic integration, such as the Stratonovich integral, which is often used in quantum field theory in physics and instead evaluates $g$ at the midpoint of the interval.

THEOREM A.4  **Itô Integrals are Martingales.** *The stochastic integral*

$$I_t = \int_0^t g_s \, dW_s \,,$$

*is a martingale.*

The intuition here is that for any given simple process approximation, $g^{(n)}$, to $g$, the Itô integral is given by $\mathfrak{I}_t^{(n)} = \sum_k g_{t_{k-1}}^{(n)} \left( W_{t_k} - W_{t_{k-1}} \right)$, and so

$$\mathbb{E}\left[\mathfrak{I}_t^{(n)}\right] = \sum_k \mathbb{E}\left[g_{t_{k-1}}^{(n)} \left( W_{t_k} - W_{t_{k-1}} \right)\right]$$

$$= \sum_k \mathbb{E}\left[\mathbb{E}[g_{t_{k-1}}^{(n)} \left( W_{t_k} - W_{t_{k-1}} \right) | \mathcal{F}_{t_{k-1}}]\right]$$

$$= \sum_k \mathbb{E}\left[g_{t_{k-1}}^{(n)} \mathbb{E}[\left( W_{t_k} - W_{t_{k-1}} \right) | \mathcal{F}_{t_{k-1}}]\right] = 0 \,.$$

The third equality follows since $g_{t_{k-1}}^{(n)}$ is $\mathcal{F}_{t_{k-1}}$ measurable, and the expected value of the increment of the Brownian motion is zero. This argument can be formalised to show that the increment of the integral, and not just the simple process approximation, has zero mean, and all that remains is to show that the integral is a strict martingale, and not just a local martingale. This last part follows from the assumption that $g \in L^2$.

Next, we touch on Itô's formula which allows us to transform Itô processes, i.e. stochastic processes which satisfy SDEs with Brownian noise terms, into other Itô processes. More specifically, it allows us to identify the SDE which the transformed process satisfies.

THEOREM A.5  **Itô's Formula.** *Let $\boldsymbol{W}_t$ denote an n-dimensional (column) vector of independent Brownian motions and suppose that the m-dimensional (column) vector-valued processes $\boldsymbol{X}_t$ satisfy the SDE:*

$$d\boldsymbol{X}_t = \boldsymbol{\mu}(t, \boldsymbol{X}_t) \, dt + \boldsymbol{\sigma}(t, \boldsymbol{X}_t) dW_t \,, \tag{A.3}$$

*where $\boldsymbol{\mu}(t, \boldsymbol{X}_t)$ is an m-dimensional (column) vector of drifts, and $\boldsymbol{\sigma}(t, \boldsymbol{X}_t)$ is an $m \times n$ matrix of volatilities. Next, introduce a new stochastic process $Y_t = f(t, \boldsymbol{X}_t)$, where $f(t, \mathbf{x})$ is twice differentiable in each $x^{(i)}$, cross-derivatives exist and $f$ is once differentiable in t. Then $Y_t$ is an Itô process satisfying the SDE*

$$dY_t = \left(\partial_t f(t, \boldsymbol{X}_t) + \boldsymbol{\mu}(t, \boldsymbol{X}_t)' \, \boldsymbol{D}f(t, \boldsymbol{X}_t) + \tfrac{1}{2} \mathrm{Tr}\, \sigma(t, \boldsymbol{X}_t)\, \sigma(t, \boldsymbol{X}_t)' \, \boldsymbol{D}^2 f(t, \boldsymbol{X}_t) \right) dt$$
$$+ \boldsymbol{D}f(t, \boldsymbol{X}_t)' \, \sigma(t, \boldsymbol{X}_t) \, dW_t \,,$$

*where for any vector or matrix $\boldsymbol{A}$, $\boldsymbol{A}'$ denotes its transpose, $\boldsymbol{D}f(t, \boldsymbol{X}_t)$ denotes the m-dimensional (column) vector of first derivatives, i.e. $(\boldsymbol{D}f(t, \boldsymbol{X}_t))_j = \partial_{x^j} f(t, \boldsymbol{X}_t)$, and $\boldsymbol{D}^2 f(t, \boldsymbol{X}_t)$ denotes the $m \times n$-dimensional matrix of mixed second derivatives, i.e. $(\boldsymbol{D}f(t, \boldsymbol{X}_t))_{jk} = \partial_{x^j x^k} f(t, \boldsymbol{X}_t)$.*

DEFINITION A.6    The **infinitesimal generator**, sometimes called simply the generator, denoted by $\mathcal{L}_t$ of a process $X_t$ acts on functions which are twice differentiable in the following manner:

$$\mathcal{L}_t f(x) = \lim_{h \downarrow 0} \frac{\mathbb{E}\left[f(X_{t+h}) \mid X_t = x\right] - f(x)}{h}.$$

The infinitesimal generator is the generalisation of a derivative of a function to make it applicable to a stochastic process. From Itô's formula in Theorem A.5, we see that the generator of an Itô process satisfying (A.3) is given by

$$\mathcal{L}_t f(\boldsymbol{x}) = \boldsymbol{\mu}(t, \boldsymbol{x})' \boldsymbol{D} f(\boldsymbol{x}) + \tfrac{1}{2} \boldsymbol{Tr}\sigma(t, \boldsymbol{x}) \, \sigma(t, \boldsymbol{x})' \, \boldsymbol{D}^2 f(\boldsymbol{x}).$$

## A.2    Jump Processes

The basic building blocks for jump processes are **counting processes**, and more specifically, **Poisson processes**.

DEFINITION A.7    **A Poisson process** $N = (N_t)_{0 \le t \le T} \in \mathbb{Z}_+$, with intensity $\lambda$, is a stochastic process which satisfies the following properties:

(i) $N_0 = 0$, a.s.,
(ii) $N_t - N_0$ has Poisson distribution with parameter $\lambda t$, i.e.

$$\mathbb{P}(N_t - N_0 = n) = e^{-\lambda t} \frac{(\lambda t)^n}{n!},$$

(iii) has independent increments, so that $N_t - N_s$ $(t > s)$ is independent of $N_v - N_u$ $(v > u)$ whenever $(s, t) \cap (u, v) = \varnothing$,
(iv) has stationary increments, so that $N_{s+t} - N_s \overset{d}{=} N_t$, for all $s, t \ge 0$.

An easy consequence of this definition is that the time between the jumps of $N$ are independent and exponentially distributed. Moreover, we have that $\mathbb{E}[N_t] = \lambda t$, so that the following proposition holds.

PROPOSITION A.8    *The compensated Poisson process* $\widehat{N} = \{\widehat{N}_t\}_{0 \le t \le T}$, *with* $\widehat{N}_t = N_t - \lambda t$, *is a martingale.*

As with Brownian motions, we can define stochastic integrals with respect to a compensated Poisson processes in such a way that the resulting object is a martingale. To this end, let $g$ be an $\mathcal{F}_t$-adapted process, where $\mathcal{F}_t$ is the natural filtration generated by a Poisson process $N$.

DEFINITION A.9    We define the stochastic integral $Y = \{Y_t\}_{0 \le t \le T}$ of $g$ with respect to the compensated Poisson process $\widehat{N}$ as follows:

$$Y_t = \int_0^t g_{s^-} \, d\widehat{N}_s = \sum_{k=1}^{N_t} g_{\tau_k^-} - \int_0^t g_s \, \lambda \, ds, \tag{A.4}$$

where $\{\tau_1, \tau_2, \dots\}$ is the collection of times at which $N$ jumps.

Note that the summation term evaluates $g$ from its left limit, and not at the time of the jump of $N$. This is a technical condition which renders the stochastic integral a martingale. If we evaluate $g$ at the time of the jump, we can easily find examples for $g$ which make the integral not a martingale. An alternate for the above integral is

$$Y_t = \int_0^t g_{s-} \, d\widehat{N}_s = \sum_{0 \le s \le t} g_{s-} \, \Delta N_s - \int_0^t g_s \, \lambda \, ds \,, \tag{A.5}$$

where the notation

$$\Delta N_s = N_s - N_{s-}$$

is the size of the jump in $N$ at time $s$ which in this case is zero or one.

THEOREM A.10    *Itô's Formula for Poisson Processes. Suppose that $Y$ is given by (A.5). Moreover, let $Z = \{Z_t\}_{0 \le t \le T}$ with $Z_t = f(t, Y_t)$ for some function $f$, once differentiable in $t$. Then,*

$$\begin{aligned} dZ_t &= (\partial_t f(t, Y_t) - \lambda \, g_t \, \partial_y f(t, Y_t)) \, dt \\ &\quad + [\, f(t, Y_{t-} + g_{t-}) - f(t, Y_{t-})\,] \, dN_t \\ &= \{\partial_t f(t, Y_t) + \lambda([\, f(t, Y_{t-} + g_{t-}) - f(t, Y_{t-})\,] - g_t \, \partial_y f(t, Y_t))\} \, dt \\ &\quad + [\, f(t, Y_{t-} + g_{t-}) - f(t, Y_{t-})\,] \, d\widehat{N}_t \,. \end{aligned} \tag{A.6}$$

The interpretation of this formula is as follows: the second term accounts for the change in $Z$ whenever a jump in $\widehat{N}$ arrives, while the first term accounts for the drift corrections.

From the above, we see that the generator $\mathcal{L}_t^Y$ of the process $Y$ in A.5 acts as follows:

$$\mathcal{L}_t^Y f(y) = \lambda([\, f(y + g_t) - f(y)\,] - g_t \, \partial_y f(y)) \,.$$

The above framework can be generalised to incorporate both diffusions and jumps. To this end, consider the sum of two stochastic integrals

$$Y_t = \int_0^t f_s \, ds + \int_0^t g_s \, dW_s + \int_0^t h_{s-} \, d\widehat{N}_s \,, \tag{A.7}$$

where, $f$, $g$ and $h$ are $\mathcal{F}_t$ adapted processes, and the filtration $\mathcal{F}$ is the natural one generated by both the diffusion $W$ and the Poisson process $N$ (assumed to be mutually independent). Each integral should be interpreted as in their individual definitions, i.e. the first term is simply a Riemann integral, the second term as in (A.1) and the third term as in (A.5). The Itô formula generalises naturally in this case.

THEOREM A.11    *Itô's Formula for Single Jumps and Diffusion. Suppose that $Y$ is given by (A.7). Moreover, let $Z = \{Z_t\}_{0 \le t \le T}$ with $Z_t = \ell(t, Y_t)$ for* *some function $\ell$, once differentiable in $t$ and twice differentiable in $y$. Then,*

$$dZ_t = \left(\partial_t + f_t\,\partial_y + \tfrac{1}{2}\,g_t^2\,\partial_{yy} - \lambda\,h_t\,\partial_y\right)\ell(t,Y_t)\,dt$$
$$+ \left[\ell(t, Y_{t-} + h_{t-}) - \ell(t, Y_{t-})\right]\,dN_t$$
$$= \left\{\left(\partial_t + f_t\,\partial_y + \tfrac{1}{2}\,g_t^2\,\partial_{yy}\right)\ell(t,Y_t)\right.$$
$$+ \lambda\left(\left[\ell(t, Y_{t-} + h_{t-}) - \ell(t, Y_{t-})\right] - h_t\,\partial_y\ell(t,Y_t)\right)\Big\}\,dt$$
$$+ \left[\ell(t, Y_{t-} + h_{t-}) - \ell(t, Y_{t-})\right]\,d\widehat{N}_t\,. \tag{A.8}$$

Again, we can identify the action of the generator $\mathcal{L}_t^Y$ of the generalised $Y$ process in (A.7) as

$$\mathcal{L}_t^Y\ell(y) = f_t\,\partial_y\ell(y) + \tfrac{1}{2}\,g_t^2\,\partial_{yy}\ell(y) + \lambda\left(\left[\ell(y + h_t) - \ell(y)\right] - h_t\,\partial_y\ell(y)\right).$$

The final generalisation which we wish to address in this section is the case of **compound Poisson processes**. A compound Poisson process $J = \{J_t\}_{0\le t\le T}$ is build out of a Poisson process $N$ (with intensity $\lambda$) and a collection of independent and identically distributed random variables $\{\varepsilon_1, \varepsilon_2, \ldots\}$ with distribution function $F$ ($\mathbb{E}[\varepsilon] < +\infty$, where $\varepsilon \sim F$), and is given by

$$J_t = \sum_{k=1}^{N_t} \varepsilon_i\,, \quad t \ge 0\,.$$

That is, the process jumps at the time of arrival of a Poisson process, and the size of the jump is independently drawn each time from the distribution function $F$.

PROPOSITION A.12 *The compensated compound Poisson process $\widehat{J} = \{\widehat{J}_t\}_{0\le t\le T}$, with $\widehat{J}_t = J_t - \mathbb{E}[\varepsilon]\,\lambda\,t$, is a martingale.*

Analogous to the Poisson case, we can define stochastic integrals with respect to compound Poisson processes as well.

DEFINITION A.13 Let $\mathcal{F}$ denote the natural filtration generated by $\widehat{J}$. We define the stochastic integral $Y = \{Y_t\}_{0\le t\le T}$ of an $\mathcal{F}$-adapted process $g$ with respect to the compensated compound Poisson process $\widehat{J}$ as follows:

$$Y_t = \int_0^t g_{s-}\,d\widehat{J}_s = \sum_{s\le t} g_{s-}\,\Delta J_s - \int_0^t g_s\,\lambda\,\mathbb{E}[\varepsilon]\,ds$$

$$= \sum_{s\le t} g_{s-}\,\varepsilon_{N_s}\,\Delta N_s - \int_0^t g_s\,\lambda\,\mathbb{E}[\varepsilon]\,ds\,, \tag{A.9}$$

where, as before, $\Delta J_t = J_t - J_{t-}$ represents the jump size of $J$ at time $t$.

In the next generalisation, we have the analog of the sum of stochastic integrals as in (A.7) by introducing three $\mathcal{F}$-adapted stochastic processes, $f$, $g$ and $h$, where $\mathcal{F}$ is the natural filtration generated by an independent Brownian motion $W$ and $\widehat{J}$, and defining the stochastic integral $Y$ as follows:

$$Y_t = \int_0^t f_s \, ds + \int_0^t g_s \, dW_s + \int_0^t h_{s-} \, d\widehat{J}_s \,, \tag{A.10}$$

where each term is interpreted appropriately.

THEOREM A.14 **Itô's Formula for Jump-Diffusion.** *Suppose that $Y$ is given by (A.10). Moreover, let $Z = \{Z_t\}_{0 \leq t \leq T}$ with $Z_t = \ell(t, Y_t)$ for some function $\ell$, once differentiable in $t$ and twice differentiable in $y$. Then,*

$$
\begin{aligned}
dZ_t &= \left( \partial_t + f_t \, \partial_y + \tfrac{1}{2} g_t^2 \, \partial_{yy} - \lambda \, \mathbb{E}[\varepsilon] \, h_t \, \partial_y \right) \ell(t, Y_t) \, dt \\
&\quad + \left[ \ell(t, Y_{t-} + \varepsilon_{N_t} h_{t-}) - \ell(t, Y_{t-}) \right] dN_t \\
&= \Big\{ \left( \partial_t + f_t \, \partial_y + \tfrac{1}{2} g_t^2 \, \partial_{yy} \right) \ell(t, Y_t) \\
&\qquad + \lambda \left( \mathbb{E} \left[ \ell(t, Y_{t-} + \varepsilon_{N_t} h_{t-}) - \ell(t, Y_{t-}) \right] - \mathbb{E}[\varepsilon] \, h_t \, \partial_y \ell(t, Y_t) \right) \Big\} \, dt \quad \text{(A.11)} \\
&\quad + \left[ \ell(t, Y_{t-} + \varepsilon_{N_t} h_{t-}) - \ell(t, Y_{t-}) \right] d\widehat{N}_t \,.
\end{aligned}
$$

For the process $Y$ in (A.10), we can see from the above that the action of the generator $\mathcal{L}_t^Y$ is as follows:

$$
\begin{aligned}
\mathcal{L}_t^Y \ell(y) &= f_t \, \partial_y \ell(y) + \tfrac{1}{2} g_t^2 \, \partial_{yy} \ell(y) \\
&\qquad + \lambda \left( \mathbb{E} \left[ \ell(t, y + \varepsilon \, h_t) - \ell(t, y) \right] - \mathbb{E}[\varepsilon] \, h_t \, \partial_y \ell(t, Y_t) \right).
\end{aligned}
$$

## A.3    Doubly Stochastic Poisson Processes

Here we consider the generalisation of jump processes which have stochastic intensity. A simple way to define such processes is by expanding filtrations. Suppose we have a counting process $N$ and we want its intensity process $\lambda = \{\lambda_t\}_{0 \leq t \leq T}$ to be 'stochastic'. One approach to defining such an object is to provide a way of computing the probability that an event arrives at time $t$ given the information we have at time $s$, i.e. to define $\mathbb{P}(N_t - N_s = n \mid \mathcal{F}_s)$, where $\mathcal{F}$ is the natural filtration generated by $(N, \lambda)$. If we have a way to compute this, then in principle we can compute all other quantities of interest. **Doubly stochastic Poisson processes**, also known as **Cox processes**, have the following property:

$$\mathbb{P}\left(N_t - N_s = n \mid \mathcal{F}_s \vee \sigma(\{\lambda_u\}_{s \leq u \leq t})\right) = \exp\left\{-\int_s^t \lambda_u \, du\right\} \frac{\left(\int_s^t \lambda_u \, du\right)^n}{n!}, \tag{A.12a}$$

so that

$$\mathbb{P}\left(N_t - N_s = 0 \mid \mathcal{F}_s\right) = \mathbb{E}\left[ \exp\left\{-\int_s^t \lambda_u \, du\right\} \frac{\left(\int_s^t \lambda_u \, du\right)^n}{n!} \,\middle|\, \mathcal{F}_s \right]. \tag{A.12b}$$

Here, $\sigma\left(\{\lambda_u\}_{s\leq u\leq t}\right)$ denotes the smallest $\sigma$-algebra generated by the intensity process $\lambda$ over the time interval $[s,t]$, and the 'join' operation $\mathcal{F}_1 \vee \mathcal{F}_2$, where $\mathcal{F}_1$ and $\mathcal{F}_2$ are two $\sigma$-algebras, represents the coarsest $\sigma$-algebra generated by taking unions of measurable sets from each individual $\sigma$-algebra. Simply put, in the above case, $\mathcal{F}_s \vee \sigma\left(\{\lambda_u\}_{s\leq u\leq t}\right)$ represents the information contained from the observations of $(N,\lambda)$ up to time $s$ together with the information on the entire path of $\lambda$ up to time $t$, but excluding the information on the $N$ process on the interval $(s,t]$. Therefore, we see from (A.12a) that the doubly stochastic Poisson process is conditionally (conditioned on $\sigma\left(\{\lambda_u\}_{s\leq u\leq t}\right)$) an inhomogeneous Poisson process with the conditionally known intensity.

The driver of the intensity process may in principle be anything ranging from an independent diffusion, an independent jump process, a combination of these, or even the counting process itself (in which case the process is known as a Hawkes process). Below are a few examples of the intensity process $\lambda$, where $W$ is an independent Brownian motion and $J$ is an independent compound Poisson process (with intensity $\lambda_J$ and i.i.d. jumps $\varepsilon \sim F$ ) with non-negative jumps:

$$d\lambda_t = \kappa(\theta - \lambda_t)\,dt + \eta\,\sqrt{\lambda_t}\,dW_t\,, \qquad\qquad \text{Feller process,} \quad \text{(A.13a)}$$

$$d\lambda_t = -\,\kappa\,\lambda_t\,dt + \gamma\,dJ_t\,, \qquad\qquad \text{Ornstein-Uhlenbeck process,} \quad \text{(A.13b)}$$

$$d\lambda_t = \kappa(\theta - \lambda_t)\,dt + \eta\,\sqrt{\lambda_t}\,dW_t + \gamma\,dJ_t\,, \qquad\qquad \text{Jump-diffusion,} \quad \text{(A.13c)}$$

$$\lambda_t = \int_0^t g(t-s)\,dN_s\,, \qquad\qquad \text{Hawkes process.} \quad \text{(A.13d)}$$

The first three processes exhibit mean-reversion. The first and third processes mean-revert to $\theta$, while the second mean-reverts to 0. With jumps, however, the mean-reversion level does not reflect the long-run behaviour. Instead we should rewrite the SDEs in terms of their compensated versions.

PROPOSITION A.15   *The compensated doubly stochastic Poisson Process* $\widehat{N} = \{\widehat{N}_t\}_{0\leq t\leq T}$, *with* $\widehat{N}_t = N_t - \int_0^t \lambda_s\,ds$, *is a martingale.*

Writing the OU (A.13b) and jump-diffusion (A.13c) in terms of their compensated versions we have

$$d\lambda_t = \kappa\left(\tfrac{\gamma\lambda_J}{\kappa}\,\mathbb{E}[\varepsilon] - \lambda_t\right)dt + \gamma\,d\widehat{J}_t\,, \qquad\qquad \text{OU process,} \quad \text{(A.14a)}$$

$$d\lambda_t = \kappa\left(\theta + \tfrac{\gamma\lambda_J}{\kappa}\,\mathbb{E}[\varepsilon] - \lambda_t\right)dt + \eta\,\sqrt{\lambda_t}\,dW_t + \gamma\,d\widehat{J}_t\,, \quad \text{Jump-diffusion.} \quad \text{(A.14b)}$$

From this expression, we see that the expected average intensities (in the long run) are $\frac{\gamma\lambda_J}{\kappa}\,\mathbb{E}[\varepsilon]$ and $\theta + \frac{\gamma\lambda_J}{\kappa}\,\mathbb{E}[\varepsilon]$, respectively. That is, the expected long run intensity is the mean-reversion level plus the jump correction term $\frac{\gamma\lambda_J}{\kappa}\,\mathbb{E}[\varepsilon]$.

DEFINITION A.16   We define the stochastic integral $Y = \{Y_t\}_{0\leq t\leq T}$ of $g$ with respect to the compensated doubly stochastic Poisson process $\widehat{N}$ as follows:

$$Y_t = \int_0^t g_{s^-}\,d\widehat{N}_s = \sum_{k=1}^{N_t} g_{\tau_k^-} - \int_0^t g_s\,\lambda_s\,ds\,, \qquad\qquad \text{(A.15)}$$

where $\{\tau_1, \tau_2, \dots\}$ is the collection of times at which $N$ jumps.

This definition is completely analogous to Definition A.9 for the simple Poisson processes. The only difference appears in the compensator, which is now a stochastic process in its own right.

THEOREM A.17 *Itô's Formula for Single doubly Stochastic Poisson processes. Let $N$ be a doubly stochastic Poisson process with intensity $\lambda$ satisfying the SDE*

$$d\lambda_t = \mu_t \, dt + \sigma_t \, dW_t + \eta_{t-} \, d\widehat{J}_t \,, \tag{A.16}$$

*where $W$ and $J$ are a Brownian motion and compound Poisson process (with intensity $\lambda_J$, i.i.d. jumps $\varepsilon \sim F$, and corresponding counting process $M$), all mutually independent and independent of $N$, and $\mu$, $\sigma$ and $\eta$ are adapted to the natural filtration generated by $N$, $W$ and $J$. Furthermore, let $Z = \{Z_t\}_{0 \le t \le T}$ with $Z_t = \ell(t, N_t, \lambda_t)$ for some function $\ell$, once differentiable in $t$ and twice differentiable in $\lambda$. Then,*

$$
\begin{aligned}
dZ_t &= \left( \partial_t + (\mu_t - \lambda_G \, \mathbb{E}[\varepsilon]) \, \partial_\lambda + \tfrac{1}{2} \sigma_t^2 \, \partial_{\lambda\lambda} - \lambda_t \, \partial_n \right) \ell(t, N_t, \lambda_t) \, dt \\
&\quad + \left[ \ell(t, N_{t-}, \lambda_{t-} + \varepsilon_{M_t}) - \ell(t, N_{t-}, \lambda_{t-}) \right] dM_t \\
&\quad + \left[ \ell(t, N_{t-} + 1, \lambda_{t-}) - \ell(t, N_{t-}, \lambda_{t-}) \right] dN_t
\end{aligned}
\tag{A.17}
$$

$$
\begin{aligned}
&= \Big\{ \left( \partial_t + \mu_t \, \partial_\lambda + \tfrac{1}{2} \sigma_t^2 \, \partial_{\lambda\lambda} \right) \ell(t, N_t, \lambda_t) \\
&\quad + \lambda_G \left( [\ell(t, N_{t-}, \lambda_{t-} + \varepsilon_{M_t}) - \ell(t, N_{t-}, \lambda_{t-})] - \mathbb{E}[\varepsilon] \partial_\lambda \ell(t, N_{t-}, \lambda_{t-}) \right) \\
&\quad + \lambda \left( [\ell(t, N_{t-} + 1, \lambda_{t-}) - \ell(t, N_{t-}, \lambda_{t-})] - \partial_n \ell(t, N_{t-}, \lambda_{t-}) \right) \Big\} \, dt \\
&\quad + \left[ \ell(t, N_{t-}, \lambda_{t-} + \varepsilon_{M_t}) - \ell(t, N_{t-}, \lambda_{t-}) \right] d\widehat{M}_t \\
&\quad + \left[ \ell(t, N_{t-} + 1, \lambda_{t-}) - \ell(t, N_{t-}, \lambda_{t-}) \right] d\widehat{N}_t \,.
\end{aligned}
\tag{A.18}
$$

From this, we see that the generator $\mathcal{L}_t^{N,\lambda}$ of the joint processes $(N, \lambda)$, where the intensity of $N$ is $\lambda$ and satisfies (A.16), acts on functions as follows:

$$
\begin{aligned}
\mathcal{L}_t^{N,\lambda} \ell(n, \lambda) &= \mu_t \, \partial_\lambda \ell(n, \lambda) + \tfrac{1}{2} \sigma_t^2 \, \partial_{\lambda\lambda} \ell(n, \lambda) \\
&\quad + \lambda_G \left( \mathbb{E}[\ell(n, \lambda + \varepsilon) - \ell(n, \lambda)] - \mathbb{E}[\varepsilon] \partial_\lambda \ell(n, \lambda) \right) \\
&\quad + \lambda \left( \mathbb{E}[\ell(n, \lambda) - \ell(n, \lambda)] - \partial_n \ell(n, \lambda) \right).
\end{aligned}
$$

The doubly stochastic Poisson process can be further generalised to the case of a doubly stochastic compound Poisson process, where jumps arrive at the arrival times of a doubly stochastic Poisson process.

## A.4    Feynman–Kac and PDEs

Certain linear partial differential equations are closely linked to stochastic processes and particularly to stochastic differential equations. The simplest example of this is the case of the heat equation:

$$\begin{cases} \partial_t h(t,x) + \frac{1}{2}\partial_{xx} h(t,x) = 0\,, \\ \qquad\qquad\qquad h(T,x) = H(x)\,, \end{cases} \tag{A.19}$$

where $H \in L^1$, i.e. $\mathbb{E}[|H(X_T)|] < +\infty$ where $X = \{X_t\}_{0 \le t \le T}$ is a Brownian motion. At first glance, there seems to be no connection whatsoever to a stochastic process or SDEs. To see the connection, suppose we introduce a Brownian motion $X$ on a probability space $(\Omega, \mathbb{P}, \mathcal{F} = \{\mathcal{F}_t\}_{0 \le t \le T})$, where $\mathcal{F}$ is the natural filtration generated by $X$, and define the stochastic process

$$f_t = \mathbb{E}[\,H(X_T)\,|\,\mathcal{F}_t\,]\,.$$

A key property in making the connection to the PDE, is the fact that such a stochastic process is a martingale because (i) for $s \le t \le T$, we have

$$\mathbb{E}\left[f_t\,|\,\mathcal{F}_s\right] = \mathbb{E}\left[\mathbb{E}[\,H(X_T)\,|\,\mathcal{F}_t\,]\,|\,\mathcal{F}_s\right] = \mathbb{E}\left[\,H(X_T)\,|\,\mathcal{F}_s\right] = f_s\,,$$

where the second equality follows from the iterated expectation property (since $\mathcal{F}_s$ is a coarser $\sigma$-algebra than $\mathcal{F}_t$), and (ii)

$$\mathbb{E}\left[|f_t|\right] = \mathbb{E}\left[\left|\mathbb{E}[\,H(X_T)\,|\,\mathcal{F}_t\,]\right|\right] \le \mathbb{E}\left[\mathbb{E}\left[|H(X_T)|\,|\,\mathcal{F}_t\right]\right] < +\infty\,,$$

where the first inequality follows from Jensen's inequality and the last from the assumption that $H$ is in $L^1$. Next, note that the process $f_t$ is in fact Markov in $X$ since

$$f_t = \mathbb{E}[\,H(X_T)\,|\,\mathcal{F}_t\,] = \mathbb{E}[\,H\,((X_T - X_t) + X_t)\,|\,\mathcal{F}_t\,]$$
$$= \mathbb{E}\left[H\left(\sqrt{T-t}\,Z + X_t\right)\,|\,\mathcal{F}_t\right] = g(t, X_t)\,,$$

where $g(t, x)$ is some function, the third equality follows from the independence of $(X_T - X_t)$ and $X_t$, and $Z$ is an standard normal random variable independent of $X_t$. If we further assume that $g$ is smooth enough, then we can use Itô's lemma to write (for any $h > 0$)

$$g(t+h, X_{t+h}) - g(t, X_t)$$
$$= \int_t^{t+h} \left\{\partial_t g(s, X_s) + \frac{1}{2}\partial_{xx}g(s, X_s)\right\} ds + \int_t^{t+h} \partial_x g(s, X_s)\,dX_s\,.$$

Since $g$ is a martingale, we have

$$0 = \mathbb{E}_{t,x}\left[g(t+h, X_{t+h}) - g(t, X_t)\right]$$
$$= \mathbb{E}_{t,x}\left[\int_t^{t+h} \left\{\partial_t g(s, X_s) + \frac{1}{2}\partial_{xx}g(s, X_s)\right\} ds + \int_t^{t+h} \partial_x g(s, X_s)\,dX_s\right]$$
$$= \mathbb{E}_{t,x}\left[\int_t^{t+h} \left\{\partial_t g(s, X_s) + \frac{1}{2}\partial_{xx}g(s, X_s)\right\} ds\right]\,.$$

Next, dividing by $h$, taking $h \downarrow 0$, and invoking the Fundamental Theorem of Calculus, we have

$$0 = \lim_{h \downarrow 0} \frac{1}{h} \mathbb{E}_{t,x} \left[ \int_t^{t+h} \left\{ \partial_t g(s, X_s) + \tfrac{1}{2} \partial_{xx} g(s, X_s) \right\} ds \right]$$

$$= \partial_t g(t, x) + \tfrac{1}{2} \partial_{xx} g(t, x),$$

and from the definition of $g$, we also have $g(T, x) = H(x)$. Hence, we see that $g(t, x)$ satisfies the PDE (A.19). This sequence of arguments is one example of a more general result.

THEOREM A.18    **Feynman–Kac.** *Let $X$ denote an Itô process satisfying the SDE*

$$dX_t = \mu(t, X_t)\, dt + \sigma(t, X_t)\, dW_t,$$

*where $W$ is a Brownian motion. Define the function*

$$f(t, x) = \mathbb{E}_{t,x} \left[ \int_t^T e^{-\int_t^s \gamma(u, X_u)\, du}\, g(s, X_s)\, ds + e^{-\int_t^T \gamma(u, X_u)\, du}\, h(X_T) \right],$$

*where $\mathbb{E}[|h(X_T)|] < +\infty$, $\mathbb{E}\left[ \int_0^T |g(s, X_s)|\, ds \right] < +\infty$, and $\int_0^T \gamma(t, X_t)\, dt$ is bounded from below a.s. . Then $f(t, x)$ satisfies the PDE*

$$\begin{cases} \partial_t f(t, x) + \mathcal{L}_t^X f(t, x) + g(t, x)\, f(t, x) = \gamma(t, x)\, f(t, x), \\ \qquad\qquad\qquad\qquad\qquad\quad f(T, x) = h(x), \end{cases} \qquad (\text{A.20})$$

*where $\mathcal{L}_t^X$ represents the infinitesimal generator of $X$, specifically,*

$$\mathcal{L}_t^X f = \mu(t, x)\, \partial_x f + \tfrac{1}{2} \sigma^2(t, x)\, \partial_{xx} f.$$

## A.5     Bibliography and Selected Readings

Baxter & Rennie (1996), Jacod & Shiryaev (1987), Shreve (2005), Øksendal (2010), Shreve (2013), Steele (2010).

# Bibliography

Abergel, Frédéric, Jean-Philippe Bouchaud, Thierry Foucault, Charles-Albert Lehalle & Mathieu Rosenbaum (2012), *Market Microstructure: Confronting many Viewpoints*, John Wiley & Sons.

Abergel, Frédéric, Marouane Anane, Anirban Chakraborti, Aymen Jedidi & Ioane Muni Toke (2015), *Limit Order Books*, Available at http://fiquant.mas.ecp.fr/limit-order-books.

Abramowitz, Milton & Irene A. Stegun (1972), *Handbook of Mathematical Functions with Formulas, Graphs and Mathematical Tables*, 10th ed., National Bureau of Standards Applied Mathematics Series, Vol. 55.

Aït-Sahalia, Yacine, Per A. Mykland & Lan Zhang (2005), 'How often to sample a continuous-time process in the presence of market microstructure noise', *Review of Financial Studies* **18**(2), 351–416.

Alfonsi, Aurélien, Alexander Schied & Alla Slynko (2012), 'Order book resilience, price manipulation, and the positive portfolio problem', *SIAM Journal on Financial Mathematics* **3**(1), 511–533.

Alfonsi, Aurélien, Antje Fruth & Alexander Schied (2010), 'Optimal execution strategies in limit order books with general shape functions', *Quantitative Finance* **10**(2), 143–157.

Almgren, Robert (2003), 'Optimal execution with nonlinear impact functions and trading-enhanced risk', *Applied Mathematical Finance* **10**(1), 1–18.

Almgren, Robert (2010), *Execution Costs*, John Wiley & Sons.

Almgren, Robert (2013), 'Execution strategies in fixed income markets', *High Frequency Trading; New Realities for Trades, Markets and Regulators*, Easley, D., M. López de Prado, and M. O'Hara (editors), Risk Books.

Almgren, Robert, Chee Thum, Emmanuel Hauptmann & Hong Li (2005), 'Direct estimation of equity market impact', *Risk* **18**, 5752.

Almgren, Robert & Neil Chriss (2000), 'Optimal execution of portfolio transactions', *Journal of Risk* **3**, 5–39.

Andersen, Torben G. & Oleg Bondarenko (2014), 'Reflecting on the VPIN dispute', *Journal of Financial Markets* **17**, 53–64.

Avellaneda, Marco & Sasha Stoikov (2008), 'High-frequency trading in a limit order book', *Quantitative Finance* **8**, 217–224.

Bandi, Federico M. & Jeffrey R. Russell (2008), 'Microstructure noise, realized variance, and optimal sampling', *The Review of Economic Studies* **75**(2), 339–369.

Bandi, Federico M. & Roberto Renò (2012), 'Time-varying leverage effects', *Journal of Econometrics* **169**(1), 94–113.

Barndorff-Nielsen, Ole E. & Neil Shephard (2004), 'Power and bipower variation with stochastic volatility and jumps', *Journal of Financial Econometrics* **2**(1), 1–37.

Barndorff-Nielsen, Ole E. & Neil Shephard (2006*a*), 'Econometrics of testing for jumps in financial economics using bipower variation', *Journal of Financial Econometrics* **4**(1), 1–30.

Barndorff-Nielsen, Ole E. & Neil Shephard (2006*b*), 'Econometrics of testing for jumps in financial economics using bipower variation', *Journal of Financial Econometrics* **4**(1), 1–30.

Bauwens, Luc & Nikolaus Hautsch (2006), 'Stochastic conditional intensity processes', *Journal of Financial Econometrics* **4**(3), 450–493.

Bauwens, Luc & Nikolaus Hautsch (2009), *Modelling Financial High Frequency Data using Point Processes*, Springer.

Baxter, Martin & Andrew Rennie (1996), *Financial Calculus: An Introduction to Derivative Pricing*, Cambridge University Press.

Bayraktar, Erhan & Michael Ludkovski (2011), 'Optimal trade execution in illiquid markets', *Mathematical Finance* **21**(4), 681–701.

Bayraktar, Erhan & Michael Ludkovski (2012), 'Liquidation in limit order books with controlled intensity', *Mathematical Finance* **24**(4), 627–650.

Bechler, Kyle & Mike Ludkovski (2014), 'Optimal execution with dynamic order flow imbalance', *arXiv preprint arXiv:1409.2618*.

Bertsekas, Dimitri P. & Steven E. Shreve (1978), *Stochastic Optimal Control*, Vol. 139 of *Mathematics in Science and Engineering*, Academic Press, New York.

Bertsimas, Dimitris & Andrew W. Lo (1998), 'Optimal control of execution costs', *Journal of Financial Markets* **1**(1), 1–50.

Biais, Bruno, Christophe Bisiere & Chester Spatt (2010), 'Imperfect competition in financial markets: An empirical study of Island and Nasdaq', *Management Science* **56**(12), 2237–2250.

Biais, Bruno, Larry Glosten & Chester Spatt (2005), 'Market microstructure: A survey of microfoundations, empirical results, and policy implications', *Journal of Financial Markets* **8**(2), 217–264.

Bialkowski, Jedrzej, Serge Darolles & Gaëlle Le Fol (2008), 'Improving VWAP strategies: A dynamic volume approach', *Journal of Banking & Finance* **32**(9), 1709–1722.

Boehmer, Ekkehart, Kingsley Y. L. Fong & Juan (Julie) Wu (2012), 'International evidence on algorithmic trading', SSRN http://ssrn.com/abstract=2022034.

Boehmer, Ekkehart, Kingsley Y.L. Fong & Juan Julie Wu (2014), International evidence on algorithmic trading, *in* 'AFA 2013 San Diego Meetings Paper'.

Bouchard, Bruno, Ngoc-Minh Dang & Charles-Albert Lehalle (2011), 'Optimal control of trading algorithms: A general impulse control approach', *SIAM Journal on Financial Mathematics* **2**, 404–438.

Brogaard, Jonathan, Terrence Hendershott & Ryan Riordan (2014), 'High-frequency trading and price discovery', *Review of Financial Studies*, DOI: 10.1093/rfs/hhu032.

Buti, Sabrina & Barbara Rindi (2013), 'Undisclosed orders and optimal submission strategies in a limit order market', *Journal of Financial Economics* **109**(3), 797–812.

Buti, Sabrina, Barbara Rindi & Ingrid M. Werner (2011*a*), 'Dark pool trading strategies', *Available at http://ssrn.com/abstract=1571416*.

Buti, Sabrina, Barbara Rindi & Ingrid M Werner (2011*b*), 'Diving into dark pools', *Available at http://ssrn.com/abstract=1630499*.

Cameron, A. Colin & Pravin K. Trivedi (2005), *Microeconometrics: Methods and Applications*, Cambridge University Press.

Carmona, Rene & Kevin Webster (2012), 'High frequency market making', *arXiv preprint arXiv:1210.5781*.

Carmona, Rene & Kevin Webster (2013), 'The self-financing equation in high frequency markets', *arXiv preprint arXiv:1312.2302*.

Cartea, Álvaro (2013), 'Derivatives pricing with marked point processes using tick-by-tick data', *Quantitative Finance* **13**(1), 111–123.

Cartea, Álvaro & Dimitrios Karyampas (2012), 'Assessing the performance of different volatility estimators: A Monte Carlo analysis', *Applied Mathematical Finance* **19**(6), 535–552.

Cartea, Álvaro & Dimitrios Karyampas (2014), 'The relationship between the volatility of returns and the number of jumps in financial markets'. Forthcoming in *Econometric Reviews*.

Cartea, Álvaro & José Penalva (2012), 'Where is the value in high frequency trading?', *The Quarterly Journal of Finance* **13**(1), 1–46.

Cartea, Álvaro, Ryan Donnelly & Sebastian Jaimungal (2013), 'Algorithmic trading with model uncertainty', *SSRN: http://ssrn.com/abstract=2310645*.

Cartea, Álvaro & Sebastian Jaimungal (2013), 'Risk metrics and fine tuning of high frequency trading strategies'. Forthcoming in *Mathematical Finance*, DOI: 10.1111/mafi.12023.

Cartea, Álvaro & Sebastian Jaimungal (2013), 'Optimal execution with limit and market orders', *SSRN eLibrary, http://ssrn.com/abstract=2397805*.

Cartea, Álvaro & Sebastian Jaimungal (2014*a*), 'A closed-form execution strategy to target VWAP', *Available at SSRN 2542314*.

Cartea, Álvaro & Sebastian Jaimungal (2014*b*), 'Incorporating order-flow into optimal execution', *Available at SSRN 2557457*.

Cartea, Álvaro, Sebastian Jaimungal & Damir Kinzebulatov (2013), 'Algorithmic trading with learning', *SSRN eLibrary, http://ssrn.com/abstract=2373196*.

Cartea, Álvaro, Sebastian Jaimungal & Jason Ricci (2014), 'Buy low, sell high: A high frequency trading perspective', *SIAM Journal on Financial Mathematics* **5**(1), 415–444.

Cartea, Álvaro & Thilo Meyer-Brandis (2010), 'How duration between trades of underlying securities affects option prices', *Review of Finance* **14**(4), 749–785.

Cebiroglu, Gökhan & Ulrich Horst (2015), 'Optimal order display in limit order markets with liquidity competition', *Available at SSRN http://ssrn.com/abstract=2549739*.

CFTC & SEC (2010), Findings regarding the market events of May 6, 2010, Report, SEC.

Chaboud, Alain, Erik Hjalmarsson, Clara Vega & Ben Chiquoine (2009), 'Rise of the Machines: Algorithmic Trading in the Foreign Exchange Market', *SSRN eLibrary*.

Chang, Fwu-Ranq (2004), *Stochastic Optimization in Continuous Time*, Cambridge University Press.

Clauset, Aaron, Cosma Rohilla Shalizi & Mark E.J. Newman (2009), 'Power-law distributions in empirical data', *SIAM review* **51**(4), 661–703.

Cohen, Samuel N. & Lukasz Szpruch (2012), 'A limit order book model for latency arbitrage', *Mathematics and Financial Economics* **6**(3), 211–227.

Colliard, Jean-Edouard & Thierry Foucault (2012), 'Trading fees and efficiency in limit order markets', *Review of Financial Studies* **25**(11), 3389–3421.

Cont, Rama, Arseniy Kukanov & Sasha Stoikov (2013), 'The price impact of order book events'. Forthcoming in *Journal of Financial Econometrics*, DOI: 10.1093/jjfinec/nbt003.

Corsi, Fulvio (2009), 'A simple approximate long-memory model of realized volatility', *Journal of Financial Econometrics* **7**(2), 174–196.

Corsi, Fulvio, Davide Pirino & Roberto Renò (2010), 'Threshold bipower variation and the impact of jumps on volatility forecasting', *Journal of Econometrics* **159**(2), 276–288.

Corsi, Fulvio & Roberto Renò (2012), 'Discrete-time volatility forecasting with persistent leverage effect and the link with continuous-time volatility modeling', *Journal of Business & Economic Statistics* **30**(3), 368–380.

Crisafi, M. Alessandra & Andrea Macrina (2014), 'Optimal execution in lit and dark pools', *arXiv preprint arXiv:1405.2023*.

Curato, Gianbiagio, Jim Gatheral & Fabrizio Lillo (2014), 'Optimal execution with nonlinear transient market impact', *Available at SSRN 2539240*.

Cvitanic, Jaksa & Andrei A. Kirilenko (2010), 'High Frequency Traders and Asset Prices', *SSRN eLibrary*.

de Jong, Frank & Barbara Rindi (2009), *The Microstructure of Financial Markets*, 1st ed., Cambridge University Press.

Ding, Shengwei, John Hanna & Terrence Hendershott (2014), 'How slow is the NBBO? a comparison with direct exchange feeds', *Financial Review* **49**(2), 313–332.

Duffie, Darrell & Larry G. Epstein (1992), 'Asset pricing with stochastic differential utility', *The Review of Financial Studies* **5(3)**, 411–436.

Easley, David, Marcos M. López de Prado & Maureen O'Hara (2012), 'Flow toxicity and liquidity in a high-frequency world', *Review of Financial Studies* **25**(5), 1457–1493.

Easley, David & Maureen O'Hara (1992), 'Time and the process of security price adjustment', *The Journal of Finance* **47**(2), 577–605.

Easley, David, Robert F. Engle, Maureen O'Hara & Liuren Wu (2008), 'Time-varying arrival rates of informed and uninformed trades', *Journal of Financial Econometrics* **6**(2), 171.

Elliott, Robert J., John Van Der Hoek & William P. Malcolm (2005), 'Pairs trading', *Quantitative Finance* **5**(3), 271–276.

Engle, Robert F. (2000), 'The econometrics of ultra-high-frequency data', *Econometrica* **68**(1), 1–22.

Fama, Eugene F. (1970), 'Efficient capital markets: A review of theory and empirical work', *The Journal of Finance* **25**(2), 383–417.

Farmer, J. Doyne, Austin Gerig, Fabrizio Lillo & Henri Waelbroeck (2013), 'How efficiency shapes market impact', *Quantitative Finance* **13**(11), 1743–1758.

Fleming, Wendell H. & Halil Mete Soner (2006), *Controlled Markov Processes and Viscosity Solutions*, Springer.

Foucault, Thierry & Albert J. Menkveld (2008), 'Competition for order flow and smart order routing systems', *The Journal of Finance* **63**(1), 119–158.

Foucault, Thierry, Ohad Kadan & Eugene Kandel (2013), 'Liquidity cycles and make/take fees in electronic markets', *The Journal of Finance* **68**(1), 299–341.

Foucault, Thierry, Ohad Kadan & Eugene Kandel (2005), 'Limit order book as a market for liquidity', *Review of Financial Studies* **18**(4), 1171–1217.

Frei, Christoph & Nicholas Westray (2013), 'Optimal execution of a VWAP order: a stochastic control approach'. Forthcoming in *Mathematical Finance*, DOI: 10.1111/mafi.12048.

Gatheral, Jim (2010), 'No-dynamic-arbitrage and market impact', *Quantitative Finance* **10**(7), 749–759.

Gatheral, Jim & Alexander Schied (2013), 'Dynamical models of market impact and algorithms for order execution', *Handbook on Systemic Risk*, Jean-Pierre Fouque, Joseph A. Langsam (editors), Cambridge University Press, 579–599.

Gatheral, Jim, Alexander Schied & Alla Slynko (2012), 'Transient linear price impact and fredholm integral equations', *Mathematical Finance* **22**(3), 445–474.

Gerig, Austin (2008), 'A theory for market impact: How order flow affects stock price', *arXiv preprint arXiv:0804.3818*.

Gerig, Austin (2012), 'High-frequency trading synchronizes prices in financial markets', *arXiv preprint arXiv:1211.1919*.

Gerig, Austin & David Michayluk (2010), 'Automated liquidity provision and the demise of traditional market making', *arXiv preprint arXiv:1007.2352*.

Glosten, Lawrence R. & Paul R. Milgrom (1985), 'Bid, ask and transaction prices in a specialist market with heterogeneously informed traders', *Journal of Financial Economics* **14**(1), 71–100.

Gould, Martin D., Mason A. Porter, Stacy Williams, Mark McDonald, Daniel J. Fenn & Sam D. Howison (2013), 'Limit order books', *Quantitative Finance* **13**(11), 1709–1742.

Graewe, Paulwin, Ulrich Horst & Eric Séré (2013), 'Smooth solutions to portfolio liquidation problems under price-sensitive market impact', *arXiv:1309.0474*.

Graewe, Paulwin, Ulrich Horst & Jinniao Qiu (2013), 'A non-Markovian liquidation problem and backward SPDEs with singular terminal conditions', *arXiv:1309.0461*.

Greene, William H. (2011), *Econometric Analysis*, Pearson Education.

Grossman, Sanford J. (1976), 'On the efficiency of competitive stock markets where trades have diverse information', *The Journal of Finance* **31**(2), 573–585.

Grossman, Sanford J. (1977), 'The existence of futures markets, noisy rational expectations and informational externalities', *The Review of Economic Studies*, **44**, 431–449.

Grossman, Sanford J. (1978), 'Further results on the informational efficiency of competitive stock markets', *Journal of Economic Theory* **18**(1), 81–101.

Grossman, Sanford J. & Joseph E. Stiglitz (1976), 'Information and competitive price systems', *The American Economic Review* **66**, 246–253.

Grossman, Sanford J. & Merton H. Miller (1988), 'Liquidity and market structure', *The Journal of Finance* **43**(3), 617–37.

Guéant, Olivier (2014), 'Execution and block trade pricing with optimal constant rate of participation', *Journal of Mathematical Finance* **4**, 255–264.

Guéant, Olivier & Charles-Albert Lehalle (2013), 'General intensity shapes in optimal liquidation'. Forthcoming in *Mathematical Finance*, DOI: 10.1111/mafi.12052.

Guéant, Olivier, Charles-Albert Lehalle & Joaquin Fernandez Tapia (2012), 'Optimal portfolio liquidation with limit orders', *SIAM Journal on Financial Mathematics* **3**(1), 740–764.

Guéant, Olivier, Charles-Albert Lehalle & Joaquin Fernandez-Tapia (2013), 'Dealing with the inventory risk: a solution to the market making problem', *Mathematics and Financial Economics* **7**(4), 477–507.

Guéant, Olivier, Jiang Pu & Guillaume Royer (2013), 'Accelerated share repurchase: pricing and execution strategy', *arXiv, http://arxiv.org/abs/1312.5617*.

Guéant, Oliver & Guillaume Royer (2014), 'VWAP execution and guaranteed VWAP', *SIAM Journal on Financial Mathematics* **5**(1), 445–471.

Guilbaud, Fabien & Huyên Pham (2013), 'Optimal high-frequency trading with limit and market orders', *Quantitative Finance* **13**(1), 79–94.

Guo, Xin, Adrien De Larrard & Zhao Ruan (2013), 'Optimal placement in a limit order book', *Preprint*.

Hagströmer, Björn & Lars L. Norden (2013), 'The diversity of high-frequency traders', *Journal of Financial Markets*, **16**(4), 741–770.

Hall, Anthony D. & Nikolaus Hautsch (2007), 'Modelling the buy and sell intensity in a limit order book market', *Journal of Financial Markets* **10**(3), 249–286.

Hansen, Peter R. & Asger Lunde (2006), 'Realized variance and market microstructure noise', *Journal of Business & Economic Statistics* **24**(2), 127–161.

Hasbrouck, Joel (1991), 'Measuring the information content of stock trades', *The Journal of Finance* **46**(1), 179–207.

Hasbrouck, Joel (1993), 'Assessing the quality of a security market: A new approach to transaction-cost measurement', *Review of Financial Studies* **6**(1), 191–212.

Hasbrouck, Joel (1995), 'One security, many markets: Determining the contributions to price discovery', *The Journal of Finance* **50**(4), 1175–1199.

Hasbrouck, Joel (2013), 'High frequency quoting: Short-term volatility in bids and offers', *SSRN eLibrary*.

Hasbrouck, Joel & Gideon Saar (2013), 'Low-latency trading', *Journal of Financial Markets* **16**(4), 646–679.

Hendershott, Terrence, Charles M. Jones & Albert J. Menkveld (2011), 'Does algorithmic trading improve liquidity?', *The Journal of Finance* **66**(1), 1–33.

Hendershott, Terrence & Ryan Riordan (2013), 'Algorithmic trading and the market for liquidity', *Journal of Financial and Quantitative Analysis* **48**(04), 1001–1024.

Hirschey, Nicholas (2013), 'Do high-frequency traders anticipate buying and selling pressure?', *SSRN eLibrary, Working paper*.

Ho, Thomas & Hans R. Stoll (1981), 'Optimal dealer pricing under transactions and return uncertainty', *Journal of Financial Economics* **9**, 47–73.

Hoffmann, Peter (2014), 'A dynamic limit order market with fast and slow traders', *Journal of Financial Economics* **113**(1), 156–169.

Horst, Ulrich & Felix Naujokat (2014), 'When to cross the spread? trading in two-sided limit order books', *SIAM Journal on Financial Mathematics* **5**(1), 278–315.

Huitema, Robert (2013), 'Optimal portfolio execution using market and limit orders', *SSRN eLibrary, http://ssrn.com/abstract=1977553*.

Humphery-Jenner, Mark L. (2011), 'Optimal VWAP trading under noisy conditions', *Journal of Banking & Finance* **35**(9), 2319–2329.

Iyer, Krishnamurthy, Ramesh Johari & Ciamac C. Moallemi (2014), 'Welfare analysis of dark pools', *Available at http://moallemi.com/ciamac/papers*.

Jacod, Jean & Albert N. Shiryaev (1987), *Limit Theorems for Stochastic Processes*, Springer.

Jaimungal, Sebastian & Damir Kinzebulatov (2013), 'Optimal Execution with a Price Limiter', *SSRN eLibrary, http://ssrn.com/abstract=2199889*.

Jaimungal, Sebastian & Mojtaba Nourian (2015), 'Mean-field game strategies for a major-minor agent optimal execution problem', *Available at SSRN: http://ssrn.com/abstract=2578733*.

Jarrow, Robert & Hao Li (2013), 'Abnormal profit opportunities and the informational advantage of high frequency trading', *The Quarterly Journal of Finance* **3**(02), 1350012.

Kharroubi, Idris & Huyên Pham (2010), 'Optimal portfolio liquidation with execution cost and risk', *SIAM Journal on Financial Mathematics* **1**(1), 897–931.

Konishi, Hizuru (2002), 'Optimal slice of a VWAP trade', *Journal of Financial Markets* **5**(2), 197–221.

Kyle, Albert S. (1985), 'Continuous auctions and insider trading', *Econometrica* **53**, 1315–1335.

Kyle, Albert S. (1989), 'Informed speculation with imperfect competition', *The Review of Economic Studies* **56**(3), 317–355.

Large, Jeremy (2007), 'Measuring the resiliency of an electronic limit order book', *Journal of Financial Markets* **10**(1), 1–25.

Laruelle, Sophie, Charles-Albert Lehalle & Gilles Pagès (2011), 'Optimal split of orders across liquidity pools: a stochastic algorithm approach', *SIAM Journal on Financial Mathematics* **2**(1), 1042–1076.

Laruelle, Sophie, Charles-Albert Lehalle & Gilles Pagès (2013), 'Optimal posting price of limit orders: learning by trading', *Mathematics and Financial Economics* **7**(3), 359–403.

Lehalle, Charles-Albert (2009), 'The impact of liquidity fragmentation on optimal trading', *Trading* **2009**(1), 80–87.

Leung, Tim & Mike Ludkovski (2011), 'Optimal timing to purchase options', *SIAM Journal on Financial Mathematics* **2**(1), 768–793.

Leung, Tim & Xin Li (2014), 'Optimal mean reversion trading with transaction costs & stop-loss exit', *International Journal of Theoretical & Applied Finance* **18**(3), 1550020.

Li, Tianhui Michael & Robert Almgren (2014), 'Option hedging with smooth market impact', preprint, NYU.

Lipton, Alexander, Umberto Pesavento & Michael G. Sotiropoulos (2013), 'Trade arrival dynamics and quote imbalance in a limit order book', *arXiv:1312.0514*.

Lorenz, Julian & Robert Almgren (2011), 'Mean-variance optimal adaptive execution', *Applied Mathematical Finance* **18**(5), 395–422.

Martínez, Miguel A., Belén Nieto, Gonzalo Rubio & Mikel Tapia (2005), 'Asset pricing and systematic liquidity risk: An empirical investigation of the Spanish stock market', *International Review of Economics & Finance* **14**(1), 81–103.

Martinez, Victor Hugo & Ioanid Rosu (2013), 'High frequency traders, news and volatility', *SSRN eLibrary, Working paper*.

Martinez, Victor Hugo & Ioanid Rosu (2014), 'Fast and slow informed trading', *SSRN eLibrary, Working paper*.

McCulloch, James & Vlad Kazakov (2012), 'Mean variance optimal VWAP trading', *SSRN eLibtrary, http://ssrn.com/abstract=1803858*.

Menkveld, Albert J. (2013), 'High frequency trading and the new market makers', *Journal of Financial Markets* **16**(4), 712–740.

Merton, Robert (1971), 'Optimum consumption and portfolio rules in a continuous-time model', *Journal of Economic Theory* **3**, 373–413.

Merton, Robert (1992), *Continuous-Time Finance*, Wiley-Blackwell.

Mitchell, Daniel, Jedrzej Pawel Bialkowski & Stathis Tompaidis (2013), 'Optimal VWAP tracking', *SSRN eLibtrary, http://ssrn.com/abstract=2333916*.

Moallemi, Ciamac C. & Mehmet Saglam (2013), 'The cost of latency in high-frequency trading', *Operations Research* **61**(5), 1070–1086.

Moallemi, Ciamac C., Mehmet Saglam & Michael Sotiropoulos (2014), 'Short-term predictability and price impact', *Available at SSRN 2463952*.

Moro, Esteban, Javier Vicente, Luis G. Moyano, Austin Gerig, J. Doyne Farmer, Gabriella Vaglica, Fabrizio Lillo & Rosario N. Mantegna (2009), 'Market impact and trading profile of hidden orders in stock markets', *Physical Review E* **80**(6), 066102.

Mykland, Per A. & Lan Zhang (2012), 'The econometrics of high frequency data'. In *Statistical Methods for Stochastic Differential Equations*, Mathieu Kessler, Alexander Lindner and Michael Sorensen (editors), CRC Press, 109–190.

Obizhaeva, Anna A. & Jiang Wang (2013), 'Optimal trading strategy and supply/demand dynamics', *Journal of Financial Markets* **16**(1), 1–32.

O'Hara, Maureen (1995), *Market Microstructure Theory*, 1st edn, Blackwell Publishers, Cambridge, Mass.

O'Hara, Maureen, Chen Yao & Mao Ye (2014), 'What's not there: odd lots and market data', *The Journal of Finance* **69**(5), 2199–2236.

Øksendal, Bernt (2010), *Stochastic Differential Equations: An Introduction with Applications*, 6th ed., Springer.

Øksendal, Bernt & Agnes Sulem (2007), *Applied Stochastic Control of Jump Diffusions*, 2nd ed., Springer.

Pascual, Roberto & David Veredas (2009), 'What pieces of limit order book information matter in explaining order choice by patient and impatient traders?', *Quantitative Finance* **9**(5), 527–545.

Pham, Hûyen (2010), *Continuous-time Stochasic Control and Optimization with Financial Applications*, Springer.

Ramsay, James & Bernard W. Silverman (2010), *Functional Data Analysis*, Springer.

Riordan, Ryan & Andreas Storkenmaier (2012), 'Latency, liquidity and price discovery', *Journal of Financial Markets* **15**(4), 416–437.

Roll, Richard (1984), 'A simple implicit measure of the effective bid-ask spread in an efficient market', *The Journal of Finance* **39**(4), 1127–1139.

Rosu, Ioanid (2009), 'A dynamic model of the limit order book', *Review of Financial Studies* **22**(11), 4601–4641.

Schied, Alexander (2013), 'Robust strategies for optimal order execution in the Almgren–Chriss framework', *Applied Mathematical Finance* **20**(3), 264–286.

Schied, Alexander & Torsten Schöneborn (2009), 'Risk aversion and the dynamics of optimal liquidation strategies in illiquid markets', *Finance and Stochastics* **13**(2), 181–204.

SEC (2010), Concept release on equity market structure, Concept Release No. 34-61358; File No. S7-02-10, SEC. 17 CFR Part 242.

SEC (2013a), Alternative trading systems: Description of ATS trading in national market system stocks, Technical report, SEC-DERA Memorandum.

SEC (2013b), Release no. 34-71057; file no. sr-chx-2013-21, Technical report, Securities and Exchage Commission.

Shreve, Steven E. (2005), *Stochastic Calculus for Finance I: the Binomial Asset Pricing Model*, Springer.

Shreve, Steven E. (2013), *Stochastic Calculus for Finance II: Continuous-Time Models*, Springer.

Steele, J. Michael (2010), *Stochastic Calculus and Financial Applications*, Springer.

Stoikov, Sasha & Mehmet Sağlam (2009), 'Option market making under inventory risk', *Review of Derivatives Research* **12**(1), 55–79.

Tourin, Agnés & Raphael Yan (2013), 'Dynamic pairs trading using the stochastic control approach', *Journal of Economic Dynamics and Control* **37**(10), 1972–1981.

Touzi, Nizar (2013), *Optimal Stochastic Control, Stochastic Target Problems, and Backward SDE*, Vol. 29 of *Fields Institute Monographs*, Springer.

Vayanos, Dimitri & Jiang Wang (2009), Liquidity and asset prices: A unified framework, Technical report, National Bureau of Economic Research.

Vives, Xavier (1996), 'Social learning and rational expectations', *European Economic Review* **40**(3), 589–601.

Yong, Jiongmin & Xun Yu Zhou (1999), *Stochastic controls: Hamiltonian systems and HJB equations*, Springer.

Zheng, Ban, Eric Moulines & Frédéric Abergel (2012), 'Price jump prediction in limit order book', *Available at http://ssrn.com/abstract=2026454*.

# Glossary

$\mathcal{A}_{t,T}$ a set of admissible strategies over the interval $[t, T]$, 102

$\mathcal{F}$ filtration, 315

$\Omega$ space of all events, 315

$\mathbb{P}$ real-world probability measure, 315

$\mathcal{L}$ Infinitesimal generator of a process, 318

$\varepsilon$-**optimal control** a control which leads to a performance criteria which is within $\varepsilon$ of the value function, 106

**Agency broker** an agent who executes trade(s) on behalf of a client, 135

**Arrival price** type of benchmark price given by the quoted price, midprice for instance, in effect at the time the order is sent to trading desk, 135

**At-The-Touch** the price and/or volume at the best bid or ask, 295

**Benchmark price** a price against which to measure the actual price of the executed shares (measure on a per share basis). Typical examples include the arrival price, TWAP and VWAP, 135

**Bonds** are contracts whereby the corporation commits to pay the holder a regular income (interest) but includes no decision rights, 4

**CBOE** Chicago Board Options Exchange – the largest options exchange in the US for options on stocks, indices, and ETFs, 43

**Closed-end Fund** a mutual fund with a fixed number of shares. These are usually issued once at inception, through an initial public offering. Closed-end mutual fund shares that are not redeemable, that is, investors cannot sell them back to the fund. Its shares, like those of ETFs, are listed and traded continuously, 5

**Colocated** also known as colocation, means that an agent's trading system is physically housed at the electronic exchange and has a direct connection to the exchanges' matching engine, 50

**Colocation** see colocated, 50

**Common stock** same as ordinary shares, 4

**Dark pool** "...systems that allow participants to enter unpriced orders to buy and sell securities, these orders are crossed at a specified time at a price derived from another market..." SEC, 176

**DPE** dynamic programming equation, 100

**DPP** dynamic programming principle, 100

**LOB** Limit Order Book. The collection of currently available buy and sell orders, their available prices and their available volumes, 9

**Locked market** the situation that occurs when the bid is equal to the ask and the quoted spread is zero, 16

**Market depth** refers to the available volume posted at different levels of the LOB. A deep market has a lot of posted volume. A thin market has little posted volume, 79

**Market Order** an aggressive order which takes liquidity from the LOB and receives the best prices currently available, 9

**Matching algorithm, price-time priority** an algorithm used by exchanges to determine which of the standing limit orders will be executed against an incoming MO. The algorithm establishes that the market order will be executed against standing limit orders at the best price based on the time at which the limit orders were posted, starting from the oldest one first, 9

**Matching algorithm, prorata** an algorithm used by exchanges to determine how the quantity demanded by an incoming market order will be shared amongst standing limit orders. The sharing rule assigns the market order proportionally based on the relative quantity of shares offered by each limit order at the best price, 10

**Microprice** the price computed as the weighted average of the bid and ask, where the weight on the ask is the volume posted at the ask relative to the total volume at the bid and ask, while the weight on the ask is the relative volume posted at the bid, 18

**Midprice** the arithmetic average of the bid and ask, 16

**Minimum tick size** the minimum price movement of an asset. Stocks in the US have mostly one cent minimum tick size, while in Europe it varies by the price of the stock. Other instruments (futures, commodities, etc.) have different tick sizes, 52

**Mutual Fund** a portfolio of securities managed to meet a particular investment objective. Mutual funds may offer active asset management or passive index tracking. There are two primary types of mutual funds: closed-end and open-end funds, 5

**OLS** Ordinary Least Squares – the standard method of linear regression analysis. It minimises the sum of squared differences between observed and fitted values, 42

**OLS, robust** a version of OLS modified to reduce any undue impact from outliers on the estimated values of the parameters. It is also used for estimating models that perform well even if the distribution is not normal, 44

**Open-end Fund** a mutual fund with a number of shares that varies daily as fund managers create new shares for investors who want to acquire them, and eliminate shares as investors want to redeem them. This process takes place once a day, after the close of trading, at the (net) value of the fund's assets (NAV), 5

**Order flow** refers to the difference between executed buy and sell volume, 43

**Ordinary shares** in its simplest form it is a claim of ownership on the company that gives the owner the right to receive an equal share of the corporation's profits, 4

**Outstanding shares** number of shares being held by its shareholders. Does not include shares that are authorised but not issued, or shares issued but held/bought back by the issuing company, 42

**Pairs trading** a trading strategy which bets on a linear combination, normally short an amount of shares in one asset and long an amount of shares in the other asset, following a predictable trajectory. When the portfolio consisting of the two assets deviates from its historical levels or where it is predicted to be, the strategy places trades that bet on the portfolio returning to its predicted level, 273

**PIN** probability of informed trading, 70

**POCV** Percentage of Cumulative Volume. This refers to an execution trading strategy which targets a fixed percentage of the total traded volume of an asset over a prespecified execution horizon, 213

**POV** Percentage of Volume. This refers to an execution trading strategy which buys/sells a fixed percentage of the traded volume of an asset over intervals of time, 213

**Preferred stock** are contracts whereby the corporation commits to pay the holder a regular income (interest) but includes no decision rights, 4

**Price impact** the impact that trading has on prices, whether they are temporary (e.g., by walking the book), or permanent (e.g., by inducing an upward pressure on prices), 70

**Price-time priority matching** see matching algorithm, price-time priority, 9

**Proprietary traders** are traders who trade for their own behalf employing their own funds and not other investors' money, 7

**Prorata matching algorithm** see matching algorithm, prorata, 10

**Quoted spread** the difference between the bid and ask prices and represents the potential cost of immediacy: the difference in price from posting a passive order at the best price versus aggressively executing an MO (and hence 'crossing the spread') at any point in time, 71

**R squared** the (adjusted) R-squared (or coefficient of determination) is a measure of how good the model fits the data, and its value is between 0 (lowest) and 1 (highest), 44

**Resilience** is the speed at which the LOB recovers after a market order walks through more than one level. Many models assume resilience is "infinite" meaning that the book recovers immediately and the prevailing fundamental price does not change, 136

**Resiliency** the speed at which quotes replenish to revert to their former levels after order flow imbalance events – such as an MO walking the book, 70

**Share turnover** number of shares traded over a period divided by the number of shares outstanding. This is used as a measure of a stock's liquidity – greater turnover implies greater liquidity, 42

**Share turnover ratio** see Share turnover, 42

**Slippage** difference between arrival price and actual price. Also known as implementation shortfall, 135

**Stub quote** a stub quote is a limit order placed very far from the price range where orders are usually executed, for example a buy limit offer at one cent for an asset trading at more than 10 dollars, 16

**Survivor function** also known as the reliability function, $S(x)$. It is the probability that a random variable exceeds a certain level $S(x) = \mathbb{P}\{X > x\}$, 93

**Sweep order (intermarket sweep order)** an intermarket sweep order is a special order type in the US that allows the sender to execute an order against all markets and execute at different prices, while bypassing the RegNMS order protection rule, 85

**Tick, tick size** the smallest step between two neighbouring price levels in the LOB, 16

**TWAP** Time Weighted Average Price. A market standard index benchmark used to measure the effectiveness of a liquidation/acquisition strategy. It equals $\frac{1}{T} \int_0^T S_u \, du$ where $S_t$ is the asset's midprice and $T$ is the time horizon, 141

**VIX** volatility index – an index which represents the market's anticipation of the future volatility published by CBOE. It is derived as a weighted average of a set of short maturity options on the S&P500 index, 43

**VWAP** Volume Weighted Average Price. It is a benchmark calculated as the volume weighted average price of trades over a given time horizon, 213

**Walking the book** (walking the LOB) the process whereby a large entering market order is executed against standing LOs at increasingly worse prices, 9

# Subject index